Military Psychology: An Introduction

Edited by
Christopher Cronin, Ph.D.
Saint Leo College

SIMON & SCHUSTER
CUSTOM PUBLISHING

ISBN 0–536–00565–6
BA 97383
Library of Congress 97-062005

SIMON & SCHUSTER CUSTOM PUBLISHING
Simon & Schuster Education Group
160 Gould Street/Needham Heights, MA 02194

Dedicated to the men and women who

serve their country.

It is hoped that psychology

eases your task.

Preface

This book provides an introduction to the field of military psychology. Military psychology is the application of psychological principles to the military environment. The history of psychological contributions to the military has been well documented. Much of the information presented in this text, in language directed to undergraduates, is available in the professional literature scattered throughout journals, books and technical reports.

This book is intended for students at the undergraduate level and is written to serve as a textbook for a college course on military psychology. It is anticipated that the majority of readers will be active duty personnel and their dependents who are pursuing college degrees on military installations throughout the world and students enrolled in ROTC programs or at the military academies. While this textbook is designed to expand the outlook of students, it also has relevance for professionals in the field. It serves as a concise introduction to the activities of military psychologists.

Rather than to act as a thorough source on the subject of military psychology, the current text strives to introduce the student to the field. Coverage has been restricted to areas which will provide the reader with an overview of the variety of activities engaged in by military psychologists.

Coverage (or lack of) of a topic in this text is not intended to indicate that one area is more significant than another. Considerations regarding the inclusion of topics in the text involved both pedagogic and practical decisions. Pedagogically, the text strives to excite the reader by reviewing areas which will have the most career relevance for the majority of students. Practically, it was necessary to limit the length of the book so as to accommodate an academic term and/or semester.

Particular attention has been paid to clarity of presentation. The material is presented in a concise, understandable manner throughout the text with numerous examples. Chapters are relatively independent to provide instructors maximum flexibility in assigning the order of chapters. Important terms are boldfaced in the margin of the text; and the Glossary defines these terms. The Appendix includes an application for the Division of Military Psychology, Division 19 of the American Psychological Association as well as a list of web sites related to psychology.

I am always interested in comments and suggestions from students, instructors and other professionals regarding the text. My address is Department of Psychology, MC-2127, P.O. Box 6665, Saint Leo College, Saint Leo, Florida, USA, 33574. My e-mail address is ccronin@saintleo.edu.

Acknowledgments

Obviously I owe a great deal of appreciation to the contributors of the text. Without their Herculean efforts the text would not exist. The contributors are all experts in the field of military psychology. The list boasts eight current or former military officers, four past presidents of the Division of Military Psychology and a combined experience of nearly 250 years in the field of military psychology. It has been an honor and privilege to have worked with so many outstanding scholars. Additionally, many thanks to our colleagues who provided feedback on the numerous manuscripts.

Thanks also goes to Dr. Marilyn Mallue whose creative insight inspired this text and also provided information regarding web sites, included in the Appendix. And appreciation is offered to the students at Saint Leo College who provided feedback on the book's chapters and to Division 19 of the American Psychological Association for permission to include a membership application. Finally, thanks to Hal Hawkins and Chris Gill of Simon & Schuster/Prentice-Hall. Their support, vision and patience made the book possible.

Christopher Cronin

Table of Contents

List of Contributors

Ronald G. Ballenger earned his Ph.D. from Ball State University and AB and Masters from Indiana University. He was for two years the staff clinical psychologist at Northeastern Center in Kendallville, Indiana. Prior to that, for four years, he was professor of counseling psychology in the Ball State University European Counseling Program and the Boston University Overseas Program. He was President of the European Branch-American Counseling Association and, later, EB-ACA Treasurer. He was selected the Outstanding Member of that organization in 1988. For over six years, Dr. Ballenger was the Clinical Director of the Community Counseling Centers in Grafen-woehr and Munich, Germany providing assessment and treatment in the area of alcohol abuse and drug use. He has presented numerous workshops on addictive behavior, child assessment and management, and various other areas, to military and civilian groups throughout Europe. Dr. Ballenger is one of two Clinical Child Psychologists with the Exceptional Family Member Program in the Wuerzburg, Germany Bavarian region. His research is in the area of alcohol use among college students and adolescents and the methods that university counseling practicums can employ to increase levels of empathy within counselors.

Paul Bartone is a research psychologist now serving as the Commander of the US Army Medical Research Unit-Europe in Heidelberg. He holds a Ph.D. in Psychology and Human Development from the University of Chicago. Dr. Bartone began working in psychology in 1974 at the Boston VA's longitudinal study of aging in healthy military veterans. He completed his Ph.D. in 1985, then entered the Army. He served as a research psychologist at the Walter Reed Army Institute of Research before going to Heidelberg. Major Bartone's research studies in USAREUR have sought to clarify the nature and sources of stress and their health effects on soldiers and their families in the forward deployed force. Of special interest are units on peacekeeping and contingency operations. Dr. Bartone is a member of many professional organizations, including the American Association for the Advancement of Science, APA, APS, American Public Health Association, Association of Military Surgeons of the US, International Society for Traumatic Stress Studies, and the European Research Group on Military and Society.

Richard Bloom is a clinical psychologist and former President of the Division of Military Psychology, American Psychological Association. He served for twenty years in the United States Air Force as a clinical and political psychologist with assignments to national-level interagency committees, policy offices and planning groups, the Joint Staff and the Air Staff. He has served in the Republic of the Philippines, Panama and Turkey. Dr. Bloom is past editor of the *Military Psychology Bulletin* and *The Military Psychologist* and has served as reviewer for Air University Review, Military Review, Journal of Personality Assessment, Contemporary Psychology and Military Psychology.

He is a Diplomate of the American Board of Professional Psychology in clinical psychology. Currently, Dr. Bloom is Associate Professor of Political and Clinical Psychology at Embry-Riddle Aeronautical University in Prescott, Arizona and Coordinator of its Terrorism, Intelligence, and Security Studies Program.

Jan Cannon-Bowers is a Senior Research Psychologist in the Science and Technology Division of the Naval Air Warfare Center Training Systems Division (NAWCTSD), Orlando, FL. She holds M.A. and Ph.D. degrees in Industrial/Organizational Psychology from the University of South Florida, Tampa, FL. As the team leader for Advanced Surface Training Research at NAWCTSD, she has been involved in a number of research projects directed toward improving training for complex environments. These have included investigation of training needs and design for multi-operator training systems, training effectiveness and transfer of training issues, tactical decision-making under stress, the impact of multi-media training formats on learning and performance, and training for knowledge-rich environments. She works closely with the Aegis community and has contributed to the development of several training systems such as the Battle Force Tactical Trainer, Joint Tactical Combat Training System, and Aegis Combat Training System.

Christopher J. Cronin earned his B.S. in psychology at the University of Wisconsin, Madison and his M.A. and Ph.D. in clinical psychology at the University of Delaware. He completed his internship at the University of California, Davis Medical Center in Sacramento, California. He taught for six years on military installations in Europe with The University of Maryland and was Director of the Counseling Center at The University of Maryland, Munich Campus in Munich, Germany. In addition, he has held academic posts in the United States and at the School of Psychology, Flinders University of South Australia in Adelaide, Australia. He is a licensed clinical psychologist and Associate Professor and Chair, Department of Psychology at Saint Leo College in St. Leo, Florida. Dr. Cronin is a member of the APA and an affiliate of Division 19, Division of Military Psychology. His research interests are in the area of harm reduction for alcohol use and the identification of youth at risk for substance misuse. His clinical work is in forensic psychology.

Mickey R. Dansby has been the Director of Research for the Defense Equal Opportunity Management Institute (DEOMI), Patrick AFB, FL., since September of 1993. Prior to that time, he was a professor at Florida Institute of Technology, consultant, independent researcher, and adjunct professor for Rollins College Brevard. He holds a Masters Degree in experimental psychology and a Ph.D. in social psychology from the University of Florida and has published numerous articles, book chapters, and reports. He has taught a wide range of courses at the graduate and undergraduate levels, including social psychology, organizational development, group/team development, management, research design, and statistics. Dr. Dansby served 20 years as an officer in the U.S. Air Force, retiring in 1991 as a lieutenant colonel. While on active duty, he held a variety of educational and research positions, including six years as a faculty member of the U.S. Air Force Academy, Director of Research and Analysis for the Air Force Leadership and Management Development Center, and Director of Research for DEOMI from 1986–1991 (before the position was converted to civil service in 1993). He founded the Military Equal Opportunity Climate Survey (MEOCS) program and many other research and consulting services at DEOMI. Dr. Dansby resides in Satellite Beach, FL, with his wife, the former Diane Marie Davis. They have two children, Julie and Scott.

W. Brad Johnson completed the Ph.D. at the Graduate School of Psychology, Fuller Theological Seminary. A Navy lieutenant from 1990–1994, Dr. Johnson served as a psychology intern at the National Naval Medical Center in Bethesda, Maryland. He was then a staff psychologist and Division Head for Psychology at the Naval Medical Clinic at Pearl Harbor Hawaii. Dr. Johnson is currently Assistant Professor of Psychology in the Graduate School of Clinical Psychology at George Fox University. An Associate Fellow of the Institute for Rational-Emotive Therapy, Dr. Johnson has published numerous articles in the area of military psychology, professional training and rational-emotive therapy. He remains on inactive reserve with the Navy.

Gerald P. Krueger received his BA degree in psychology from the University of Dayton, Ohio in 1966. He obtained an MA in engineering psychology in 1975, and a Ph.D. in experimental and engineering psychology in 1977 from the Johns Hopkins University in Baltimore, Maryland. As an active duty military research psychologist for over 25 years, Dr. Krueger specialized in conducting research on soldier and aviator performance in stressful working conditions, and during deployments to harsh environments. Many of his research endeavors were on the study of sustained military performance. Dr. Krueger served as the chairman of the U.S. Department of Defense Human Factors Engineering Technical Advisory Group (DOD-HFE-TAG), 1989–90, and as the chairman of the DOD-HFE-TAG's international subgroup on Sustained/Continuous Operations from 1985 to 1989. As a Colonel in the U.S. Army Medical Service Corps, Colonel Krueger served as the commander of the U.S. Army Research Institute of Environmental Medicine (USARIEM) at Natick, Massachusetts from 1990 to 1994. Col Krueger retired from the U.S. Army in 1994. He served as the President of Military Psychology, the American Psychological Association's (APA) Division 19, in 1995 and '96. Representing the APA, Dr. Krueger testified in support of tri-service behavioral sciences research budgets at hearings of the U.S. Senate Defense Appropriations Subcommittee in June, 1995 and in June, 1996. Dr. Krueger is now a Principal Research Scientist and Ergonomist at Star Mountain, Inc., a human resources consultant firm in Alexandria, Virginia.

Alan W. Lau earned his B.A. and M.A. in psychology at American University and his Ph.D. in applied psychology at the University of Utah. He has served as a Research Psychologist at the Army Research Institute, the Navy Personnel Research & Development Center and the U.S. Public Health Service, and as Scientific Officer at the Office of Naval Research. He is a member of the American Psychological Association, American Psychological Society, American Management Association and the Academy of Management. He has authored over 40 articles and technical reports. He recently concluded a four-year longitudinal research program funded by ARI at the Virginia Military Institute on leadership development and effectiveness. Dr. Lau currently teaches in the Graduate Management Program with The University of Maryland on military installations in Europe.

Stephanie C. Payne is a doctoral student in the Industrial/Organizational Psychology program at George Mason University, Fairfax, VA. She received her B.A. degree in psychology from St. Leo College, St. Leo, FL and her M.S. degree in Industrial/Organizational Psychology from the University of Central Florida, Orlando, FL. She was a research assistant in the team training lab at the Naval Air Warfare Center Training Systems Division, Orlando, FL. Some of the research projects she has contributed to include identifying air traffic controller team competencies, developing a performance measure and debriefing guide to assess teamwork, and evaluating leader briefing skills. Her research interests include team training, individual differences, self-regulation, and job design.

Eduardo Salas is a Senior Research Psychologist and the head of the Training Technology Development Branch, Science and Technology Division, of the Naval Air Warfare Center Training Systems Division. He received his Ph.D. degree (1984) in Industrial/Organizational Psychology from Old Dominion University. He is a Fellow of the American Psychological Association and the Human Factors and Ergonomics Society. His research interests include team training and performance, training effectiveness, tactical decision making under stress, team decision making, human performance measurement and modeling, and learning strategies for teams. He has co-authored over one-hundred journal articles and book chapters and has co-edited four books.

John Sexton has been a Marine Corps and Navy officer since his graduation from the U.S. Naval Academy in 1972. He earned a Masters degree in counseling psychology from the University of Northern Colorado, and a Masters and Ph.D. in clinical psychology from the California School of Professional Psychology. Since 1983, he has served as a Navy clinical psychologist in varied assignments around the world. In 1994, Dr. Sexton was one of the first two graduates of the Department of Defense Psychopharmacology Demonstration Project in which he completed approximately 75% of the medical student didactic courses at the Uniformed Services University of Health Science's Medical School, and a one year residency at Walter Reed Army Medical Center. Commander Sexton was assigned to the Naval Medical Center, Portsmouth, Virginia; and in February 1995, he became the first independent practitioner prescribing psychologist.

Kimberly Smith-Jentsch is a Research Psychologist with the Naval Air Warfare Center Training Systems Division, Orlando, FL. She received her Ph.D. in Industrial/Organizational Psychology from the University of South Florida in 1994. She manages programs of research which investigate training and performance measurement issues particularly for teams which operate in complex environments such as air traffic control towers, navy warships, and aircraft cockpits. Specifically, she has conducted research in the areas of team training needs analysis, team competency models, human performance measurement, training strategies to enhance team self-correction, leadership and team member assertiveness, transfer of training climate, pre-training experiences and their impact on training effectiveness, on-the-job training, and continuous learning environments.

Brian K. Waters is Program Manager of the Manpower Analysis Program of the Human Resources Research Organization (HumRRO). He joined HumRRO in 1980, after retiring from the Air Force, where he taught and was Director of Evaluation at the Air War College, was a research and development manager and researcher with the Air Force Human Resources Laboratory, and was a navigator. He holds a Ph.D. and M.S. in Educational Measurement and Testing from Florida State University, and an M.B.A. from Southern Illinois University. He is a Fellow of the American Psychological Association (APA) and a former President of the Division of Military Psychology of APA. He has authored over 100 journal articles, books, and professional papers, primarily dealing with the selection, classification, and testing of military and civilian personnel.

Martin F. Wiskoff is currently a Director/Senior Scientist for BDM International Inc., in Monterey, CA. He was previously a Senior Scientist for the Defense Personnel Security Research Center in Monterey, CA and Director of the Manpower and Personnel Laboratory at the Navy Personnel Research and Development Center, San Diego, CA. Dr. Wiskoff is a Fellow and past President of the Division of Military Psychology of the American Psychological Association. He is also the founder and editor of *Military Psychology*, a journal published by the Division of Military Psychology.

Defense Department Photo

1

Introduction to Military Psychology

Christopher Cronin, Ph.D.

* Introduction * Definition of Military Psychology * Brief Historical Overview
of Military Psychology * Domain of Military Psychology * Overview of the
Book * Getting Started as a Professional: Division of Military Psychology
* Summary * References

Introduction

This text is about an area of psychology which enjoys a long history of setting precedents in the field of behavioral science. This chapter begins with a definition of military psychology and a brief historical overview of the relationship between psychology and the military. It then reviews the specific areas of research and practice which constitute the field of military psychology. The discussion previews areas covered in the current text while also touching upon aspects of military psychology not covered in this book. The subsequent chapters are described briefly and the chapter ends with some suggestions on how one can immediately start to become an active professional in the field of military psychology.

Definition of Military Psychology

Most branches of psychology are defined by either the area of study or a common set of techniques. For example, clinical psychology focuses on the diagnosis and treatment of behavior disorders (area of study) whereas comparative psychology uses different species to understand behavior (common set of techniques). However, it is the area of application, the military environment, which differentiates military psychology from the other specialties in the field of psychology (Driskell & Olmstead, 1989). *Military psychology* is defined as the study and application of psychological principles and methods to the military environment (Mangelsdorff & Gal, 1991; Taylor & Alluisi, 1994).

    Military psychology is a microcosm of all psychology disciplines and one will find that work in military psychology affects almost all aspects of the military setting. Indeed, it is difficult to imagine any function in which military personnel are involved that has not benefitted from the application of psychological research. Take a moment to briefly consider the wide range of activities in which our armed forces are involved.

clinical psychology

Military psychology

1

Combat readiness

Perhaps you imagined troops deployed on a military operation. Currently, U.S. troops are stationed all over the globe. Combat readiness is the most salient aspect of the Armed Forces and is usually the first image which comes to mind. After all, the military's primary mission is defense of the nation. Similarly, the primary objective of military psychology is to assist the Armed Services in achieving their mission. Anything which does not contribute to combat readiness is of secondary importance (Driskell & Olmstead, 1989).

However, as every service member knows, maintaining combat readiness starts well before troops are sent to the field. And it is no surprise that psychology has played a vital role in *every* aspect of military life. For example, military psychologists are involved in the selection and classification of recruits, the identification and training of leaders, helping service members and families cope with the stress of military life, identifying and treating emotional problems among members of the military community and maintaining morale and a sense of esprit de corps among an active duty force of approximately 1.44 million people.

Psychologists also play a vital role in optimizing human performance under extreme environmental conditions, developing high-tech human-machine weapon systems, waging information warfare, negotiating with terrorists and conducting research in military laboratories throughout the world. This is not an exhaustive list of the application of psychology to the military mission, but rather some illustrative examples. The remaining chapters in this text will serve to inform you of the numerous contributions of psychology to the military.

The astute reader will quickly recognize that this relatively small textbook cannot do justice to an area as broad as military psychology. Rather, this book will acquaint the student of psychology to the many exciting opportunities found in the application of psychological principles to military problems. This introduction to the field of military psychology may hopefully encourage students to pursue further study, and perhaps even a career, in military psychology.

A final comment regarding the definition of military psychology is necessary. Civilian and uniformed military psychologists work in a variety of settings such as colleges, universities, troop units, Veterans Administration (now Veterans Affairs) and military hospitals, military installations, defense contract companies, military academies, and consultant groups, to name a few. As mentioned above, military psychology is defined as the application of psychological principles to the military environment, regardless of who is involved or where the work is conducted.

Brief Historical Overview of Military Psychology

This section will briefly trace the development of military psychology in the United States. Attention will focus on how the relationship has been mutually beneficial to both the Armed Services and the discipline of psychology.

The founding of Wilhelm Wundt's experimental psychology laboratory in Leipzig, Germany in 1878 is identified as the beginning of modern experimental psychology. Wundt defined psychology as the study of consciousness. Germany was the first country to award degrees in psychology and Wundt trained many of the early psychologists, including leading American psychologists such as James Cattell and Edward Titchener.

Psychology is a relatively young science and it was only recently that the American Psychological Association (APA), the world's largest association of psychologists with over 125,000 members, celebrated its centennial anniversary.[1] Modern American psychology is marked by the founding of the APA in 1892 (Leahey, 1997). The APA was founded to promote the *science* of psychology. However, it was not long before the early psychologists stressed the personal and social benefits of their discipline.

American Psychological Association

The following discussion will explain how the military played an influential role in the development of psychology as an applied discipline. The practice of psychology in the late 1800s was very different than the way the public currently perceives the field. Initially, psychologists were academics who taught and conducted research in the 'ivory towers' of the universities. Indeed, one requirement for membership in the APA was a published study beyond the dissertation. Though there were scattered efforts to make psychology practical, neither the public nor psychologists saw psychology as an applied profession. The shift from principally an academic discipline to an applied profession can be attributed in large part to psychology's relationship with the military (Yerkes, 1919; McGuire, 1990; Leahey, 1997).

Scholars date the beginning of military psychology in the United States to the beginning of World War I (Taylor & Alluisi, 1994; McGuire, 1990). In 1917, Robert Yerkes, as president of the APA, organized the efforts of psychologists to aid in the war effort. In a letter to the APA Council, he wrote, "It is obviously desirable that the psychologists of the country act unitedly in the interests of defense. Our knowledge and our methods are of importance to the military service of our country, and it is our duty to cooperate to the fullest extent and immediately toward the increased efficiency of our Army and Navy" (Yerkes, 1918). This 'call to arms' by Yerkes resulted in twelve APA committees tasked with applying the scientific principles of psychology to the war effort. Thus started the current eighty year tradition of military psychology.

Initially, psychologists found their niche in the selection and classification of military recruits. Prior to the war, the military had been relatively small. America's entry into World War I and the concomitant influx of high numbers of men into the army created the need for an economical and efficient method of classifying new soldiers and identifying potential officer candidates.

Yerkes, an experimental psychologist, led a committee which focused on the psychological examination of recruits and the identification of individuals deemed mentally unfit for military service. These efforts led directly to the development of the *Army Alpha* and *Army Beta* group mental tests. These tests are the predecessors of the present *Armed Services Vocational Aptitude Battery* (ASVAB), familiar to virtually every reader of this book, and currently administered to over 1.5 million individuals annually.

Armed Services Vocational Aptitude Battery

Concurrently, Walter Dill Scott, an industrial psychologist, led a committee to develop tests for job classification. Scott's committee had developed proficiency tests for 83 military jobs (Leahey, 1997). By the end of the First World War, psychologists had tested nearly 4,675,000 men.

The war effort also led to psychologists serving in military hospitals. In 1918, the Surgeon General authorized the first billet (duty assignment) of a psychologist to Walter Reed General Hospital (McGuire, 1990). Originally, psychologists played an educational role in the medical setting, training hospital staff and surveying patients.

billet

However, the need to economically screen large numbers of recruits for indications of mental abnormality during World War I prompted the development of a structured measure of **psychopathology** psychopathology (Dubois, 1970). This initial work with patients laid the groundwork for psychology's entry into the domain of psychiatry, the diagnosis, treatment and prevention of **psychiatry** mental disorders.

Prior to the war, psychologists were academics who worked quietly in university settings with little recognition from society. During the war, psychology literally touched the lives of millions. Due to the success of military psychology, the field had transformed itself from an obscure academic discipline into an applied profession ready to tackle a variety of practical problems.

The new relationship between psychology and the military simmered during the years after the First World War. However, the field of psychology had changed as a result of this relationship and the peacetime years saw a fortifying of psychology's role in military applications and applied settings. American psychologists were busy developing and refining group mental tests, helping industry to achieve worker satisfaction and offering advice to both **Applied psychology** advertisers and parents. Applied psychology, though existent prior to the war, had come of age in its post-war fame.

When the United States entered into the Second World War, psychology was poised to make significant contributions. Buoyed by its newfound confidence and recognition as a respected profession, psychology ambitiously addressed the military's, and society's, problems. Once again, prior to the United States direct involvement in the global conflict, the APA started planning for its contribution to the war effort. Several months prior to the Japanese attack on Pearl Harbor, the *Psychological Bulletin* devoted an entire issue to "Military Psychology" (Leahey, 1997). The war effort provided a significant boost to military psychology. In 1943, fully half of all pages of the *Psychological Bulletin* were devoted to military psychology and 25% of the nation's psychologists were engaged in military psychology (Driskell & Olmstead, 1989).

World War II gave psychologists an opportunity to further refine the selection and classification techniques they had developed during the First World War. By the end of WW II, over 13,000,000 persons had been administered the Army General Classification Test (Taylor & Alluisi, 1994).

In addition to the selection, classification and assignment efforts for manpower management, psychologists were creating new opportunities. Experimental psychologists, applying **human factors** their skills to military problems, developed a new branch of psychology called human factors **engineering** engineering and engineering psychology (Taylor & Alluisi, 1994). *Human factors engineering* aims to optimize human performance by designing (weapon) systems which can account for the strengths and weaknesses of the humans who operate and maintain the systems (Chatelier & Alluisi, 1991).

Psychologists, initially involved in assisting engineers to optimally place controls in aircraft cockpits during WW II, have since become involved in the design of instrument displays, consoles, system controls, protective clothing, gunsights, combat information centers, communications equipment, etc. The effects on performance of harsh environmental conditions such as extreme heat and cold are a focus of interest, along with other adverse conditions such as extreme sleep deprivation and the prolonged use of chemical protective clothing.

With the rapid advances in computer technology, a major Research and Development (R&D) thrust has been the development of simulated combat environments networking numerous ground vehicle, aircraft, artillery, tactical operations centers, and command and control simulators (such as the earlier flight simulators) on one area network (Taylor & Alluisi, 1994). This sort of virtual world allows for the use of real troops training in and testing weapon systems in "virtually" any environment (physical as well as climatic) under simulated combat conditions. Clearly, military psychology's contribution to human factors engineering has come a long way since the early 1940s and undoubtedly will remain an integral part of human-machine system development.

The growth in applied psychology propelled by the war effort also had a significant impact on clinical psychology. After WW I and prior to the end of WW II, clinical psychologists were principally involved in test administration and interpretation, not only in military settings but also in schools, industry, prisons, and as private consultants. However, by the end of WW II, it became clear that there was an overwhelming need for the provision of psychological services which could not be met by psychiatrists alone. Over half of the nearly 75,000 hospitalized veterans at the end of the war were hospitalized for psychiatric reasons.

During WW I psychologists had become involved in the identification of psycho-pathology using structured assessment instruments and had duty assignments in hospitals as early as 1918. As members of the medical team, they performed diagnostic duties. But due to the overwhelming need for psychotherapy, psychologists also began to serve as therapists at the end of the Second World War (Leahey, 1997). This transition from diagnostician to therapist marks the beginning of modern clinical psychology.

The Veterans Administration (VA) hospitals, tasked with the responsibility of providing psychological services to veterans, has played a major role in defining modern day clinical psychology. In 1946, the VA set up training programs at major universities to turn out clinical psychologists whose job would be therapy as well as diagnosis (Leahey, 1997). The VA continues to train clinical psychologists, providing APA approved pre-doctoral internships and post-doctoral fellowships at numerous facilities.

APA approved internships

fellowships

And the military continues to have a strong influence on the training and professional development of clinical psychologists. Currently there are over 400 authorized billets for clinical psychologists among the Navy, Air Force and Army; all three branches have clinical psychology internship programs and post-doctoral fellowship training. The Uniformed Services University of the Health Sciences (USUHS) recently started a Ph.D. training program in Military Clinical Psychology. Also, military clinical psychologists are leading the way in the right for prescription privileges, the authorization to prescribe psychotropic medications to patients. This last issue, prescription privileges, is covered in detail in Chapter 12 of this text.

Uniformed Services University of the Health Sciences

prescription privileges

psychotropic

The rapid expansion of the military during World War II also focused attention on human behavior within organizations. The need to identify and train leaders was of paramount importance to a military structure which relied heavily on small units to engage the enemy. The selection, training and evaluation of leaders was established as an area of research for military psychologists.

A primary task of leaders is to develop cohesion in their unit. Many of the functions performed in the military are conducted in units or teams. How well individuals function as a

cohesion

team is often critical to mission success. The study of individual and team behavior has become a crucial area of interest to military psychologists.

The success of psychology's contributions during the two war efforts legitimized the specialty of military psychology. With regards to current research and applied work, the military environment can serve as an excellent "laboratory" which enables military psychology to be on the cutting edge of the behavioral sciences.

Although far from comprehensive, this brief history has introduced you to psychology's early contributions to the military dating back to World War I. Those initial years demonstrated that the field of psychology had a lot to offer to the military environment. Currently, psychology plays a role in almost every facet of the Armed Services' mission and, without exception, every service member benefits from the work of military psychologists.

Just as importantly, every psychologist owes a debt of gratitude to the military for the significant role it has played in the advancement of the profession. The military has helped psychology to develop as both an applied and a scientific discipline. Today, the Department of Defense (DoD) is the largest employer of psychologists.

The interested reader can consult a number of historical (Meier, 1943; Bray, 1948; Flanagan, 1952; Geldard, 1962) and contemporary (Gal & Mangelsdorff, 1991; McGuire, 1990; Taylor & Alluisi, 1994; Driskell & Olmstead, 1989) publications to further explore the development of military psychology. Other contributors to this text have also taken the opportunity to review the historical development of military psychology highlighting slightly different areas.

Domain of Military Psychology

Military psychology encompasses a broad domain of research and practice in the field of psychology. Essentially all of the areas studied in military psychology are also represented in the civilian world. It is the unique application of these principles to the military environment which defines the area of military psychology. In their comprehensive text, *Handbook of Military Psychology*, Gal and Mangelsdorff (1991) identified six principal areas of military psychology:

(1) *selection, classification and assignment,*
(2) *human factors,*
(3) *environmental factors,*
(4) *leadership,*
(5) *individual and group behavior,*
(6) *clinical and consultative/organizational psychology.*

Additionally, they incorporated a special topics section covering such areas as propaganda, hostage negotiation, women's role in the military and prisoners of war. Wiskoff (1997) identifies two additional specialties in his chapter on careers in military psychology;

(1) *training and education,*
(2) *manpower management decision making support.*

The above eight areas (the six proposed by Gal and Mangelsdorff (1991) and the two proposed by Wiskoff (1997)) comprised the domain of military psychology. Due to the breadth of topics covered, it is easy to see why military psychology has frequently been called a microcosm of psychology. The *Handbook of Military Psychology* consists of 39 chapters by 62 authors from seven different countries, testimony to the scope and diversity of the field. The uniqueness of military psychology and what sets it apart from the other specialties in the field is the application of scientific findings in these eight diverse areas to the solution of practical problems in the military environment.

Overview of the Book

This text, a primer on military psychology, will acquaint you with the following areas:

(1) psychological research in the military setting,
(2) selection and classification,
(3) leadership training,
(4) human performance under adverse conditions,
(5) stress in the military setting,
(6) team performance,
(7) information warfare,
(8) cultural diversity and gender issues,
(9) clinical/counseling psychology, and
(10) careers in military psychology.

Note that several of these areas directly correspond to one or more of the eight specialties listed above. I encourage the reader to identify which of the eight domains of military psychology are associated with the chapters of this text. The remainder of this chapter will briefly describe each of the following chapters.

Chapter 2, by Dr. Gerald Krueger, introduces the reader to the wide range of research activities in which military psychologists are involved. This chapter also identifies the major military research labs and familiarizes the reader with the Department of Defense's (DoD) budget categories (6.1–6.5). These categories, part of DoD's Program for Research, Development, Test & Evaluation (RDT&E) indicate the type of R&D funded (Driskell & Olmstead, 1989). After completing this chapter, you will be able to identify, *(1) the types of research conducted in military psychology, (2) the major military research laboratories, and (3) career opportunities for military research psychologists.*

Chapter 3, by Dr. Brian Waters, introduces the reader to the mainstay of military psychology, personnel selection and assignment. Dr. Waters first discusses issues germane to both military and civilian selection decisions: setting standards, establishment of cut scores on selection tests and validation of the selection procedure. The chapter moves on to cover military personnel selection at both the enlisted and officer level. Dr. Waters then covers the area of classification, the process of finding the right person for the right job. Again, the chapter treats enlisted and officer classification procedures separately. This chapter will undoubtedly remind many readers who are associated with the military of the time they completed the ASVAB, the three hour aptitude battery all recruits complete. Additionally, prospective recruits will gain insight into the selection and classification process for enlisted and officer positions.

In Chapter 4, by Dr. Al Lau, you will be introduced to theoretical and applied areas of *leadership training*. Areas covered include an overview of major leadership theory and research, leadership training programs in the military and suggestions garnered from the empirical data base on how you can improve your own leadership effectiveness. This chapter is spiced with anecdotal incidents which will help you to understand and apply the information presented to your own situation. After completing this chapter you will be able to, *(1) define effective leadership, distinguishing between transactional and transformational leadership components, (2) discuss the major leadership theories and research, (3) be familiar with some of the major leadership training programs in the military, and (4) identify current issues which will influence leadership research and theory in the future.*

In Chapter 5, Teams and Teamwork in the Military, you are introduced to a concept crucial to mission success in the military, *teamwork*. As every member of the armed forces realizes, they are all members of a team which must work together to ensure the desired outcome of most, if not all, assignments. It is not surprising that psychologists have focused their attention on ways to improve selection, training and maintenance of high performing teams. The authors, Drs. Eduardo Salas, Janis Cannon-Bowers, Kimberly Smith-Jentsch and doctoral candidate, Stephanie Church-Payne, provide an overview of what military psychologists have learned about teams and teamwork in the military. It seems fitting that the chapter on teamwork is the only co-authored chapter in the book.

Chapter 5 starts with a rationale for the importance of teamwork. The authors then define the concepts of *team* and *teamwork* using examples drawn from the military setting. The last two sections of this chapter review the issues involved in *measuring team performance* and *team training*. Once you finish this chapter, you will be able to answer the five questions the authors pose at the beginning of their chapter: (1) why is teamwork important in the military, (2) what is a team, (3) what comprises teamwork, (4) how is team performance measured, and (5) how are teams trained.

Chapter 6, also by Dr. Krueger, introduces the reader to the varied types of environmental conditions in which service personnel must perform. In addition to focusing on human-machine interface, psychologists have also investigated ways to optimize human performance under *adverse environmental conditions*. Military personnel are frequently assigned to harsh environments, required to perform under conditions of extreme *temperatures, high altitude, noise, motion including acceleration, vibration and motion induced sickness, toxic gases, radiation, use of chemical protective clothing and conditions of sleep deprivation*. Dr. Krueger also covers *countermeasures* developed to optimize human performance under these various environmental conditions.

Major Paul Bartone discusses the ubiquitous matter of *stress* and the role it plays in the military setting in Chapter 7. First, Dr. Bartone defines what is meant by the term stress and offers a psychological framework for considering stress in the military setting. After a brief historical overview with comments on *military psychiatry*, Dr. Bartone introduces the reader to *deployable field psychological research teams* in the U.S. Army. This section is brought to life by current examples related to recent military operations including Operation Just Cause (Panama), the Gulf War, Operation Restore Hope (Somalia), Operation Provide Promise (Croatia), Operation Uphold Democracy (Haiti), Operation Vigilant Warrior (Kuwait), and Operation Joint Endeavor (Bosnia).

countermeasures

Based on current empirical findings, Dr. Bartone proposes five general dimensions (*isolation, ambiguity, powerlessness, boredom and threat/danger*) as a way of categorizing stress in military operations. The author then turns our attention to stress in the military family and military community and examples of *stress-resistance resources.*

In Chapter 8, Dr. Richard Bloom discusses the often clandestine world of information warfare (IW), perhaps better known to many readers as the field of propaganda. Dr. Bloom first introduces the complexities of IW by challenging readers to question their own assumptions regarding the accuracy of information and the identification of 'real' motives, the feasibility of research on IW and the potential dangers inherent in IW, both scientific and real.

information warfare

Dr. Bloom provides a brief historical perspective on IW and then furnishes the reader with a list of reasons for and advantages of IW, along with current theoretical perspectives. The reader is then treated to a number of present-day examples in which IW has been utilized, including the Mideast, former Yugoslavia, Israel, Africa, China, and Iraq. Chapter 8 also explores IW as an applied social science, covering areas such as IW typology, effectiveness, process and training. Dr. Bloom predicts future trends for the field of IW and the IW practitioner and concludes his chapter on a personal note.

Dr. Mickey Dansby's contribution, Chapter 9, covers the rather timely areas of *cultural diversity* and *gender issues.* Dr. Dansby reviews the history of racial integration in the military and evaluates the success of programs to insure diversity and eliminate discrimination. The area of gender issues is also covered, including a detailed discussion on *sexual harassment.* He concludes his chapter by highlighting other areas of research regarding gender issues (such as pregnancy and single parenthood) and then discusses future considerations regarding diversity issues.

Chapters 10 through 12 address issues related to clinical/counseling military psychology. I wish to offer an explanation for this apparent bias toward the clinical aspects of military psychology. First, experience and surveys indicate that the majority of psychology majors who plan to pursue a career in psychology typically indicate a preference for a clinical or counseling degree. A major goal of this text is to be relevant for students who may be considering careers in military psychology. Secondly, other excellent descriptions of military psychology have omitted information on the clinical specialty (e.g., Taylor & Alluisi, 1995; Driskell & Olmstead, 1989). Perhaps these three chapters will help to fill this information gap.

Chapter 10, by Dr. Brad Johnson, discusses in detail the training opportunities for clinical psychologists in the military. He walks the reader through the pre-doctoral military *internship*, covering the application and selection process, the officer *indoctrination process*, structure of the military internships and the unique training opportunities and stresses to be found in these programs. Dr. Johnson then provides an overview of the practice opportunities and duty assignments for clinical psychologists. The chapter ends with a discussion on the particular ethical dilemmas facing military clinical psychologists.

Chapter 11, by Dr. Ronald Ballenger, discusses the duties of psychologists in several of the military's mental health programs. After World War II, the composition of the military changed due largely to an increase in professional soldiers, individuals who stayed in the military after their initial service obligation. The services went from consisting essentially of single men to being composed of married men with families. For example, today nearly two-thirds of the U.S. Army's 495,000 soldiers are married and support 474,000 children. An illustrative example of the impact of married life on a military installation is Fort Hood, Texas, a base with 41,500 soldiers and 71,650 dependents.

The military responded to this change in demographics by establishing mental health clinics and outpatient facilities to help service personnel and their families adapt to the military and deal with the stresses of problems in living (Mangelsdorff & Gal, 1991). As you will see in Chapter 7 by Dr. Bartone, these types of programs are successful in increasing retention.

In Chapter 11, Dr. Ballenger reviews several of these programs including the Army's *Alcohol and Drug Abuse Prevention and Control Program* (ADAPCP), the *Exceptional Family Member Program* (EFMP) and the *Neuropsychiatry and Mental Health Department*. Frequently, professionals speak in a language all their own (jargon) when among colleagues, and psychologists are no exception. This occupational hazard tends to be compounded among military psychologists due to the opportunity to use both psychological and military jargon (Driskell & Olmstead, 1989). The military is famous for its fondness of acronyms and Chapter 11 will give you a sample of their liberal use. Similar to Chapter 10, this chapter also includes case examples and concludes with another perspective on the ethical dilemmas confronting clinical psychologists working in a military setting.

Chapter 12, the last of the clinical triad, introduces the reader to the DoD's *Psychopharmacology Demonstration Project* (PDP). The author, Commander John Sexton, a clinical psychologist in the Navy, is one of the first graduates of the *PDP* and is one of only a handful of psychologists who have *prescription privileges*, which is the authorization to prescribe *psychotropic medications*. Currently, prescription privileges for psychologists is a frequently debated topic in the profession and serves as an excellent example of military psychologists' role as pioneers in the field. Dr. Sexton addresses the issues involved regarding prescription privileges for psychologists, reviews the history of the debate and gives a detailed overview of the training he received in the PDP. His chapter concludes with a look toward the future of prescription privileges for both military and civilian psychologists.

Chapter 13 will help you to explore the options of a career in military psychology. Dr. Martin Wiskoff will inform you about career opportunities, salary ranges, professional growth and the type of academic preparation required for a career as a military psychologist. Dr. Wiskoff concludes his chapter on a personal note, describing some of his own experiences as a military psychologist.

Chapter 14, by this author, tackles the rather ambitious task of predicting how military psychology will develop in the near future. The chapter identifies four trends which will play a role in the future development of military psychology: *(1) rapid advancements in technology, (2) political changes such as the end of the Cold War and expansion of NATO, (3) budget restrictions and the military drawdown, and (4) changes in personnel demographics.* Chapter 14 ends with some comments on recent developments regarding the profession of military psychology and optimistic predictions for the specialty's continued growth.

Getting Started as a Professional: Division of Military Psychology

As a student of psychology, you are encouraged to get involved in the profession. There are several ways to do this. First, I highly recommend becoming a student affiliate of the **Division of Military Psychology, Division 19** of the American Psychological Association. As mentioned above, the APA is the world's largest association of psychologists with over 125,000 members. The Division of Military Psychology was one of the original 19 charter divisions founded in 1946. Today, the APA is comprised of 49 Divisions which represent the many specialized interests of psychologists.

Division of Military Psychology

In the Appendix is an application for the Division of Military Psychology. Before you read any further, I encourage you to complete the application and mail it out today so that you can start to benefit from membership even before your course on military psychology is finished.

As a student affiliate of Division 19, you will receive the quarterly journal, *Military Psychology* and the biannual newsletter, *The Military Psychologist. Military Psychology* is a peer reviewed journal which publishes behavioral science research articles having military applications in the areas of clinical and health psychology, training and human factors, manpower and personnel, social and organization systems, and testing and measurement.

The newsletter will inform you of what is going on within the division and other news of relevance to military psychologists. In particular, you will know about conferences and special events of interest to military psychologists and will also develop a feel for the profession and a familiarity of the other professionals in military psychology.

Students should also consider getting involved in conducting and presenting research. I urge you to seek out a mentor (perhaps your course instructor) who will help you to formulate and conduct research. There are numerous opportunities for students to present research at state, regional, national and international conferences. This is an excellent way to acquaint yourself with the field and goes a long way to improving one's chances for acceptance into graduate school, where you can pursue the master's degree or the Doctor of Philosophy (Ph.D.) in psychology or Doctor of Psychology (Psy.D.) degree.

Finally, in the Appendix is a list of web sites which will assist you in gathering more information on psychology. Some of these sites are specifically about military psychology and a few are general sites which will link you to a world of psychologically related information.

There are also several good books on various aspects of military psychology (e.g., Geldard, 1962; McGuire, 1990) and an excellent reference, the *Handbook of Military Psychology* (Gal & Mangelsdorff, 1991), has recently been published. The *Handbook of Military Psychology* is an encyclopedic source and the reader is referred to that publication several times throughout this text. Other useful sources of information regarding military psychology are the references at the end of each chapter in the text.

Summary

This introductory chapter has provided you with an overview of the field of military psychology and a frame of reference for the rest of the text. Military psychology is defined as the study and application of psychological principles and methods to the military environment. The early days of American military psychology date back to WW I. The president of the APA, Robert Yerkes, rallied the scientific discipline of psychology to assist in the war effort as the United States entered into World War I. Initially involved in testing, termed psychometrics, the work of these early military psychologists eventually found them in the hospital setting.

psychometrics

The brief interlude in the growth of military psychology which occurred during peacetime ended with the United States' entry into World War II. By the 1940s, psychologists had firmly established their role in selection and classification and were continuing to carve out their niche in the medical corp. The beginning of World War II saw psychology's contribution to ergonomics, an applied science which strives to increase job performance through an understanding of the characteristics of people and the jobs they perform.

ergonomics

Eight areas were identified as comprising the domain of military psychology: (1) selection, classification and assignment, (2) human factors, (3) environmental factors, (4) leadership, (5) individual and group behavior, (6) clinical and consultative/organizational psychology, (7) training and education, (8) manpower management decision making support.

The remaining chapters were previewed and suggestions for involvement in the field were presented. The remainder of the book will help to expand your knowledge of the field and hopefully spark an interest in and appreciation for the men and women who serve their nation as military psychologists.

References

Adler, H. E. & Rieber, R. W. (Eds.) (1995). Aspects of the history of psychology in America: 1892–1992. Washington, DC: American Psychological Association.

Bray, C. W. (1948). *Psychology and military efficiency*. Princeton, N.J.: Princeton University Press.

Chatelier, P. & Alluisi, E. (1991). Introduction to Section 2. In R. Gal & A. D. Mangelsdorff (Eds.). *Handbook of military psychology*. New York: John Wiley & Sons.

Driskell, J. E. & Olmstead, B. (1989). Psychology and the military. *American Psychologist*, 44(1), 43–54.

Dubois, P. H. (1970). *A history of psychological testing*. Boston: Allyn & Bacon.

Flanagan, J. C. (Ed.). (1952). *Psychology in the world emergency*. Pittsburgh, PA.: University of Pittsburgh Press.

Gal, R. & Mangelsdorff, A. D. (Eds.) (1991). *Handbook of military psychology*. New York: John Wiley & Sons.

Geldard, F. A. (Ed.) (1962). *Defence psychology: Proceedings of a symposium held in Paris, 1960*. New York: Pergamon Press.

Leahey, T. H. (1997). A history of psychology: Main currents in psychological thought. (4th ed.) Upper Saddle River, N.J.: Prentice-Hall.

Mangelsdorff, A. D. & Gal, R. (1991). Overview of military psychology. In R. Gal & A. D. Mangelsdorff (Eds.). *Handbook of military psychology*. New York: John Wiley & Sons.

McGuire, F. L. (1990). *Psychology aweigh! A history of clinical psychology in the United States Navy, 1900–1988*. Washington, DC: American Psychological Association.

Meier, N. C. (1943). *Military psychology*. New York: Harper.

Street, W. R. (1994). A chronology of noteworthy events in American psychology. Washington, DC: American Psychological Association.

Taylor, H. L. & Alluisi, E. A. (1994). Military Psychology, pp. 191–201. In V. S. Ramachandran (Ed.) Encyclopedia of human behavior. (vol. 3, J-P). New York: Academic Press.

Wiskoff, M. F. (1997). Defense of the Nation: Military Psychologists. In: R. J. Sternberg (Ed.) Career Paths in Psychology: Where your degree can take you. Washington, DC: American Psychological Association.

Yerkes, R. M. (1918). Psychology in relation to the war. *Psychological Review*, 25, 85–115.

Yerkes, R. M. (1919). Report of the Psychology Committee of the National Research Council. *Psychological Review*, 26, 83–149.

Endnotes

[1]For a thorough history of American psychology, see Adler and Rieber (1995) and Street (1994).

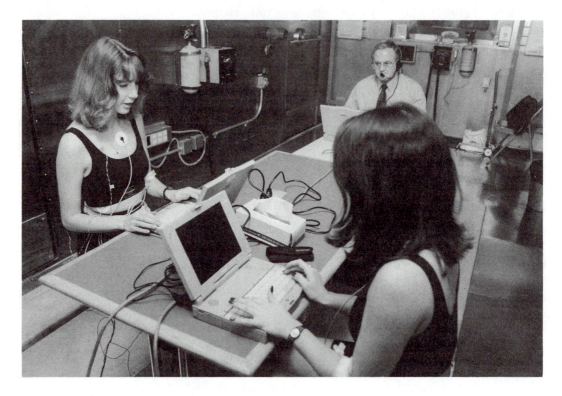

Photo courtesy of L. E. Banderet, Ph.D.

2

Psychological Research
in the Military Setting

Gerald P. Krueger, Ph.D.

* Introduction * Early History of Military Psychological Research * Post
WW II Man-Machine System Experiments * Military Labs and Psychological
Research: Then and Now * U.S. Army Labs * U.S. Navy Labs * U.S. Air
Force Labs * Military Research Funding * The Making of Military Research
Psychologists * Projection for the Future * Summary * References

Introduction

I have enjoyed an exciting, rewarding 30+ year career in military psychological research in a variety of research settings spread over 25 years as an active duty U.S. Army research psychologist, and more than five years as a government defense contractor. I therefore was delighted to be asked to prepare this chapter.

Psychological research in military settings had its modern beginnings in World War I, but more especially in the early days of World War II. An introduction to this broad topic requires considerable narrowing of what can be written in a short treatise. This chapter is intended to be a taste of history, a statement of the current status and opportunities in military psychological research, and a projection for the future.

Military psychology. Some behavioral scientists raise issue with whether or not *military psychology* is really a separate brand of psychology, or merely the practice of applying psychological principles to the lives of military personnel, their families, leaders, and government decision makers who affect military personnel in their unique occupations. Probably it is a little of both. The American Psychological Association's Division 19: Military Psychology, has an international membership of over 500 civilian and uniformed military psychologists. Those members who do research (probably about half the members) have diverse backgrounds reflecting a wide diversity of specialties ranging from industrial-organizational psychology, to tests and measurements specialties, and to experimental and applied engineering psychologists. Collectively, these psychologists perform research on problems of military populations, work in governmental laboratories, universities, schools of medicine, as defense contractors, or as members of consultant groups. Uniting them is an occupation that applies specialized research techniques and methodologies to solving human problems unique to military environments.

Psychological research in military settings includes work on: a) recruitment, selection, placement, training and retention of military personnel; b) prediction and enhancement of combatant performance in harsh environments; c) human engineering design of complex weapon systems for effective use by soldiers, sailors, airmen and marines; d) training procedures to mold well-honed military teams by maximizing specialized differential skills to permit success on tough missions; e) soldier coping mechanisms for deployment to foreign lands, or to carry out extended hours of work, or to work in austere conditions; f) troop abilities to adjust to countless intricacies of military life-style; g) collection and interpretation of large amounts of behavioral and psychological data to assist military leaders and civilian authorities in making smart decisions and informed policies that affect millions of military members and their families; and h) providing advice on integrating people of diverse ethnic and social backgrounds into the workplace (Krueger, 1996). For a treatise on careers in military psychology, see Wiskoff's Chapter 13 in this text.

Early history of military psychological research. Modern day history stakes the beginning of U.S. military psychology in 1917, when a committee of psychologists under direction of Robert M. Yerkes assisted in classification of 1.5 million recruits to make administrative decisions about World War I (WW I) duty assignments or admission to officer training. Their efforts resulted in the country's first group of intelligence tests which were extensively used in WW I. Thereafter, these tests were widely adapted as group intelligence tests in civilian applications (American Psychological Association, 1983). In conjunction with the military personnel selection and classification programs the research on psychological tests and measurements during WW I is usually identified as some of the first American *psychological research in military settings.*

In the early days of World War II (WW II) research psychologists expanded the U.S. Army and Navy's selection and training programs. Additionally, the interests which initially focused on personnel classification quickly broadened during 1942–43 to include proficiency testing, development of operator manuals, and alternative ways of training military personnel on complex systems of men and machines. For example, members of the Applied Psychology Panel of the National Defense Research Committee initiated studies of the control of naval gunfire to improve training of Navy rangefinder personnel throughout the war. In 1944, jointly with the Armored Medical Research Laboratory, they investigated the sources of errors in Army field artillery to develop several experimental models of new gunsight scales. In 1944, these psychologists also began to conduct research leading to improvements in design of gunsights for the flexible gunnery equipment of B-29 aircraft (Bray, 1948).

In WW II, the inclusion of radar capability on U.S. naval vessels forced integration of new technology into the arrangement of associated communications equipment, procedures, components and interfaces with the older combat elements of ships, thereby bringing about
human factors studies | elements of what came to be known as "human factors" studies (Parsons, 1972). An extensive WW II Applied Psychology Panel program to select and train radar operators and Morse code communicators spawned a series of shipboard radar-associated investigations by researchers at the Systems Research Laboratory of Harvard University, and at a dedicated ground-based laboratory at Beavertail Point, Jamestown, Rhode Island. Research focused on improving the

Navy's Combat Information Centers (CICs) in which man-machine complexes of radar and other information were viewed on various display scopes, evaluated by operator personnel, and distributed for weapons and battle direction (Parsons, 1972). These on shipboard studies were transferred to the Johns Hopkins University in 1948 and eventually were sponsored there by the Office of Naval Research (ONR) until 1958.

The membership of the American Psychological Association (APA) established functional divisions loosely formulated into subject matter groupings in 1946. *Military Psychology*, APA's Division 19, was one of the original nineteen divisions organized at that time and initially was made up of hundreds of psychologists who played important roles for the U.S. military during WW II, thereby solidifying roots of military psychology as a specialty field. It is estimated that during WW II, more than two thousand uniformed and civilian psychologists worked on the war effort (Parsons, 1972).

In the early 1940s and carrying through the war years into the 1950s, development of sophisticated technologies and complex weapon system operator stations (e.g. those for high performance aircraft, radar, sonar, submarines, shipboard command and control stations, etc.) necessitated more attention to human factors engineering design issues, especially concerning human information processing and cognitive task performance during high operational work- **cognitive** loads. Applied experimental psychologists (later known as engineering psychologists) **engineering** organized together as APA's Division 21, now called the Division of Applied Experimental **psychologist** and Engineering Psychologists. These psychologists attacked problems of determining the capabilities and limitations of man in operation of complex systems of men and machines— a concentration on "man-machine interface" with goals of making equipment easier to operate, to maximize performance and productivity, and to minimize errors and accidents— thus the budding young field of human engineering took shape.

Military psychological research laboratories, established adjacent to engineering design offices, made numerous design contributions that proved to be valuable in making equipment easier and safer to operate. These psychological research laboratories became permanent establishments within the U.S. Department of Defense and at several research universities and defense contractor facilities that participated heavily in support of the U.S. war effort in WW II.

One of the earliest of the several laboratories that took on a military psychological research mission was the US Army Quartermaster Climatic Research Laboratory formed at Lawrence, Massachusetts in April 1943 as a field installation charged to conduct research on environmental protection for soldiers. Throughout WW II, Psychologists in that laboratory worked on multidisciplinary research programs. Over ensuing years the lab was assumed into the Natick Army Laboratories, now a part of the Army Soldier Systems Command at Natick, Massachusetts, and the mission expanded to studying design of individual equipment, uniforms, clothing, helmets, protective goggles, protective clothing, parachutes, field feeding systems, etc.

During WW II, the U.S. Army Air Force had research psychologists working on human engineering design of high performance aircraft in several locations, but especially at Wright Field (now Wright-Patterson Air Force Base) near Dayton, Ohio. In 1947, a sizable portion of these research psychologists formalized their laboratory work for the newly established U.S. Air Force. An enormous, growing aerospace industry employed numerous human factors psychologists in aerospace developmental research efforts over the next 50 years.

Beginning in 1947, the Office of Naval Research employed a large contingent of researchers at the Johns Hopkins University in Baltimore, Maryland. Many of those psychologists conducted studies for shipboard command and control (e.g. naval combat information centers), and formulated many of the early human engineering design principles which now appear in our handbooks (Chapanis, Garner & Morgan, 1949).

Post WW II man-machine system experiments. Parsons (1972) described 43 different research programs and over 160 *man-machine system experiments* conducted from 1948 through the 1960s. Almost all of these studies were multidisciplinary in nature, and most were directed at military applications. All of them benefitted from considerable involvement by research psychologists. The studies described by Parsons were either conducted in military research centers at government facilities, or in university and private industrial research foundations working on government contract or on a federal funding grant basis. The many behavioral scientists employed undoubtedly would have described their work as applied military psychology.

For Parsons, *man-machine system experiments* were large scale experiments in which human subjects interacted with machines, and each other, in complex system settings based to a considerable extent on simulation. Many elaborate laboratory facilities were created for that purpose. In some cases knowledge was sought concerning the manned operation of particular systems or system components; in others the aim was to discover how human beings function in such environments (Parsons, 1972). Examples of some of these programs and experiments include such diverse studies as those of: a) radar-aided military terminal air traffic control, at Ohio State University; b) land-based air defense and interception, at Columbia University, and at the Massachusetts Institute of Technology Lincoln Labs; c) shipboard Combat Information Centers for fleet air defense, surveillance and interception, at the Naval Research Laboratory Chesapeake Bay Annex; d) battlefield surveillance, at the University of Michigan; e) Army field operations, tactics, organization, devices and equipment, at the Combat Development Experimentation Center at Fort Ord, California; and f) hypothetical command center for nuclear war, at the System Development Corporation. For a complete treatise, see Parsons (1972).

There were other psychological research programs under military sponsorship after World War II as well. One of the more prominent and impactful of these was that of the Human Resources Research Office (HumRRO) of the George Washington University, established in 1951 to conduct studies for the U.S. Army on training methods. For over twenty years this highly successful federal captured contract research program accomplished much of the Army's training research in field offices throughout the country. The mission was eventually brought in-house to a government lab, as it was turned over to the Army Research Institute (ARI) established in 1972. From 1972 to 1975 the HumRRO program was phased out as an Army only program (Uhlaner, 1977). The HumRRO still functions today in Alexandria, Virginia, as a principal federal government contract agency but with a more diversified, smaller customer base.

Military Labs and Psychological Research: Then and Now

Since World War II, each of the military services has sponsored numerous multidisciplinary research programs and isolated projects within government owned and operated laboratories, or via extramural grant and research contract funding at universities and corporate research centers. The amount of clearly identifiable military psychology invoked in these programs and studies varied widely, because so many of them were multidisciplinary programs in which research psychologists played key roles as members of research teams. Often it was the military psychologists whose expertise in experimental design and statistics determined the direction taken by many research programs. Since the goals of large military research projects often revolved around predictions of equipment operator performance, it was the military psychologists who were best equipped to help research teams account for the trickiest of human performance issues: a) people are different (individual differences), b) people are motivated differently (motivation), and c) people improve through practice (learning and skill development). Careful attention to experimental design issues for performance studies with people became the forte of applied psychologists.

With continually changing national priorities, adjustments in military service research budgets, frequent organizational restructuring of missions, consolidations, downsizing, and resultant laboratory name changes, it would be difficult to cover the complete history in this chapter. Many name changes and effective dates of organizational restructuring prior to 1975 are outlined in Uhlaner (1977). Presented here is a synopsis of today's U.S. military in-house research laboratories that conduct or sponsor psychological research in military settings in 1997.

U.S. Army Labs

The U.S. Army Human Research Engineering Directorate (HRED) was founded as the better known Human Engineering Laboratory (HEL) at the Aberdeen Proving Ground, Maryland in 1951. Now as HRED, it is a part of the Army Research Laboratory. From the beginning, the HEL devoted about one-third of its efforts to basic (6.1) psychological research on human sensory input and handling of information; about one-third of laboratory efforts were field experiments measuring performance of soldiers operating developmental hardware systems; and about one-third of the work was devoted to ensuring appropriate application of human engineering design guidance for development and testing of thousands of Army materiel systems. Positioning of human engineering specialists in numerous field offices at most Army development and test agencies, and at Army training and combat doctrine development centers, ensured most major materiel development programs benefited from the HEL's work. HEL's behavioral scientists and human factors specialists worked on a diversity of Army systems ranging from designing the interface of field telephones and computerized battlefield communication switching systems, to determining display symbology for missile site control stations, to designing pilot controls for advanced development helicopters, and determining ammunition loading procedures for armored tanks and self-propelled artillery systems (USAHEL's Milestones, 1990).

Division of Neuropsychiatry at the Walter Reed Army Institute of Research (WRAIR). The WRAIR formalized a medical psychology research unit in Washington, DC in

1951. This unit eventually took on the title: Division of Neuropsychiatry (Div. NP) and traditionally employed about 20% of the WRAIR staff. Historically, about one-third of a sizable Div. NP program dedicated studies of basic (6.1) neuroscience (including a mix of animal and human experiments) to elucidating mechanisms of human stress reaction and responses to a wide array of stresses encountered by military personnel. About one-third of the WRAIR psychology studies (6.2) were devoted to psychophysiological assessments of soldier performance in sustained duration missions that necessitate significant loss of sleep and rest. Seminal work on this and related research topics established the WRAIR as a center of sustained performance and sleep loss expertise in the 1950s and 1960s (e.g. Williams, Lubin, & Goodnow, 1959; Krueger, 1986; 1991); and in the 1980s and 90s its staff directed significant research to explorations of stimulant drugs and sleep aids for military use in the field (Newhouse et al., 1992). The remaining one-third of Div. NP's work, employing social experimental psychologists and social anthropologists, historically examined more holistic stresses facing soldiers deployed to various regions of the world, and pioneered research in stress syndromes associated with the harshness of military occupations and lifestyle.

The U.S. Army Research Institute of Environmental Medicine (USARIEM) was opened by the Army Surgeon General at Natick, Massachusetts in 1961. Collocated with the Natick Army Laboratories, the USARIEM examines soldier performance as it is affected by the influences of harsh environmental stressors of high heat, extreme cold and high terrestrial altitude. Over time, the USARIEM also has become the world's leading military research laboratory on physical fitness, biomechanics, biophysics of protective clothing, effects of military diet and nutrition, and general, preventive occupational and environmental medicine. Research psychologists have been an integral part of those research programs every step of the way (Krueger, Cardinal & Stephens, 1992).

The U.S. Army Aeromedical Research Laboratory (USAARL), opened by the Army Medical Department at Fort Rucker, Alabama in 1963 with a predominant mission to accomplish aviation medicine research for the growing Army air force, which during and after the Vietnam conflict included thousands of helicopter aviators. In the mid-1970s the USAARL established an Aviation Psychology Division to study pilot workload and per-formance issues. Research topics included inflight and simulator assessments of helicopter pilot performance under various stressful flight conditions, including: extended flight operations involving aviator fatigue; extensive flight with night vision goggles; flight while wearing chemical protective clothing; with various drugs such as chemical warfare agent prophylaxis compounds, antihistamines, and stimulants. The USAARL also broadened its behavioral science studies to involve assessments of health hazards associated with operation of large Army weapons and material systems (Krueger, 1983). These later studies included multidisciplinary studies of the mechanical threats to the body by impact, shock, acceleration and vibration, acoustical noise, and visual enhancement via night vision systems (USAARL's annotated bibliographies).

The U.S. Army Research Institute (ARI) for the Behavioral and Social Sciences was activated at Alexandria, Virginia by the Army Deputy Chief of Staff for Personnel in 1972 as a consolidation and replacement for three other Army manpower, motivation, training and behavior systems research centers. ARI took on research missions of personnel selection, classification, assignment placement, retention, etc. and extensive studies of leadership, all directed to promote maximum levels of military and civilian personnel performance throughout the Army, to support effective operations of current and future combat and tactical systems, and to maintain the proficiency of groups of individuals working as teams. As the Army's one-stop psychological research shop, ARI's research programs were responsive to requirements as far reaching as providing behavioral science support to the Department of the Army Personnel Staff, to the Army Reserve Commands, and to Department of the Army civilian personnel management offices. Over the years ARI programs provided important data, information and recommendations for decision making on Army-wide personnel policies that affected all soldiers and often their dependent families as well.

ARI was staffed with the country's largest single concentration of military research psychologists (numbering over 400 personnel in the 1980s), and became the developing agency for the social sciences in the U.S. Army. From time to time, ARI staffed 15–20 field offices at various Army training centers around the country so as to produce research directly applicable to specific training needs at those centers. About 1990, much of ARI's research support for developmental materiel systems was transferred to the HRED at Aberdeen Proving Ground.

The U.S. Army Medical Research Institute of Chemical Defense (USAMRICD) was established by the Army Medical Research and Development Command (circa 1979) at the Edgewood Area of Aberdeen Proving Ground, Maryland as the Army transitioned from development of offensive chemical weapons to devoting significant amounts of research toward medical defense against chemical battlefield threats. As a key part of multi-faceted medical research programs, these research psychologists conduct numerous programs, experiments using either animal or human models, to develop pretreatment, prophylaxis, and treatment drugs, as well as medical antidotes for those who might be exposed to battlefield chemical threats. The research psychologists who work at USAMRICD tend to be physiological or pharmacological psychologists, neuroscientists, and comparative behavioral biologists.

U.S. Navy Labs

Naval Health Research Center (NHRC). In 1974 the Navy Medical Neuropsychiatric Research Unit (NPRU) merged with other Navy assets to become the Naval Health Research Center (NHRC) in San Diego, California. The NHRC currently fulfills the U.S. Navy's mission to support fleet operational readiness through research, development, test and evaluation of the biomedical and psychological aspects of Navy and Marine Corps personnel health and performance. Numerous uniformed and Navy civilian behavioral scientists and military psychologists conduct research in settings on-board ships and submarines, with the Navy and Marines in training and in the field, and during laboratory studies of performance in operation of naval equipment and systems, or while subjected to various stressors unique to naval and marine operations. Research on coping mechanisms, sleep management strategies, performance while wearing chemical protective clothing, or during arduous special forces (i.e. Navy Seals) training, and many other military psychology topics have been, and continue to be researched at NHRC (NHRC, 1996).

Naval Aerospace Medical Research Lab (NAMRL). Although NAMRL has undergone some mission changes and mergers over the years, naval aviator performance and aviation medicine have been studied by Navy psychologists at NAMRL in Pensacola, Florida since the 1950s. The topics of research for Navy military psychologists at NAMRL have included: longitudinal studies of health and fitness of Naval aviators; aerospace motion sickness; visual performance under high G force stresses; or during use of vision enhancement systems; sustained aviation operations and crew work-rest scheduling; as well as other issues of biomedical safety and health.

Naval Medical Research Institute (NAMRI). Most of the U.S. Navy's basic (6.1) research on neurosciences and psychological performance traditionally has been conducted at NAMRI in Bethesda, Maryland. Topics studied at NAMRI are as diverse as personnel performance in hypobaric conditions (high aerospace altitudes), and in hyperbaric (under-water diving) atmospheric mixes, or during exposure to extreme cold such as might be experienced in aircraft emergency bail-out into the North Atlantic ocean.

U.S. Naval Submarine Medical Research Laboratory (NSMRL) at Groton, Connecticut, employs several research psychologists to conduct performance experiments with submarine personnel in conjunction with the overall submarine medicine research mission of the Laboratory. Studies range from assessing performance in varying submarine atmospheric breathing mixes to examining confinement and personnel readiness factors for unique undersea missions.

Navy Command, Control & Ocean Surveillance Center (NCCOSC). U.S. Naval materiel systems development missions are presently undergoing some restructuring and reorganization, but some military psychologists are positioned at NCCOSC in San Diego, California to work on human engineering of new weapons and large materiel systems for the shipborne Navy. New technologies, such as embedded virtual reality systems, digital pre-processing and display of command and control information, and resultant effects on rapid decision making are studied at NCCOSC's Research Development, Test and Evaluation Center (NRaD). Aviation systems and air warfare developmental work accomplished for the Naval Air Systems Command recently relocated to Patuxent River, Maryland.

Naval Personnel Research and Development Center (NPRDC). The Navy's personnel research center, concentrating on personnel selection, training, job placement, retention etc. remains at San Diego, but it too is undergoing some change and flux at this writing. For the past twenty years NPRDC has been involved in many assessments and re-configurations of Navy personnel systems.

U.S. Air Force Labs

U.S. Air Force Armstrong Laboratory (USAF-AL). In a merger to enact a *superlab* structure (circa 1990–91) the U.S. Air Force merged several of its research laboratories under the umbrella organization, the Armstrong Laboratory at Brooks Air Force Base, San Antonio, Texas. Under this superlab arrangement, military research psychologists work not only at Brooks AFB, but also at the Human Resources Research Directorate at Williams Air Force Base, near Meza, Arizona; and at the Crew Systems and Human Engineering Directorate at Wright-Patterson Air Force Base, Dayton, Ohio. Research psychologists and human engineering specialists work on numerous Air Force materiel development programs, and conduct both in-house and field oriented research on aviator and ground airmen performance under conditions uniquely within the Air Force mission. Multidisciplinary studies are very much in evidence here too, and behavioral scientists are key members of research teams studying a variety of human response variables involving human systems interface, biomedical safety and health, and manpower, personnel and training system applications.

Military Research Funding

Funding sources. Psychological research in military settings throughout the world is normally funded by centralized governmental agencies such as a defense ministry, or in a few cases, by a foreign intelligence agency. In the United States, the Department of Defense and its individual service Departments of Army, Navy, Air Force, and U.S. Marines fund scientific research via a Department of Defense research program, entitled Research, Development, Test and Evaluation (RDT&E) Program No. 6. RDT&E under Program 6 is predominately configured along a materiel life cycle management continuum ranging from basic research, through applications, engineering design and development testing, to procurement and fielding of materiel systems. That linear continuum of a defense product development management model occasionally muddies issucs of *where and when* behavioral science research is to be managed, or applied to system design within the materiel life cycle management model. Behavioral science almost always struggles for acknowledgment of its cost effectiveness and recognition of its true worth to system design since it is easily (mis)perceived as tangential to materiel development priorities.

> Research, Development, Test and Evaluation

Annually, Programs 6.1: Basic Research, and 6.2: Applied Research, fund most of the behavioral and psychological research in military settings. These two programs together historically provided 65–75% of the nation's funding for military behavioral sciences research. The funds are usually specifically directed to support work accomplished by in-house governmental research laboratories, or to extramural contracts and grants to universities, private research foundations, defense contracting companies, or to individual consultant firms. The remaining 25–35% of behavioral science funding is generally for applications work in Program 6.3: Advanced Materiel Development, and to Program 6.4: Engineering Development programs, where efforts often are directed to manufacture and procurement of very specific materiel product lines. Thus 6.3 and 6.4 work typically involves making human engineering adjustments, frequently late in the design cycle or in the form of product fix-it retrofits of new weapon systems containing complex operator stations built with sophisticated new technologies (e.g. high performance aircraft, naval ships, global telecommunications, tele-robotic systems, artificial intelligence, embedded training technologies etc.).

> 6.1: Basic Research
> 6.2: Applied Research

> Program 6.3: Advanced Materiel Development

> Program 6.4: Engineering Development

Each of the U.S. military service departments centralizes behavioral science funding in management offices established to administer contracts and grants for 6.1 basic science research projects extramurally. The Office of Naval Research (ONR) in Arlington, Virginia administers behavioral science programs for the Navy and the Marines. The U.S. Air Force Office of Scientific Research (AFOSR) at Bolling Air Force Base, Washington, DC manages the Air Force 6.1 program for both the Air Force labs (i.e. Armstrong Lab) and for the university and research contracts and grants recipients. Administration of the Army's 6.1 human resources research program is split four ways: a) research on personnel selection, training, placement, and retention is administered by the Army Research Institute (ARI) for Social and Behavioral Research in Alexandria, Virginia; b) research in human engineering and materiel applications is managed by the Army Materiel Command's Human Research and Engineering Directorate at Aberdeen Proving Ground, Maryland; c) predominately medical programs are managed by the U.S. Army Medical Research and Materiel Command (USA-MRMC) at Fort Detrick, Frederick, Maryland; and d) a small portion of basic behavioral research (university grants) is administered by the Army Research Office (ARO) at Durham, North Carolina.

Program 6.1 basic psychological research. Basic psychological research programs typically search for *basic mechanisms* to explain how physiological or psychological systems work—particularly on issues unique to military settings. Most of the animal model research accomplished (in-house or extramurally) for the military is managed under these programs. Examples of such animal model work might entail determining how cognitive and memory processes are affected by administration of a psychoactive drug being investigated as a prophylaxis for chemical warfare nerve agent exposure (at USAMRICD); or determining how exposure to high intensity, low frequency acoustical noise affects temporary hearing threshold shifts and therefore response to caution warning signals (at USAARL). Examples of 6.1 research with human test participants might include a positron emission tomography (PET) examination of brain glucose utilization in sleep deprived subjects to explain the fatigue effects of sustained operations on operator performance (at WRAIR); or examination of effects of restricted fluid intake (hypohydration) and exposure to cold weather on cognitive performance of military tasks (at USARIEM).

University 6.1 research support of military labs. For about thirty years after World War II the U.S. Congress provided substantial 6.1 basic research funding with only loose control strings or requirements attached. It was almost an unstated national policy that the Congress would allocate a fixed percentage (roughly 9 to 12 % per year) of military research funds to 6.1 research, intending for the military departments to pass along many of the funds to research universities throughout the country. Sometimes Congress actually specified what percentage of the 6.1 budget was to be allocated to university-based research programs, and often gave specific priority guidance for allocation to specific programs or specific universities. It was an acknowledged governmental mechanism to sponsor university research which supported graduate student development programs to produce our nation's young scientists and to train many future government service oriented specialists to fulfill national research needs in a diversity of specialty areas.

Since WW II, many of the nation's research psychologists benefited at one time or another from U.S. Defense Department funding, either by having received financial support during their graduate school training, or by obtaining financial support for research programs in their laboratories. The most reputable and productive university researchers annually could count on obtaining research grants or contracts from the military sponsoring agencies to pursue research on topics largely of their own choosing, provided the work was generally within the stated overall mission needs of the sponsoring agency. Whenever defense spending tightened, Congress and the military sponsoring agencies tightened the ground rules, and restricted grants and contracts to those university or private foundation programs that demonstrated clear military application and best payoff for the proposed basic research. As overall research funding declined in the late 1980s through the late 1990s, so has traditional training support for basic scientists in topical areas of interest to the military declined as well. With that, fewer behavioral scientists are now being trained with federal government financial assistance.

University research for the military labs. Scientific staffs have been *downsized* in numbers during the entire decade of the 1990s; and therefore the dependence of military lab science programs upon university research has increased. Thus, the long standing relationship between the government's military lab programs and their augmentation by funding university-based research programs to meet specific research requirements has become even closer. The concept is: the military lab staffs establish close communication and coordination with line military leaders, especially the world regional Commanders-in-Chiefs (CINCs), and with the commanders of training and personnel staffing centers in each service, so as to gain a clear understanding of what their research needs are—usually envisioning needs with several years lead time to permit research to be done to meet them. In-house military lab programs are more actively conforming their infrastructures and their business procedures to more clearly obtain answers to applied questions from field CINCs.

In situations where military labs do not have the resources to do all the work in-house, they are to contract for specific research capabilities as an augmentation of their own staffing, essentially hiring extended research hands at universities and private research centers to obtain answers to those field oriented questions. In effect, the in-house lab scientists are to blend the extramural research into their overall programs; and thus in-house government researchers become the conduit between University scientists and the military commanders in the field to provide research-based solutions to practical problems of modern military forces.

Of course the government labs continue to be involved in their own in-house programs to establish advanced development technologies and to push the state of the art of military psychology. These programs may have less obvious immediate application to today's practical problems, but offer potential payoff for future military forces. Extramural research contracts and grants, funded from either 6.1 or 6.2, and less often from 6.3 funds, are administered for these programs as well.

In-house 6.2 exploratory research funding. Prioritization and allocation of 6.2 funding has generally been within the control of the military laboratory commands to sponsor core in-house research programs. In-house psychological research sponsorship is often a combination of 6.1 and 6.2 funding. With these core funds, the laboratories pay the salaries of civil service employees, research staffs of scientists and technicians, the administrative and overhead costs, and the costs associated with accomplishment of research programs, i.e. purchase of instrumentation and equipment, data collection and analysis costs, and other associated research costs. Salaries of uniformed scientists and technicians assigned to the labs usually come from centralized service personnel budgets. Thus, having a higher percentage of uniformed laboratory scientists enhances the in-house research personnel talent base to broaden the research capability, but also, because uniformed personnel are funded centrally, helps the labs meet budget requirements as well.

The Making of Military Research Psychologists

Researcher background. What does it take to make a good research military psychologist? Most military psychologists (uniformed, government civil servant, or contractors and university grant recipients) in the United States have excellent training, usually to the Ph.D. level at reputable universities known for their research programs. Some of the personal features that accompany successful military research psychologists include: a) a good foundation in experimental design and statistics; b) plenty of practical experience; c) being a good generalist and able to work as part of a multidisciplinary team of scientists; and d) capable of studying many variables simultaneously, especially in field research where some variables are difficult to control, or reliably measure (Krueger, 1983).

The government laboratory specialist is not much different than a university professor, and has to struggle for funding priorities, access to instrumentation and equipment too. However, once the budget request is won, the government researcher frequently has easier access to much more elaborate and expensive equipment; and probably opportunities for continuation of funding year to year are easier than they are for a non-government researcher looking toward the government labs for next year's funding.

Research opportunities. Studies in military psychology range from research on basic mechanisms, to carefully controlled laboratory experiments using military participants on many of the topics in the specific laboratory programs listed earlier in this chapter. They also include conducting grandiose field studies of man-machine system experiments of the type described by Parsons (1972). In the realms of military psychological research there is ample room for many divergent research orientations. A large variety of research opportunities exist in the military community. Those who wish to work on basic mechanisms often do research on animal models. Those who wish to rigorously design laboratory experiments with the latest state-of-the art equipment and instrumentation, and those who like the challenges of field research will find those opportunities in U.S. military labs.

Answering tricky questions regarding predictions of performance of military personnel who will use the next generation of high-tech equipment and hardware, or who will fight battles or wars with new doctrine, present exciting challenges to the heartiest of researchers. Conducting large experiments in military field environments presents some of the greatest challenges. Often the military psychologist must piggy-back his/her work with that of other researchers and field training staffs such as proponents of new tactical doctrine and/or training technologies. As an example, Army behavioral scientists in the 1980s devoted considerable field test time to the study of troops wearing chemical protective clothing in the field where bulkiness of the clothing interferes with tactual and sensorimotor response, the facial protective gas mask impedes head movements or restricts vision, and the impermeability of the uniforms make them a micro-encapsulation for those who wear them thereby threatening heat stress in warm environments (Taylor & Orlansky, 1993; Krueger & Banderet, 1997). Large field studies with groups of participants numbering in the hundreds (simultaneously) frequently encounter numerous *glitches* attributed to data collection failures and human response expectations. The soundness of experimental design is often taxed in military field studies, where many uncontrollable variables (e.g. sizes of groups, the weather, maintenance status of operational equipment, missing, data etc.) sometimes relegate weeks of unrepeatable *indelicate experiments* hard work to the category of indelicate experiments described by Sinaiko (1965). Indeed, the challenges of military field experiments are great, and of course for those who pursue them, rewarding as well.

Social psychological research studies, and survey research with very large *Ns* are also carried out within the military context (e.g. studies of leadership, group cohesion, group behavior, etc.). A recent example of this work is the ARI sponsored extensive psychological studies on: Reserve component soldiers as peacekeepers (Phelps & Farr, 1996).

Military test participants. Employing military personnel as subject-participants in research studies often facilitates collecting data from the desired numbers of highly selected subjects, ranging in some cases from a handful of experienced test pilots, to a few hundred participants in some tests, or to collecting survey data on thousands of uniformed personnel. Sometimes, because the numbers of active duty personnel are decreasing, recruitment of suitable participants for military studies can be exceedingly difficult. Soldiers all have jobs to do; and their commanders are reluctant to permit them to take time away from their regular jobs to participate in behavioral science tests unless they are offered a chance at obtaining some relevant training value in return for their participation. The more relevant the research is to practical operations, the more likely military participants will be available in sufficient numbers to conduct a particular research project.

Sometimes the specialty skills of military forces who serve as test participants make for rather unique population samples with which to conduct research. Both the caliber of subjects, and their motivational levels to perform well, can be very high, e.g. test pilots, combat veterans, highly trained specialists in any of a hundred specializations, highly motivated troops like the special forces, the marines, or airborne parachute units. The research subjects tend to be healthy young men and lately, more women, all who usually strive to perform at their best. Opportunities to conduct performance research with a military population constitutes a *definite plus* in an applied psychology career.

Connections to overseas colleagues. One of the things that distinguishes psychological research in military settings is the growing international network of psychologists involved in military research around the globe. Some countries, like the U.S., deploy uniformed military personnel to work in other countries for extended periods of time, or provide 2–3 year assignments of civil service employees as scientific liaisons to labs of our allies. Military psychologists internationally, commonly link together to collaborate on joint projects, or to share resources, instrumentation, and data with one another for their work. Countries with formal alliances agreeing to cooperate and help one another often sponsor periodic professional meetings and conferences at which researchers present technical papers on their work and intermingle with their counterparts from other countries. NATO, for example, sponsors a variety of cooperative research panels, boards, international meetings and exchange forums. U.S. military psychologists participate in these to the fullest. Most participants talk highly of the usefulness of such international connections.

Publishing military psychology. Psychological research accomplished in military settings is generally published in either open literature referred journals and book chapters, or in the form of governmental laboratory technical reports, or both. Many labs produce a laboratory technical report enroute to, or in lieu of having the research published in the open literature. The format of a lab report is controlled by the individual laboratory and therefore permits any length document desired, allowing for detailed descriptions of procedures and analyzed data. Many open literature publications are re-printed with a laboratory technical report cover on them as well.

Most laboratory technical reports are submitted to the Defense Technical Information Center (DTIC) at Fort Belvoir, Virginia for permanent storage in a state of the art document repository from which defense contractors can easily retrieve them. DTIC has hundreds of thousands of government laboratory documents on file dating back to about 1950. In turn, DTIC renders many of the documents publicly accessible by making them available through the Department of Commerce's National Technical Information Service (NTIS). Some military laboratories periodically produce annotated bibliographic listings of their published reports; these bibliographies can usually be obtained by writing the scientific information section of the laboratories and asking for them.

Occasionally some military laboratory research may be classified for military security reasons, but not much of the psychological research fits into the classified category. Medical research labs, true to the Hippocratic oath, and preferring to stay within the *helping mankind* mode, conduct or sponsor medical-psychophysiological research which is normally published in the open scientific literature.

Much of the U.S. military lab psychological research is published in refereed journals, especially those produced by the American Psychological Association. Reports of research may be published in a variety of specialty journals appropriate to their content. Some journals which publish military psychological research include: *Military Psychology; Journal of Applied Psychology; Journal of Experimental Psychology; Human Factors; Ergonomics; Aviation, Space and Environmental Medicine; Aviation Psychology; Military Medicine;* and *Work and Stress.*

Projection for the Future

After the fall of the Berlin Wall in 1989, and the demise of the Soviet Union as we have known it for over forty years, the military services of the so-called western allies have all been significantly downsized, restructured, and re-prioritized for centralized governmental funding. The state of flux and change these reorganizations brought about is seemingly about to stabilize for a time. As organizational and mission responsibility are now more stable, military laboratories are refining research objectives and programs. Military research lab trimming and consolidating has attempted to preserve resources needed for the year 2000 and beyond.

In the future we will likely witness more cross-service collaborative studies among the Air Force and Army, or Marines and Army, etc., or even additional tri-service sponsorship of research programs. Our allies all experienced similar cutbacks in resources, and more multinational experiments and joint research programs are likely to be the wave of the future. There are already many international research programs, but it is easy to foresee these will become more common during the next ten years. With the easy omni-access to electronic telecommunications, the age of international electronic communications via internet will make international collaborations much more likely. Joint multi-national studies are often more difficult to coordinate and carry out, but the rewards of getting to know one's allies help to maintain working relationships which often pay off when joint military maneuvers are called for to meet operational commitments. Military psychologists will continue to be in the forefront of joint military service and multinational research programs.

Summary

This chapter presents a short history of psychological research in U.S. military settings, describes where and how this research is conducted, and highlights opportunities for participation in those research programs. For additional coverage of this topic the Handbook of Military Psychology (Gal & Mangelsdorff, 1991) provides ample research results from many military research programs only mentioned here. Wiskoff (1997; this text, Chapter 13) describes more detail regarding career opportunities for military research psychologists.

References

American Psychological Association (1983). *Military Psychology: An Overview*. Washington, DC: American Psychological Association, Division 19.

Bray, C. W. (1948). *Psychology and military proficiency: A history of the Applied Psychology Panel of the National Defense Research Committee*. Princeton, NJ: Princeton University Press.

Chapanis, A., Garner, W. R., & Morgan, C. T. (1949). *Applied experimental psychology: Human factors in engineering design*. New York: Wiley.

Gal, R., & Mangelsdorff, A. D. (Eds.) (1991). *Handbook of military psychology*. New York: Wiley & Sons.

Krueger, G. P. (1983). *The role of the behavioral scientists in assessing the health hazards of developmental weapon systems*. In: A.W. Schopper & U.V. Nowak (Eds.) Proceedings of the Army Medical Department Behavioral Sciences R&D Conference, Fort Rucker, AL: U.S. Army Aeromedical Research Laboratory.

Krueger, G. P. (1986). Publications of the Department of Human Behavioral Biology— 1958–1986, Walter Reed Army Institute of Research. (WRAIR-BB Tech. Rept. No. 86–1). Washington, DC: Walter Reed Army Institute of Research.

Krueger, G. P. (1991). Sustained military performance in continuous operations: Combatant fatigue, rest and sleep needs (Chapter 14, pp. 255–277). In: R. Gal & D. Mangelsdorff (Eds.) *Handbook of military psychology*. New York: Wiley & Sons.

Krueger, G. P. (1996). Division Focus: Division 19 Military Psychology. In *Psychological Science Agenda*. Washington, DC: American Psychological Association (May/June 1996; page 7).

Krueger, G. P., Cardinal, D. T., & Stephens, M. E. (1992). *Publications and technical reports of the United States Army Research Institute of Environmental Medicine*. Natick, MA: U.S. Army Research Institute of Environmental Medicine.

Krueger, G. P., & Banderet, L. E. (1997). The effects of chemical protective clothing on military performance: A review of the issues. *Military Psychology*, 9, (4), 255–286.

Naval Health Research Center (1996). *Bibliography of scientific publications, 1962–1996*. San Diego, CA: Naval Health Research Center.

Newhouse, P. A., Penetar, D. M., Fertig, J. B., Thorne, D. R., Sing, H. C., Thomas, M. L., Cochran, J. C., & Belenky, G. L. (1992). Stimulant drug effects on performance and behavior after prolonged sleep deprivation: A comparison of amphetamine, nicotine, and deprenyl. *Military Psychology*, 4, 207–233.

Parsons, H. M. (1972). *Man-machine system experiments*. Baltimore, MD: The Johns Hopkins Press.

Phelps, R. H. & Farr, B. J. (1996). *Reserve component soldiers as peacekeepers*. Washington, DC: U.S. Army Research Institute for the Behavioral and Social Sciences.

Sinaiko, J. W. (1965). The indelicate experiment. In J. Spiegel, & D. E. Walker (Eds.). *Second Congress on the Information System Sciences*. Washington, DC: Spartan Books.

Taylor, H. L., & Orlansky, J. (1993). The effects of wearing protective chemical warfare combat clothing on human performance. *Aviation, Space, and Environmental Medicine*, 64 (3; Section II), A1-A-41.

Uhlaner, J. E. (1977). *The research psychologist in the Army—1917 to 1977*. (ARI Res. Rept. No 1155). Alexandria, VA: U.S. Army Research Institute for the Behavioral and Social Sciences.

United States Army Aeromedical Research Laboratory (1991). *Annotated bibliography of USAARL technical and letter reports Volume 1: June 1963—September 1987*. Fort Rucker, AL: U.S. Army Aeromedical Research Laboratory.

United States Army Aeromedical Research Laboratory (1996). *Annotated bibliography of USAARL technical and letter reports Volume 2: October 1987–September 1995*. Fort Rucker, AL: U.S. Army Aeromedical Research Laboratory.

United States Army Human Engineering Laboratory (1990). MILESTONES: An annotated bibliography of publications of the U.S. Army Human Engineering Laboratory. Aberdeen Proving Ground, MD; USAHEL.

Williams, H. L., Lubin, A., & Goodnow, J. J. (1959). Impaired performance with acute sleep loss. *Psychological Monographs* No. 484, 73, (14) 1–26.

Wiskoff, M. F. (1997). Defense of the Nation: Military Psychologists. In: R. J. Sternberg (Ed.) *Career paths in psychology: Where your degree can take you* (pp. 245–268). Washington, DC: American Psychological Association.

Wiskoff, M. F. (1998). Careers in Military Psychology. In: C. Cronin (Ed.) (1998) *Military Psychology: An Introduction* (pp. 257–267). Needham, MA: Simon & Schuster.

Defense Department Photo

3

Personnel Selection and Classification in the Military

Brian K. Waters, Ph.D.

* Introduction * Personnel Selection Concepts * Civilian Selection and Classification * Setting Cut Scores * Selection System Validation in the Civilian World * Military and Civilian Selection: A Comparison * Military Personnel Selection * Enlisted Selection Process * The Officer Selection Process * Military Selection Summary * Military Personnel Classification * Enlisted Classification * Officer Classification * Military Classification * Summary * Conclusion * References

Introduction

A primary function of military psychologists is the research and development into methods for recruiting, selecting, classifying, training, and assigning hundreds of thousands of new service members each year. This chapter focuses on two of those functions: selecting new service members from applicants, and classifying those selected into one of well over 1,000 jobs within the military.

Minimum selection standards for United States enlisted service have evolved since World War I. The basic objectives of 'quality' standards, summarized in a DoD report to Congress in 1981, state:

> *Proper enlistment screening and job placement are prerequisites for efficiencies in training, retention of skilled personnel, and mission performance. Any deficiencies in the selection and classification system lead to increased training times and cost, dissatisfied personnel with concomitant decreases in morale, productivity, and retention, and critical shortages of skills caused by failure to achieve optimal assignment of available manpower into the various occupations.*[1]

selection

Personnel Selection Concepts[2]

Efficiently choosing and assigning employees from a pool of available applicants is the goal of every organization's personnel selection system. Conceptually, military selection and

pool

classification are not unlike their equivalent functions in the civilian employment world. The specific details are different, however, since the general populations of applicants, the context of the employment situation, and the goals and missions of the military and civilian organizations are quite different.

This chapter begins its analyses of the selection processes within the military by first examining the topics in a more general setting, i.e., outside the military. It discusses the conceptual basis for selection decisions with regard to setting standards, establishing cut scores on selection tests, and validating the selection procedures.

In the world of civilian employment selection, rarely, if ever, are all applicants for a job accepted. Instead, there is an attempt to develop a systematic method for choosing people for employment. Just as business and industry seek the 'best qualified' employees available (however that may be defined), so too the Military Services seek to enlist the most capable recruits. The military's task is complicated by a predominantly young and inexperienced applicant pool. Furthermore, the military must select recruits for not one, but many diverse jobs simultaneously. To help the reader understand the distinctive aspects of military selection, this section briefly describes the civilian environment of personnel selection.

Civilian Selection and Classification

The theory of personnel selection identifies two types of selection errors: rejecting a candidate who would have succeeded if hired (Type I error), or accepting a candidate who fails to successfully perform on the job (Type II error). Civilian selection assumes an abundance of applicants and attempts to minimize the second type of error, which is more costly and detrimental to an organization (Howell, 1976, p. 136). The company gathers data on the applicants and uses a series of selection techniques in addition to cognitive test scores to narrow the field of candidates, based upon:

- absence or presence of criteria (e.g., job experience and education);
- minimum levels of competencies in skill areas (e.g., clerical skills); and
- analysis of the 'fit' between the applicant and the organization.

Entry into the civilian work force involves determining the most suitable match between job candidate characteristics and personnel requirements. Unlike military selection, which is primarily based on scores from one test battery, the Armed Services Vocational Aptitude Battery (ASVAB), civilian personnel selection typically employs one or more of a wide variety of techniques, instruments, and selection processes, including assessment centers, employment applications, biographical inventories, interviews, references, work samples, and aptitude aptitude tests. Most often, a combination of techniques is used during the selection process. Initially, background data from application forms, references, biodata, or resumes may be used to screen out a large proportion of applicants. These techniques are relatively cost-effective, efficient, and quick. Testing, interviews, work samples, and even assessment centers (for higher level positions) may be incorporated into the final selection decision.

In contrast, aptitude testing in the military is seen as an efficient and useful approach, primarily because of the large numbers of applicants lacking work experience and the wide variety of jobs offered in the Services. Also, since testing programs require sophisticated and extensive data bases, the military is better equipped to use this selection method than most civilian employers. As a point of fact, over the past three decades the use of aptitude tests within the civilian sector has decreased. A 1963 Bureau of National Affairs study (Lawshe & Balma, 1966) showed that virtually all large companies (with 1,000 or more employees) and 80 percent of small companies administered pre-employment tests. Less than 10 years later, only 55 percent of companies were using tests. Fine (1975) noted that this drop could be attributed to the fact that many employers discontinued using unvalidated tests, which did not meet the increasing demands imposed by Federal legislation in defining legal uses and validity of tests. Increasingly, the process used to screen out candidates must be legally defensible in terms of reliability, validity, fairness, and impact on subgroups of applicants. Professional and Federal agency standards and guidelines have been developed to aid both applicants and employers in understanding how the process does and should work (APA, 1980, 1985; *Federal Register* 43, pp. 38290–38315, 1978). Thus, aptitude testing is not the central focus of civilian selection decisions as it is in the military.

Setting Cut Scores

A 'cut score' is (APA, 1985, p. 43), "a specified point on a scale at or above which candidates pass or are accepted and below which candidates fail or are rejected." When cut scores are used, they reduce the operational scale to just two points—pass or fail. Expectancy tables or probability analyses are rarely used to arrive at cut scores in the civilian sector. A fully dependable numerical basis for a critical decision is seldom, if ever, available. The justification traditionally used for cut scores is that they be based on a reasonable rationale (Ofansko, 1985). A cut score depends upon the selection ratio, which is the number of persons selected divided by the number of applicants. When the selection ratio is low, cut scores can be set relatively high; conversely, when this ratio is high, it forces the institution to set the cut score relatively low. The selection ratio thus reflects the demand/supply relationship in the job market. When demand is strong (i.e., many applicants for each opening), the employer selects the 'best' applicants from a large pool of qualified candidates. When less job seekers are available, the employer is forced to take the best of a smaller pool or to leave the opening unfilled.

cut score

selection ratio

As shown in Figure 3–1, regardless of the specific measures used by an organization to gather data on its applicants, the 'scores' are generally combined, either implicitly or explicitly, into an 'accept-reject' or 'pass-fail' categorization. The setting of the cut score is a complex, imprecise, and primarily judgment-based process.

The conceptual relationship between predictor and criterion variables in a selection situation is shown in Figure 3–1. All applicants to the right of cut score **E-E'** 'pass' the selection criteria (qualify). All applicants above **F-F'** would succeed if selected. The objective of the selection system is to maximize the proportion of individuals in quadrants A and D, while minimizing those in quadrants B and C, thus providing a maximally efficient system. An individual in B would pass the test for a job or educational program, but then fail; a person in quadrant C would be screened out, but would have succeeded if selected. By moving **E-E'** to the right, for example, to **G-G'** (raising the cut score), an employer or school could theoretically eliminate the selection of probable failures. Such a policy would, however, drastically reduce the number of potential selectees and screen out large numbers of applicants who would have succeeded if selected. The eternal dilemma of selection policy is trying simultaneously to minimize costs to both the institution and the applicant. In essence, cut scores rest more on professional judgment call than on technical procedures (Ofsanko, 1985).

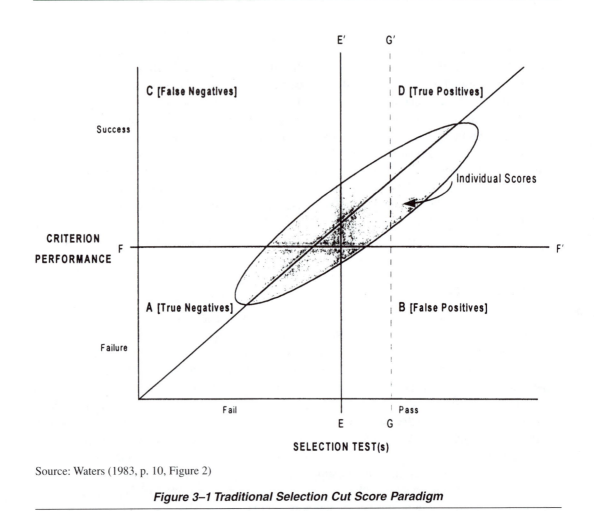

Source: Waters (1983, p. 10, Figure 2)

Figure 3–1 Traditional Selection Cut Score Paradigm

validity   Figure 3–1 also visually depicts the concept of validity, which is shown as the dispersion pattern of scores. The wider the spread of scores from the straight diagonal line, the greater the loss of accuracy in prediction and selection (i.e., the more wrong decisions). The validity of the measures has a direct impact on the effectiveness of the selection system, since less validity (more dispersion from the diagonal) results in less accurate qualified/not qualified decisions. The more precisely that the domain of an actual job is captured by the measurement system, the narrower the dispersion pattern will be.

Selection System Validation in the Civilian World

With the exception of major corporations which have a large number of applicants and jobs to be filled, predictive validation of civilian selection systems is not done routinely. It would be virtually impossible for a typical employer to validate its cut scores and selection processes empirically. In general, corporations use time-tested methods which have worked for them, although requirements for the documentation of successes and failures in selection are increasing. A careful analysis of test content and job requirements (often supported by job/task analysis studies) provides a defensible basis for most industrial selection systems.

Currently, firms are sensitive to the possibility of litigation over their selection practices, and issues surrounding adverse impact are salient in the personnel field. Although equal employment opportunity pressures are a driving force within current civilian employment selection, the utility of these practices is also a focal point. Thus, there is a need for industrial psychologists to develop more fair, tailored, and valid selection techniques.

Established professional standards and guidelines (APA, 1980, 1985; *Federal Register 43,* pp. 38290–38315, 1978) recognize the problems small firms typically face in validating their selection procedures and have tried to balance purely scientific validation requirements with practical limitations. The work in validity generalization (e.g., Hunter, 1980, 1984; Schmidt & Hunter, 1977) shows promise for reducing the burden of validating all selection procedures in all work environments. The essence of this research is that validity coefficients generalize beyond the specific situations from which they were gathered, permitting an employer to 'generalize' validities established for similar groups of jobs to his or her organization's selection program.

Military and Civilian Selection: A Comparison

Despite differences in the choice and frequency of use for various selection techniques, military and civilian selection processes are functionally quite similar. Both use an instrument(s) that is correlated to real world output—job or training performance—as a measure of an applicant's suitability for the organization (e.g., Army, Navy, AT&T, or IBM) and the job (military specialty or civilian job).

The Services are able to pay far closer attention to selection empirically 'proven' validity than the average civilian employer. Indeed, experience of the last four decades suggests that, although there are high-scoring applicants who prove ineffective and low-scoring persons who perform well, an individual's military enlistment test scores (on the average) bear a strong relationship to her or his likelihood of successful training and job performance.

A review of current selection practices suggests that the military can claim more progress in validation than can the civilian sector. Validation studies of the military selection process clearly outnumber those of the civilian process. The civilian sector suffers from an inherent lack of uniformity and cooperation among organizations in general, and a reluctance to share information on selection in particular. While military policy makers have the opportunity to survey military selection techniques across time and across the Services, civilian personnel managers must rely on limited published materials, i.e., those from local, state, and federal governments, and their own personnel selection experiences and studies.

Military Personnel Selection[3]

Military personnel selection has two basic missions—one for selecting enlisted personnel, and a second, more limited one, for selecting officers. Military psychologists in DoD and Service personnel policy management offices, the Services' recruiting commands, the Joint-Service U.S. Military Entrance Processing Command (USMEPCOM), the Service personnel research laboratories, the Service academies and ROTC, and the DoD Testing Center in Monterey, California all play a part in selecting new military personnel. In fiscal year (FY) 1995, for example, 218,891 enlisted and 31,819 officers joined the active and Reserve Military Services from civilian life.

Enlisted Selection Process

The demand for new enlisted recruits by the Army, Navy, Marine Corps, and Air Force is great, running from over a quarter of a million young men and women annually during peacetime to several million during mobilizations. The pool from which the military draws its recruits is young (generally ranging in age from 18 through the early 20s). Most often they arrive as recent high school graduates without advanced education (i.e., college) or, unlike civilian employment selection, without much previous employment or technical training history. Further, the military needs recruits who must often withstand physical demands well beyond comparable civilian jobs, and who may be required to risk their lives to accomplish their missions. Thus, military selection is not limited only to cognitive assessment, but also assesses the physical, medical, educational, and attitudinal characteristics of potential recruits. The scope of the requirements for military selection is unmatched in any other psychological measurement application. The lives of millions of American youth and our national security rest, in part, upon an effective selection system in the U.S. military.

cognitive assessment

The Armed Services Vocational Aptitude Battery (ASVAB) is given to all applicants for enlistment. Since military applicants, for the most part, do not have job histories of any length, the Services must depend upon *indicators* of potential job performance. In choosing over 200,000 active duty recruits per year from about twice that many applicants, the Services 'screen' on the basis of aptitude, education, English language skills, medical, physical fitness, moral character, age, and citizenship standards.

screen

It is primarily the first two standards, aptitude and education, that are used to define applicant capability or quality. The aptitude and education 'quality' surrogates are used in lieu of evaluating past work experience, which rarely exists for military applicants. Each Service sets aptitude standards that consist of minimum score requirements on composites of test scores from the ASVAB. The Armed Forces Qualification Test (AFQT), a composite of verbal and math test scores from the ASVAB, is the primary enlistment aptitude screen and gauge of recruit quality.

Armed Forces
Qualification Test

Education standards refer to requiring lower minimum aptitude test scores for high school diploma graduates than holders of high school equivalency certificates or for non-high school graduates. The rationale for this policy is based upon the differential military performance (in performance and attrition rates) of the three education groups. That is, members of the latter two education groups are approximately twice as likely to leave service before completing their initial period of enlistment obligation as do high school diploma graduates. Higher aptitude standards for the latter two groups are used to accept only the most promising of them from the statistically less successful and thus less preferred group of applicants (Ofsanko, 1985).

Aptitude test scores are the single, most important determinant of who gets in and who does not. Of those applicants rejected for service, about one half fail for aptitude test score reasons. Physical and moral character standards (e.g., arrest records) screen out about 12 and 1 percent of military examinees, respectively (Waters, 1983). Using fiscal year 1985 as an example, 510,802 members of the examinee pool of 695,995 (73.4 percent) met or exceeded the minimum criteria. Over 300,000 actually entered service, 43.3 percent of the examinees. Thus, about two-fifths of the individuals testing for enlistment into the military actually entered. The rest were screened out either by their own choice (a little less than one-third) or by Service selection standards (just over one-fourth).

MANPOWER POOL

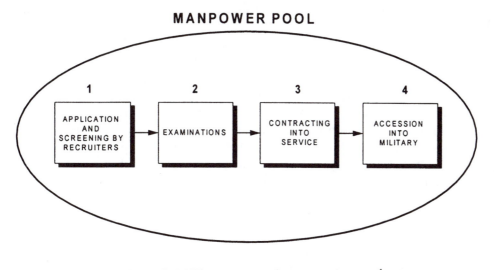

Figure 3–2 Military Enlisted Selection Process[4]

Once enlistment occurs, various combinations of ASVAB scores serve as determinants for entry into specific military jobs. Aptitude and education standards, therefore, are the most visible, analyzed, and criticized of the enlistment selection standards.

To understand the selection procedures used by the Military Services, the reader should understand how the process works and its terminology. Figure 3–2 provides a simplified model of this system.

Members of the potential military enlisted manpower pool (predominantly 18–23 year olds) enter the process by interacting with Service recruiters, who provide initial screening of applicants. Recruiters verify citizenship, age, juvenile or criminal offense background, education status, and other information—Step 1 of Figure 3–2.

Service recruiters frequently use an enlistment screening test to predict applicant scores on the full AFQT. On the basis of the examinee's score on this test, a recruiter can estimate his or her likelihood of passing the AFQT or qualifying for special bonuses or job assignments. No data are available on the proportion of applicants who are screened out at the recruiter level. It is generally assumed that this proportion is low. Waters et al. (1987) estimate (on the basis of discussions with Service recruiting managers) that about 10 percent of applicants desiring to take the ASVAB are screened out at this stage during periods of low to average youth unemployment and approximately 20 percent during periods of high unemployment. A subset of the applicants, termed examinees, formally enters the selection system—Step 2 of Figure 3–2, by taking the ASVAB at one of nearly 800 testing locations across the United States and overseas.

For the most part, a Service recruiter interests a potential recruit in a Service, not in a specific job within that Service. That function is then left to a Service career counselor (classifier) at a Military Entrance Processing Station (MEPS). MEPSs are located in 65 locations around the country, where military applicants can take the ASVAB, get medical and physical testing, and be processed for enlistment. Each MEPS has numerous remote mobile examining team sites that provide ASVAB testing in local post offices and other distributed locations throughout the geographic area served by the MEPS.

A Service career counselor has access to a computer data file that contains results from the examinee's ASVAB tests, physical examination, educational records, and other collected data. A counselor also has access to Service current and future (near-term) vacancies in technical schools and jobs. By considering the occupational interests and background of the examinee, and 'selling' specialty training slots of highest priority to the Service for which the applicant qualifies, a job-person match is made. Ideally, the assignment meets the require- ments of both the Service and the individual. Once a contract—step 3—is agreed upon, the recruit either enters active duty and basic training immediately, or, more often, joins the Reserves as a member of the Delayed Entry Program (DEP) for up to one full year prior to active duty.

job-person match

Figure 3–2 also suggests the difficulty of the military recruiting mission, and the criticality of reliable and valid selection measures to minimize the false-positive and false- negative decisions displayed earlier in Figure 3–1.

false-positive
false-negative

Table 3–1 shows the ASVAB tests and what they measure. AFQT, as noted earlier, is derived from four of the ASVAB tests (shown in Table 3–1), measures verbal and quantitative abilities. It has been shown to predict recruits' military training and job performance.

The Officer Selection Process[5]

Officer selection is a somewhat less rigid process than that of enlisted selection. The potential candidates are older, may have more job performance history, and officers are required to be college graduates. Thus, they have better shown their academic ability and perseverance in completing their education. Unlike enlisted selection, however, there is no single aptitude test used across Services to identify the best qualified officer candidates. Even within each Service, there are multiple commissioning paths (e.g., Service Academy, Reserve Officer Training Corps, Officer Candidate School, or direct commissioning), whose selection processes vary greatly. Each path has its own selection methods, measures, and standards.

The numerical requirements for officer accessions are only about one-seventh of the number of enlisted recruits needed annually. The college graduate standard effectively places the onus on the educational community to provide the screening (cognitive as well as behavioral) function which the ASVAB provides for the enlisted force. Certainly, selection testing plays an important part in who goes to college (e.g., the Scholastic Assessment Test [SAT] or American College Test [ACT] college admission tests), however the Services demand much more than just academic ability of their officers. Graduating from college requires technical knowledge (e.g., engineering or scientific skills), plus demonstration of less specifically-defined characteristics such as perseverance, maturity, leadership and manage- ment skills, and logical thinking.

Across all of the Services, aptitude test scores are only part of their officer selection processes. The so-called 'whole-person' is assessed, including leadership potential, technical skills, character, letters of recommendation, school grades, and interview performance. Selections are made by boards of senior officers and personnel specialists. Potential costs of making an errant selection can be enormous, currently up to a quarter of a million dollars at the academies, not even taking into account the operational costs and consequences of having

whole-person

Table 3-1

**Armed Services Vocational Aptitude Battery (ASVAB) Tests:
Description, Number of Questions, and Testing Time[a]**

ASVAB Test Title and Abbreviation	Description	Number of Questions	Testing Time (Minutes)
Arithmetic Reasoning (AR)	Measures ability to solve arithmetic word problems	30	36
Word Knowledge (WK)	Measures ability to select the correct meaning of words presented in context and to identify the best synonym for a given word	35	11
Mathematics Knowledge (MK)	Measures knowledge of high school mathematics principles	25	24
Paragraph Comprehension (PC)	Measures ability to obtain information from written passages	15	13
General Science (GS)	Measures knowledge of physical and biological sciences	25	11
Mechanical Comprehension (MC)	Measures knowledge of mechanical and physical principles & ability to visualize how illustrated objects work	25	19
Electronics Information (EI)	Measures knowledge of electricity and electronics	20	9
Auto and Shop Information (AS)	Measures knowledge of automobiles, tools and shop terminology and practices	25	11
Coding Speed (CS)	Measures ability to use a key in assigning code numbers to words in a speeded context	84	7
Numerical Operations (NO)	Measures ability to perform arithmetic computations in a speeded context	50	3
Totals		**334**	**144[b]**

[a]Source: Eitelberg, M. J. (1988). *Manpower for Military Occupations.* Washington, DC: Office of the Assistant Secretary of Defense (Force Management and Personnel), p. 68.

[b] Administrative time is 36 minutes, for a total testing and administrative time of 3 hours.

a 'bad' officer enter the system. The complex, expensive, and unique systems used by each of the Services to identify its officers appropriately reflect the importance of reliable and valid decisions in deciding who becomes their 'brass.'

At the end of FY 1995, there were 221,518 officers on active duty and 134,984 in the Selected Reserves. During that year, 15,746 new officers entered active duty and there were 6,073 new Reservists (DoD, 1996).[6] Table 3–2 shows officer selection tests and academic criteria used by the Services during 1987–88.

Table 3–2

Aptitude Tests and Academic Criteria Used to Screen Officer Candidates, by Program and Service, 1987–88[7]

Program	Army	Navy	Marine Corps	Air Force
Service Academy	SAT/ACT HS Rank	SAT/ACT HS Rank	—[a]	SAT/ACT HS Rank HS Activities
ROTC Scholarship	SAT/ACT HS Rank College GPA	SAT/ACT HS Rank HS GPA	—[a]	SAT/ACT HS RANK College GPA AFOQT
ROTC Non-Scholarship	OSB 3 & 4	Varies by Unit	—[a]	AFPQT SAT/ACT College GPA
OCS/OTS and Other	OSB 1 & 2 GT	OAR	SAT/ACT EL	AFOQT College GPA
Aviation OCS	N/A	AQT-FAR	SAT/ACT EL AQT-FAR	N/A

[a]Same as Navy. Up to 16% of Naval Academy and naval ROTC graduates may be commissioned as Marine Corps officers.

Abbreviations

SAT	Scholastic Assessment Test	**ASVAB**	Armed Services Vocational Aptitude Battery
ACT	American College Test	**GT**	General Technical Composite of ASVAB
HS	High School	**OAR**	Officer Aptitude Rating
GPA	Grade Point Average	**AQT-FAR**	Aviation Qualification Test-Flight Aptitude Rating
AFOQT	Air Force Officer Qualifying Test	**OTS**	Officer Training School
OSB	Officer Selection Battery	**EL**	Electronic Composite from ASVAB
OCS	Officer Candidate School	**N/A**	Not Applicable
ROTC	Reserve Officer Training Corps		

Military Selection Summary

In many ways, military personnel selection is similar to the more general subject of selection in the civilian world. DoD, however, has paved the way for much of the development of testing technology throughout this century. Since relatively small improvements in the technology yield huge payoffs in a system as large as the U.S. military, its selection processes will continue to be at the leading edge of the testing profession into and throughout the 21st century. Military psychologists will lead that development in the United States.

Military Personnel Classification

As with military selection, quite different procedures are used between enlisted and officer job development.

Enlisted Classification

Once the recruits for each of the Military Services have been selected, the Services need a system for matching them to Service jobs. The system is termed a Service's classification traditions. It is beyond the scope of this chapter to describe each system in detail, however a description of the elements in the systems will be discussed.[8]

classification

Setting Classification Standards

Military manpower representatives from each of the Services continually monitor and document performance in technical training, periodically conduct validation studies, and adjust classification standards accordingly. In addition, training schools and operational field units may also make recommendations on classification policies.

The Services take into consideration the relative difficulty of job tasks, as well as a number of training school criteria, including academic training attrition, final course grades, and school graduation rates. They use an integrated system for setting and modifying classification. If a 'problem' arises, such as a sudden increase in academic failures in a specific job training school, course curricula would be reviewed and considered for revision in lieu of, or in addition to, a classification standards change. Current composite scores of recent successful and unsuccessful trainees would be analyzed, and school grades over time would be tracked to identify the probable cause of the problem. Standards would be revised only after concurrence from affected staff agencies, recruiting, and policy personnel, based upon a full review of training performance and the quality of recruits.

JOB-PERSON MATCH

Just as meeting minimum selection standards does not guarantee entry into the military, meeting minimum classification standards for a job does not guarantee that a recruit will be assigned to that particular specialty. The essence of the development of a Service classification system is completing a job-person match for each job (termed military occupational specialty [MOS] in the Army and Marine Corps, Rating in the Navy, or Air Force Specialty Code [AFSC] in the Air Force). The jobs are analyzed via lists of the specific tasks involved, in terms of what the knowledges, skills, and abilities (KSAs) which the incumbent must possess to accomplish the job successfully. The jobs are then clustered into sets of jobs which have similar task-characteristics and needed person-characteristics. Composite scores, which are combinations of test scores from the ASVAB which have been shown to be predictive of success in the individual jobs, are then developed and validated. A detailed description of the Services' classification processes and composites can be found in Eitelberg (1988).[9]

military occupational specialty

Rating

Air Force Specialty Code

knowledges, skills, abilities (KSAs)

As an example, in the Air Force four clusters of AFSCs are used: **M**echanical, **A**dministrative, **G**eneral, and **E**lectronic. Three (of 32) AFSCs included in the Administrative cluster are administration specialist, personnel specialist, and postal specialist. The **A** composite is made up of scores from the Word Knowledge, Paragraph Comprehension,

cluster

Composite scores

Coding Speed, and Numerical Operations test scores from a recruit's ASVAB record. These tests measure verbal abilities and attention to detail under speeded conditions, which relate to job performance in administrative jobs. A new recruit scoring high on the **A** composite would likely be offered an administrative AFSC, with the availability of a training seat in the AFSC's technical training school and the desires of the recruit taken into consideration.

Service laboratory psychologists conduct occupational surveys of each of their Service's jobs on a regular basis to ensure the jobs have not changed since the composites were developed. Included in these studies is the statistical analysis of the relationships between composite scores and performance in the jobs by incumbents. This process is known as validation.

validation

After the Service and the new recruit settle on a job for the individual, the Service personnel representative then makes the assignment for the recruit to a specific technical school, followed by the recruit reporting to his/her unit. Although the terms are often confused, 'classification' refers to the job-person matching process, while 'assignment' refers to the placing of an individual in a specific location and unit.

assignment

Officer Classification

Classification of the officer corps' is similar to enlisted classification, except that more emphasis is placed on individual officer education, experience, and preferences. According to Rosenthal & Colot (1988), for example, an officer's military specialty hinges more on his or her major in college and personal preferences than on other variables:

> *In contrast to enlisted personnel, there are few jobs (notably pilots and navigators) which depend on aptitude test scores to help make assignment decisions.*[10]

Thus, many 'classification decisions' are made on officer candidates in the selection process wherein certain college majors (i.e., science and engineering) have greater priority in acceptance into commissioning programs. In fact, many initial jobs are assigned dependent upon a new officer's class standing in his or her commissioning program. A list of available assignments may be posted, with graduating officers selecting their specialty in "order of merit," based upon their preferences.

Military Classification Summary

The ultimate, unreachable goal for military classification would be to place every new service person in the job in which he or she would have the highest potential for success, considering Service needs, the person's skills, knowledges, abilities, interests, and other characteristics. This process is becoming more technical in all of the Services. Complex computer models are being used by each Service to try to approach optimum job-person matching.

Conclusion

Service personnel psychologists will be at the forefront of these research and development efforts. Military psychologists will need statistical, psychometric, operations research, economic, personnel management, computer, and modeling skills to keep up with the technical developments of the coming millennium. The opportunity exists for an exciting, challenging, and productive career in military psychology through selection and classification work towards the improvement of the Services' greatest resource—their people.

References

American Psychological Association. (1980). *Principles for the Validation and Use of Personnel Selection Procedures*. Division of Industrial-Organizational Psychology. Berkeley, CA: Author.

American Psychological Association. (1985). *Standards for Educational and Psychological Testing*. Washington, DC: Author.

Department of Defense. (1981). *Department of Defense Efforts to Develop Quality Standards for Enlistment. A Report to the House and Senate Committees on Armed Services*. Washington, DC: Office of the Assistant Secretary of Defense (Manpower, Reserve Affairs, and Logistics).

Department of Defense. (1996). *Population Representation in the U.S. Military: Fiscal Year 1995*. Washington, DC: Office of the Assistant Secretary of Defense (Force Management Policy).

Eitelberg, M. J. (1988). *Manpower for Military Occupations*. Washington, DC: Office of the Assistant Secretary of Defense (Force Management and Personnel).

Eitelberg, M. J., Laurence, J. H., & Brown, D. C. (1992). Becoming Brass: Issues in the Testing, Recruiting, and Selection of American Military Officers. In Gifford, B. R. & Wing, L. C., *Test Policy in Defense: Lessons from the Military for Education, Training, and Employment*. Boston: Kluwer Academic Publishers.

Eitelberg, M. J., Laurence, J. H., Waters, B. K., with Perelman, L. S. (1984). *Screening for Service: Aptitude and Education Criteria for Military Entry*. Washington, DC: Office of Assistant Secretary of Defense (Manpower, Installations and Logistics).

Howell, W. (1976). *Essentials of Industrial and Organizational Psychology*. Homewood, IL: Dorsey Press.

Lawshe, C. H., & Balma, M. (1966). *Principles of Personnel Testing*. New York, NY: McGraw-Hill.

Ofsanko, F. (1985). Setting Ability Test Passing Scores. Paper presented at the 93rd Annual Meeting of the American Psychological Association, Los Angeles, CA.

Rosenthal, D. B., & Colot, P. (1988). *Initial Job Assignments of Military Officers*. Alexandria, VA: Human Resources Research Organization.

Sands, W. A., Waters, B. K., & McBride, J. R. (Eds.) (1997). *Computerized Adaptive Testing: From Inquiry to Operations*. Washington, DC: American Psychological Association.

Waters, B. K. (1997). Army Alpha to CAT-ASVAB: Four-Score Years of Military Personnel Selection and Classification Testing. In Dillon, R. (Ed.) *Handbook on Testing*. Westwood, CT: Greenwood Press.

Waters, B. K., Laurence, J. H., & Camara, W. J. (1987). *Personnel Enlistment and Classification Procedures in the U.S. Military*. Washington, DC: National Research Council.

Endnotes

[1]Department of Defense. (1981). *Department of Defense Efforts to Develop Quality Standards for Enlistment. A Report to the House and Senate Committees on Armed Services* (Washington, DC: Office of the Assistant Secretary of Defense [Manpower, Reserve Affairs, and Logistics], p. 5.

[2]This section draws heavily from Waters, B. K., Laurence, J. H., & Camara, W. J. (1987). *Personnel Enlistment and Classification Procedures in the U.S. Military*. Washington, DC: National Research Council.

[3]A detailed description of military selection processes can be found in Eitelberg, M. J., Laurence, J. H., & Waters, B. K., with Perelman, L. S. (1984*). Screening for Service: Aptitude and Education Criteria for Military Entry*. Washington DC: Office of the Assistant Secretary of Defense (Manpower, Installations, and Logistics).

[4]From Waters, B. K., Laurence, J. H., & Camara, W. J. (1987). *Personnel Enlistment and Classification Procedures in the U.S. Military*. Washington, DC: National Research Council.

[5]This section has been drawn heavily from Eitelberg, M. J., Laurence, J. H., & Brown, D. C. (1992). Becoming Brass: Issues in the Testing, Recruiting, and Selection of American Military Officers. In Gifford, B. R. & Wing, L. C., *Test Policy in Defense: Lessons from the Military for Education, Training, and Employment*. Boston: Kluwer Academic Publishers.

[6]Department of Defense. (1996). *Population Representation in the U.S. Military: Fiscal Year 1995*. Washington, DC: Office of the Assistant Secretary of Defense (Force Management Policy).

[7]From Eitelberg M. J., Laurence, J. H., & Brown D. C. (1992). Becoming Brass: Issues in the Testing, Recruiting, and Selection of American Military Officers. In Gifford, B. R. & Wing, L. C., *Test Policy in Defense: Lessons from the Military for Education, Training, and Employment*. Boston: Kluwer Academic Publishers.

[8]For a more detailed description of the Services' procedures, see Waters, B. K., Laurence, J. H., & Camara, W. J. (1987). *Personnel Enlistment and Classification Procedures in the U.S. Military*. Washington, DC: National Research Council.

[9]Eitelberg, M. J. (1988). *Manpower for Military Occupations*. Washington, DC: Office of the Assistant Secretary of Defense (Force Management and Personnel).

[10]Rosenthal, D. B., & Colot, P. (1988). *Initial Job Assignments of Military Officers*. Alexandria, VA: Human Resources Research Organization, pps. 40–41.

4

Military Leadership

Al Lau, Ph.D.

* Introduction * What is Leadership? * Review of Leadership Theory and
Research * Contingency Theories of Leadership * Transformational
Leadership Theory * Building Leadership Skills
* Challenges of the Future * References

Introduction

The effective military leader today is characterized by a high degree of flexibility, initiative, and ability to lead in complex and ambiguous circumstances. A broad range of capabilities is required. Not only are high standards of ethical and organizational proficiency demanded, but military leaders also have to be effective in their dealings with local populations and different cultures. In the twenty-first century, the military services will be required to do more with less. The increasing level of sophistication in military hardware requires the military leader to empower subordinates and to take on more complex tasks with fewer resources in such areas as Bosnia and Herzogovina (Lau, Atwater, Avolio, & Bass, 1993). These new leadership requirements have been influenced by: (a) the increased pressure to do more with less—the impact of the downsizing of the military; (b) the trend toward greater power sharing and decentralization which has resulted in the increased importance of team building skills; (c) the accelerating rate of technological change and the need to provide continual on-the-job training and development to followers; (d) the increased educational level and higher expectations that members of the military have; and (e) a commitment to quality and customer service as reflected by the official adoption of Total Quality Management (TQM) across the military services. All of these changes demand a new approach to leadership.

> Total Quality
> Management
> •
> leadership

What is leadership? Do you consider yourself a leader? Is it possible to improve your leadership effectiveness or are leaders born? What kind of leadership training is being conducted in the United States military today? What changes in society have and will continue to impact future leaders? These are some of the questions that will be pursued in this chapter.

The major purposes of this chapter are to provide: (a) a definition of leadership and to distinguish its transactional and transformational components; (b) an overview of major leadership theory and research starting with leadership traits, leadership behaviors, situational leadership theories, and ending with the recent focus on transformational leadership; (c) an overview of some of the current leadership training programs in military settings; (d) suggestions from the research literature that you may be able to use in order to improve your own leadership effectiveness; and (e) to identify some current issues which will influence leadership in the future. Throughout the chapter, there is an emphasis on research findings from the military sector.

It is unfortunate that much of the leadership literature that you may have encountered is difficult to use in developing concepts that will improve your leadership performance. As pointed out by Mintzberg (1982), this is because much of the literature is written primarily for other leadership researchers and includes complex statistical methods designed to impress one's colleagues and journal editors, and is uninterpretable to individuals who are trying to become more effective leaders. This literature, however, can be useful and will be reviewed along with my own research to provide some useful concepts to bridge the gap between research and practice. An additional problem is that the vast majority of leadership studies conducted in the military sector appear in technical reports and journals that are not found in many base libraries. This useful literature is also not available for improving current military leadership training and development programs. In this chapter an attempt is made to provide the reader with an overview of some of this military literature.

What Is Leadership?

Stogdill (1974, p. 259) concluded that "there are almost as many definitions of leadership as there are persons who have attempted to define the concept," and Bennis (1959, p. 259) stated that:

> "Of all the hazy and confounding areas in social psychology, leadership theory
> undoubtedly contends for the top nomination. And, ironically, probably more
> has been written and less known about leadership than about any other topic in
> the behavioral sciences."

A student of mine once remarked that leadership was difficult to measure or define, but he easily recognized it when he saw it. Despite these difficulties, the following definitions are provided:

Directing and coordinating the work of group members (Fiedler, 1967).

The process by which an agent induces a subordinate to behave in a desired manner (Bennis, 1959).

To approach followers with an eye to exchanging one thing for another; jobs for votes, or subsidies for campaign contributions (Burns, 1978).

The Air Force defines leadership as the art of influencing and directing people to accomplish the mission, and the Army as the process of influencing others to accomplish the mission by providing purpose, direction, and motivation. The common theme in these definitions is that of a transaction or exchange relationship involving the clarification of task requirements and the application of contingent rewards by the leader to influence others to accomplish group goals. These definitions do not explain how some leaders function as role models or provide visions which motivate followers to perform at peak levels of performance.

Most leadership training programs, both in the civilian and military sectors, emphasize a transactional approach. A shortcoming of this approach is that leaders may lack the time, skills, or resources to diagnose follower needs and/or to deliver rewards such as promotion, recognition, or praise.

Recently there has been a focus on leaders who function as role models and develop visions which motivate followers to make self-sacrifices to achieve the organization's mission and to function at peak levels of performance. Transformational leadership is:

Transformational Leadership

A process in which leaders and followers raise one another to higher levels of morality and motivation (Burns, 1978).

Transforming followers, creating visions of the goals that may be attained, and articulating for the followers the ways to attain these goals (Bass, 1985; Tichy & Devanna, 1986).

To achieve understanding and commitment of subordinates for the accomplishment of purposes, goals, and objectives envisioned by the leader, beyond that which is possible through the use of authority alone (Department of the Army, 1987).

A complete definition of leadership should include both the concept of a transaction between leaders and followers as well as the concept of transforming followers by raising their awareness of the importance of achieving valued outcomes. Transformational leaders have followers who identify with them and arouse and inspire them with a vision of what can be accomplished through extra personal effort. Perhaps some examples from a recently completed study of leadership development and performance at Virginia Military Institute (Atwater, Dionne, Avolio, Camobreco, & Lau, 1996) will clarify this distinction. These were taken from leadership incidents collected from freshmen ("rats").

Transformational Incidents:

"My roommate and I messed up in parade one time. He was willing to come up on Sunday night to help us out. He took time out from his own study time to do this. . . . It made me want to improve."

"He took me aside during rat challenge (physical training) and remarked how impressed he was at my motivation. It was a sincere talk, and even though he is a very rough cadre member, he does go out of his way to give positive advice. . . . I felt like I was accomplishing something, instead of always being yelled at. It was the first remarks that made me realize that some people are here to make you a better man, instead of just run you out."

"During physical training, he is highly motivational. Competes on obstacle course and does motivational sit-ups, push-ups, and pull-ups with us. Strives for excellence and expects only that. Leads by example—Our company works together to try and outdo him and that's his purpose. . . . I want to be like him next year."

Transactional Incidents:

"He told me that I had the potential to be a good cadet. He said I should be the company leader. . . . I was motivated, challenged and full of pride and self-confidence."

"He said he was proud of our unit's performance."

A number of individuals have made a distinction between management and leadership, and it is interesting to note that the U.S. Navy quality approach is titled Total Quality

Leadership (TQL), not Total Quality Management. Bennis and Nanus (1985), for example, stated that managers do things right and leaders do the right things. Based on the study of Mintzberg (1973), which focused on the nature of managerial work, and my research on managers (Lau, Newman, & Broedling, 1980), I propose to use the terms manager and leader as being complementary—both are necessary for organizational success.

Review of Leadership Theory and Research

Traits of Leaders

traits One of the earliest approaches to studying leadership was the trait approach. Traits can be defined as personal characteristics or attributes of individuals. This approach focused on traits such as physical characteristics, intelligence, personality, and values that consistently differentiated leaders from followers or differentiated effective from less effective leaders. If traits were demonstrated to be consistently related to leadership emergence and effectiveness, it would be possible to select "natural" leaders.

Early reviews of the literature indicated that there was little evidence to support the trait approach (Stogdill, 1974). In recent years, however, this approach has been more productive and, after a thorough review of the research literature, Bass (1990) concluded that good leaders tend to have the following ten leadership traits:

A strong desire to achieve.
A desire to influence others for the common good.
A high energy level.
Persistence.
Task competence.
Good interpersonal skills.
Self-confidence.
A willingness to act (decisiveness).
A tolerance for stress.
A high degree of flexibility.

Due to the recent development of more reliable measures of traits and better measures of leadership emergence and effectiveness, the following conclusions can be drawn:

1. Intelligence is related (correlations center around .50) to leadership emergence, but relationships with leader effectiveness are much weaker. It is possible for leaders to be too intelligent.
2. Personality traits that appear to be related to leader effectiveness include, for example, self-esteem, dominance, self-confidence, achievement orientation, dependability, sociability, and tolerance of ambiguity (Hughes, Ginnett, & Curphy, 1993). A leadership training manual used in U.S. Air Force training states, for example, that effective leaders have six traits that are related to effectiveness—integrity, loyalty, commitment, energy, decisiveness, and selflessness. Reliable measurement of some of these traits, however, would appear to be difficult.

3. Traits are more strongly related to leadership emergence than to leader effectiveness. Traits such as intelligence and personality may be a necessary condition for leadership emergence, but do not guarantee that leaders will be effective (Hughes, et al., 1993). A variety of studies in the military sector, however, have examined and have reported significant relationships between some traits and leadership effectiveness. Mael and Schwartz (1991) found that emotional stability, dominance, and energy level were related to leadership performance at West Point. Atwater and Yammarino (1993) found that personality traits such as conformity and self-discipline predicted leadership as measured by supervisory ratings in a sample of midshipmen leaders at the U.S. Naval Academy. Atwater, et al. (1996) reported that physical fitness, self-esteem, and the ability to tolerate stress predicted leader effectiveness in a longitudinal study conducted at Virginia Military Institute (VMI).

4. The relationship between traits and emergence and/or effectiveness is strongly influenced by situational factors such as the education, experience, and expectations of followers as well as by group characteristics such as size, norms, and cohesiveness. Research has generally shown that there is no single list of traits that determine whether someone will be an effective leader, that leaders may have very different traits and still be effective, and that traits vary in their importance depending on the situation. It has been found, for example, that personality traits have significant relationships with effectiveness only when the leader has some discretion over which behaviors to engage in a given situation, but weak relationships when the situation dictates the behaviors the leader must engage in.

Behavior of Leaders

Another approach to studying leadership was to examine the behaviors that consistently differentiated leaders from followers or effective from less effective leaders. If this behavioral approach is valid, it would be possible to train effective leaders. It is far easier to change one's behavior than to attempt to change one's personality!

behavioral approach

Initial research in this area was conducted at Ohio State University (Fleishman, 1973) with similar findings from the University of Michigan (Likert, 1961). The studies at Ohio State found that leadership behaviors could be described by two independent dimensions named consideration and initiation of structure. *Consideration* involved friendly and supportive behaviors toward followers, showing appreciation, and caring about their personal welfare. *Initiation of structure* involved behaviors that focused on direction and control of task accomplishment. Some behavioral items that came from the surveys developed at Ohio State included:

Consideration:
    Is friendly and approachable.
    Trusts members to exercise good judgment.

Initiation of structure:
    Encourages increased production.
    Asks the members to work harder.

Research showed that consideration generally led to leadership success as measured by follower satisfaction but that relationships with effectiveness were inconsistent. This suggests that there is no universal set of leader behaviors associated with leadership effectiveness. While many leadership training programs emphasize the value of team management, participative management, and democratic management, this "one best way" approach may not always be the most effective leadership style. This is because the behavioral approach failed to consider follower and situational factors that influenced the leader behavior—leader effectiveness relationship. Other problems that confronted the behavioral approach were the over-reliance on surveys, the difficulty in determining cause and effect relationships (does considerate behavior cause follower performance or does follower performance cause considerate behavior?), and the conclusion that only two dimensions could describe leadership behavior.

Recent research suggests there is another way of looking at leadership behaviors in addition to surveys, and that leadership involves more than the two components of consideration and initiation of structure. There are numerous aspects to leadership such as decision making, resource allocation, and negotiation which have received less attention. Relatively few studies have investigated the behavioral requirements of managerial/leadership positions to identify what managers actually *do*. One particularly important study of managers was conducted by Mintzberg (1973). Based on observation of managers, Mintzberg developed ten basic roles within three areas that were common to managerial jobs:

1. Interpersonal—Figurehead, liaison, leader.
2. Informational—Monitor, disseminator, spokesman.
3. Decisional—Entrepreneur, disturbance handler, resource allocator, negotiator.

In a study of U.S. Navy civilian Senior Executive Service executives using a multi-method approach (surveys, interviews, observation, and work activity diaries), Lau, Newman, and Broedling (1980) identified four major factors that described executive jobs. These were: Leadership and supervision; information gathering and dissemination; technical problem solving; and executive decision making, planning, and resource allocation. This more comprehensive description of managerial/leadership jobs could be used, they argued, to improve executive selection, development, and performance appraisal processes. Schneider (1985, p. 359), however, in his review of the organizational behavior literature concluded that "apparently only Lau and his associates (e.g. Lau & Pavett, 1980) and McCall and his associates (e.g., McCall, et al., 1978) have recently pursued the development of measures of managerial activities that pattern the classical management functions."

Follow-up studies indicated that managerial jobs in both the public and private sectors were similar in terms of the relative importance of the roles identified by Mintzberg, with the roles of leader and resource allocator being rated as the most important by both groups (Lau & Pavett, 1980). Later research found that the importance of these roles varied by hierarchical level (Pavett & Lau, 1983). While these studies help us to understand what managers and leaders do, they do not tell us what makes a leader successful. In a study of the relationship between managerial roles and job performance as measured by performance appraisal ratings, Pavett and Lau (1982) found a negative relationship between the rated importance of the leadership role (which involves behaviors that are similar to the consideration dimension described earlier) and performance appraisals, but a positive relationship with the importance of the liaison role (establishing external contacts and developing work relationships with managers from other departments). In spite of the high importance placed on the leader role, leadership was inversely related to job performance! Since McCall & Segrist (1980) also found an inverse relationship between the importance of the leader role with promotion rate, one has to accept the possibility that the emphasis on leadership style may in some cases be contrary to successful leadership. Luthans, Hodgetts, and Rosenkrantz (1988) also used direct observation to study managers, and to identify what *successful* managers did (those managers who were promoted) versus what *effective* managers did (those who were highly rated by followers). They found that successful managers spent more time in interacting with outsiders and in socializing/politicking (networking) activities than did effective managers who spent more time in the leadership role. These studies indicate that leadership effectiveness is influenced by more than a focus on leader-follower interactions. More research, however, is needed to identify the behaviors of effective and successful leaders.

Another technique of measuring leadership behavior involves the collection of critical incidents where followers describe the specific behaviors of leaders. At West Point, followers were asked to describe, in writing, examples of either good or bad leadership observed in summer training (Adams, Prince, Instone, & Rice, 1984). Effective leadership was categorized into twelve components such as motivating (She motivated, cheered us on); role model (He always set an example by his personal appearance and conduct); contingent sanctions (She was proud of our unit's performance); and downward communication (He met with the squad beforehand to tell us what to expect). The authors suggest that these behavioral descriptions could be very useful in training, development, and counseling programs at the academy. A study by Yukl and Van Fleet (1982) also used critical incidents of especially effective and ineffective leadership behaviors collected from members of the Corps of Cadets at Texas A & M University to derive a definition of leader behavior. They found that four categories of leadership behavior were related to performance: performance emphasis, inspiration, role clarification, and maintenance of discipline.

In a review of the behavioral approach to studying leadership, the following conclusions can be drawn:

1. Behavioral descriptions of leadership need to include more than two or three dimensions. As described by Mintzberg and the studies using a critical incident technique, leadership involves a much more complex set of dimensions than those originally identified in the studies at Ohio State and the University of Michigan. Yukl, Wall, and Lepsinger (1989), for example, have developed a comprehensive survey called the Managerial Practices Survey which includes measures of leadership (motivating, recognizing and rewarding) as well as measures of information handling and processing, decision making, and networking. In general, these studies have identified a common set of leadership/ management behavioral dimensions.

2. A reliance on self-report surveys is not sufficient to describe leadership behaviors. One needs to use a multi-method approach and supplement surveys with interviews, observation, and other methods to adequately describe leadership. In

addition to self-reports, it is also important to collect leadership behavior data from peers, subordinates, and superiors.

organizational
culture

3. Leadership behavior—performance relationships are strongly influenced by follower experience and expectations as well as by other situational factors such as time constraints, task requirements, and organizational culture. There is probably no one best way to lead that is applicable across all situations. For example, is it appropriate to lead a U.S. Marine infantry unit in the same way one would lead a Marine aviation unit?

Contingency (Situational) Theories of Leadership

The first approach to studying leadership focused on the leader's traits and the second focused on the leader's behaviors. It was noted that both approaches neglected to consider the influence of situational factors such as the characteristics of followers, the nature of group factors such as size and cohesiveness, or more general characteristics of the situation such as task complexity or organizational culture. An additional factor that must be considered is that the level of the leadership position (e.g., NCO versus officer) influences the appropriateness of various leadership behaviors or styles.

Research generally did not show strong relationships between leadership behaviors and leadership performance. While leader traits and behaviors are important components that relate to effectiveness, the characteristics of followers and the characteristics of the leadership situation must also be considered. A good example of this is provided below (West Point Associates, 1988, p. 233). In reading this example, ask yourself which leadership style would have more been appropriate and why.

"His unit's mission was highly complex. All his noncommissioned officers (NCO's) were good, but new at the task. They didn't know quite how to proceed. Neither did the lieutenant—but he was afraid to ask anyone for fear of looking incompetent. Throughout, the NCO's complained because of lack of guidance. Somehow the unit stumbled through the assignment and completed it in a barely satisfactory manner. The company commander called the lieutenant aside and counseled him to be more assertive in running his platoon. "Give the people lots of your attention and guidance," he suggested, "they need it!" The lieutenant was glad that was over."

"His next mission was a very simple one—post support, which included sending groups of soldiers under the supervision of a squad leader to cut grass, rake leaves, move bleachers, pick up litter and the like. Wanting to make a good impression, this time the lieutenant outlined specific detailed instructions and assignments for the platoon sergeant. He corrected the squad leaders for making a routine decision on their own. When inspecting their detail members, he issued additional instructions, continually emphasized the need to meet the deadline (the work was already progressing ahead of schedule), and closely supervised the task to insure strict adherence to policies. The result stirred dissatisfaction and complaints among his sergeants. The lieutenant could not understand why. He was only doing what his boss had suggested."

contingency
theories

Contingency theories share three major similarities (Hughes, et al., 1993). First, there has been a considerable amount of research generated that attempts to support the validity of each theory. Second, each theory assumes that leaders can be trained to diagnose important situational factors and trained to select the most appropriate leadership style for that situation. Third, leadership effectiveness is maximized when leaders make their behaviors contingent on a correct diagnosis of important follower and situational factors.

The major contingency theories include the normative decision model (Vroom & Yetton, 1973; Vroom & Jago, 1988); situational leadership theory (Hersey & Blanchard, 1984) ; the contingency model (Fiedler, 1967); and path-goal theory (House & Dessler, 1974). These theories consider a limited and fairly general number of relevant leadership behaviors: task-orientation and relationship-orientation. The major difference between these four theories is the specific situational and follower characteristics identified that influence leadership effectiveness.

Normative decision model. This theory focuses on the optimal level of participation that followers should have in the decision making process. Decision making is widely recognized as one component of effective leadership. Decisions are influenced by two major factors: (a) the importance of making a quality decision; and (b) the importance of followers accepting that decision. Other factors that should be considered in this decision process include time pressures and whether leaders and/or followers have sufficient information to make a quality decision. The various leadership behaviors are as follows: autocratic—the leader makes the decision by herself; consultative—the leader shares the problem with his followers as a group then makes the decision himself; or group—the leader shares the problem and consensus is reached on a solution that is supported by both the leader and the group. With relatively unimportant problems where acceptance can be taken for granted (e.g., the location of office furniture), an autocratic decision is appropriate. With important problems where acceptance is important and the leader does not have sufficient information to make a high quality decision, a group decision would be appropriate. With the introduction of Total Quality Management across the services, one should expect to see a trend away from autocratic to more group-based decision making.

Situational leadership theory. The major principle behind this theory is that the appropriate leadership behavior is contingent on a diagnosis of the maturity level of followers and on the leader's ability to be flexible in his or her leadership style. Follower maturity is defined as having two components: (a) job maturity, the amount of task-relevant knowledge, experience, skill, and ability that followers have; and (b) psychological maturity, the self-confidence, commitment, and motivation possessed by followers. Leadership behaviors are described as: (a) task-oriented, where leaders focus on telling people what to do and how and when to do it; and (b) relationship-oriented, where leaders listen, encourage, and facilitate. Various combinations of these two leadership behaviors result in the following: *telling* (high task, low relationship), *selling* (high task, high relationship), *participating* (low task, high relationship), and *delegating* (low task, low relationship). When followers have low levels of group maturity (e.g., the new and inexperienced recruit who lacks self-confidence), telling him or her what to do and being very directive is the most effective leadership style. When followers have high levels of group maturity, the most effective leadership styles would be participation or delegation. In order to avoid micromanagement, Hersey and Blanchard (1984) suggest a number of developmental interventions that leaders can employ to move followers to higher levels of group maturity.

Although situational leadership theory has not received strong empirical support and neglects other important situational factors, it is very appealing to students of leadership and

Situational
Leadership
Theory

represents a commonsense kind of approach to teaching and improving leadership effectiveness. A leader should lead competent and motivated followers through the use of participation and/or delegation, and unskilled and unmotivated followers through direction and coaching. The key factors in improving leadership effectiveness are a diagnosis of follower maturity levels and the ability to be flexible in one's leadership style. Situational leadership theory has received a strong emphasis in the military setting. The Airman Leadership School which is designed for Sergeants (E-4), for example, includes instruction on how the effective use of the situational leadership model enhances a quality Air Force environment.

The contingency model. Some researchers consider this model as being more of a trait than a behavioral approach to impacting leadership effectiveness. Fiedler (1967) and Fiedler and Chemer (1982) consider leadership style to be relatively fixed and difficult to change through training, and that effectiveness is primarily determined by selecting the right kind of leader for a particular situation or training leaders to change the situation to fit their leadership style. Fiedler's least preferred co-worker (LPC) scale asks individuals to think of the single co-worker with whom he or she has had the greatest difficulty in working with. The individual then rates this LPC on 16 bipolar adjectives (e.g., friendly-unfriendly, open-guarded). Based on the LPC score, individuals are categorized into two groups: low LPC leaders who are primarily motivated by the task, and high LPC leaders who are primarily motivated by establishing and maintaining close relationships with followers. The most effective leadership style is determined by situational favorability (a combination of leader-member relations, task structure, and position power). The low LPC leader is most effective when situational favorability is high (good leader-member relations, a structured task, and high position power) or when favorability is low (poor leader-member relations, an unstructured task, and low position power). The high LPC leader is most effective when situational favorability is moderate. Although the contingency model is more complex than described and there is continued controversy over the meaning of LPC scores there is considerable empirical evidence that the model successfully predicts and improves leadership effectiveness. In fact, *Leader Match*, an instruction book based on Fiedler's model, has been widely used in training U.S. Army leaders.

Path-goal theory. This theory was developed to explain how the behavior of the leader impacts follower satisfaction and motivation. Path-goal theory is a good example of a transactional approach to leadership. As described by House (1971, p. 324):

"The motivational function of the leader consists of increasing personal payoffs to subordinates for work-goal attainment and making the path to these payoffs easier to travel by clarifying it, reducing roadblocks and pitfalls, and increasing the opportunities for personal satisfaction en route."

Four leadership behaviors are defined: supportive, directive, participative, and achievement-oriented. The first three behaviors should be familiar by now, and the fourth, achievement-oriented, refers to behaviors involving the setting of challenging goals and the emphasis on high performance standards. The appropriate leadership style is dependent on two major situational factors—task characteristics (e.g., task complexity) and follower characteristics (e.g., ability and personality). As with the other contingency theories, path-goal theory argues that leaders should first diagnose the situation and then select the leadership behavior most appropriate to that situation.

Margin notes: least preferred co-worker

Leader Match

In a review of these contingency approaches to studying leadership, the following conclusions can be drawn:

1. These theories correctly state that leadership effectiveness is dependent on characteristics of the leader, follower and the situation. Although there is not enough research evidence to conclude that one situational theory is significantly better than another or that any one theory has been demonstrated to be successfully validated, there is enough evidence to suggest that leadership performance can be enhanced by paying attention to situational factors. It is not appropriate to adopt one leadership style to cover all situations or all followers.
2. Key leadership skills involve the ability to correctly diagnose relevant situational factors and the ability to be flexible in changing one's leadership style to maximize performance. These skills are trainable, but one needs to recognize that leaders may find it difficult to apply "if—then" contingency theories because of time pressures or other factors.

Transformational Leadership Theory

When one really thinks about leaders, one often thinks about those leaders who have had strong effects in business, politics, or the military. These leaders might include, for example, Mahatma Gandhi, Adolph Hitler, Winston Churchill, Martin Luther King, Jr., John Kennedy, Lee Iacocca, Norman Schwartzkopf, and Colin Powell. These leaders share a number of common characteristics. First, they have the ability to create a common vision which they articulate and get followers to identify with. Bennis and Nanus (1985) argue that the lack of a clear vision is a major reason for the declining effectiveness of many organizations. Second, transformational leaders build trust in their leadership by exhibiting high levels of self-confidence and personal example. Third, they create an emotional involvement with their followers which results in feelings of empowerment and in a strong identification with the leader and his or her vision. An example of transformational leadership is provided by Ted Kennedy who, in his eulogy for Robert Kennedy, quoted:

"Some men see things as they are and say, why? I dream things that never were and say, why not?"

This focus on charismatic leaders is especially relevant today in light of the pressures on business and military leaders to create visions and to reinvent business and/or the military (Tichy & Devanna, 1986). Since the transactional leadership theories described earlier do not seem to help us understand the effect that transformational leaders have on followers, a recent approach to the study of leadership—which makes a distinction between transactional and transformational leadership—has received increased attention.

Bass (1985) states that a major component of transactional leadership involves contingent rewards (the exchange of praise and recognition for effort), but that transformational leaders: (a) raise the level of awareness of followers about the importance of achieving valued outcomes; (b) get followers to transcend their own self-interest for the sake of the team or organization; and (c) expand followers' portfolios of needs by raising their awareness to improve themselves and what they are attempting to accomplish. According to Bass, the major components of transformational leadership are charisma, individualized consideration,

Transactional
Leadership

intellectual stimulation, and inspirational motivation. These components are described below and some items taken from the Multifactor Leadership Questionnaire (MLQ) used in the leadership study at VMI (Atwater, et al., 1996) are provided:

Charisma—leaders become a source of admiration by followers, often functioning as their role models. ("I am ready to trust him to overcome any obstacle").

Individualized consideration—leaders assess each follower's needs, motivation, and capabilities. ("He treats followers as individuals rather than just members of a group").

Intellectual stimulation—leaders stimulate their followers to view the world from new perspectives and question old assumptions. ("He gets his followers to look at problems from many different angles").

Inspirational motivation—leaders build followers' confidence to overcome obstacles and take on greater personal challenges. ("He encourages his followers to try their best").

Numerous studies have supported the contention that transformational leadership is related to effectiveness in military settings and that it augments the effect of transactional leadership. These studies suggest that *both* types of leadership impact performance—transformational leadership does not rule out the need for good transactional leadership. Yammarino and Bass (1988), in a study conducted with a sample of U.S. Naval Academy graduates, found that transformational leadership ratings collected from subordinates were highly related to fitness report scores and recommendations for early promotion. Avolio, Atwater, and Lau (1993) found that VMI cadets rated as transformational by their peers received higher scores in Army ROTC summer camp conducted at Fort Bragg than cadets rated as less transformational. Atwater, et al., (1996), in a longitudinal leadership study conducted at VMI, reported that subordinate ratings of transformational leadership were related to peer ratings and rank in the Corps of Cadets. Curphy (1992) investigated the effects of transactional and transformational leadership on squadron performance at the U.S. Air Force Academy, finding that both components had a positive effect on squadron performance measures which required interdependent effort among subordinates. Clover (1989) also found that transformational leadership ratings from subordinates impacted team performance at the U.S. Air Force Academy.

The following conclusions can be drawn concerning transformational leadership:

1. Transactional leadership results in acceptable levels of follower performance but transformational leadership leads to performance beyond that which results from a simple exchange of rewards for effort. This effect has been demonstrated in a number of empirical studies conducted in military settings.

2. Transformational leadership can occur and has been documented in many teams, group, or organizational settings. In the study by Yukl & Van Fleet (1982), for example, inspirational and/or transformational leadership was mentioned quite often by ROTC members, and contributed to unit effectiveness. Examples included instilling pride, setting an example by one's own behavior, encouraging, and complimenting good performance. Given the importance of this type of leadership, it is important to develop military leadership training programs which go beyond instruction in transactional leadership.

3. Although women may still have fewer opportunities than men to emerge as leaders, they are just as effective when they do. A study conducted at West Point clearly demonstrated this. (Rice, Instone, & Adams, 1984) There is also evidence

that women may, in some situations, make better managers because they are more transformational than their male counterparts (Rosener, 1990; Bass & Avolio, 1992).

Building Leadership Skills

The purpose of this section is to provide a sample of research-based suggestions on how individuals can improve their leadership effectiveness. Several general principles are important. First, it is important to recognize that leadership performance is improved through education and experience. In order to become a better leader, one must manage time effectively, seek out leadership opportunities, take action, observe what happened, and then reflect on that experience, asking yourself how you could have handled that situation more effectively. You must be willing to experiment and to learn something from your successes and failures. Feedback from followers, peers and/or superiors is also important if you want to improve your leadership performance. Second, it is important to recognize that leadership style and performance is influenced by personal traits, attitudes, values and past experience. But leadership style and performance is also influenced by the ability and motivation of one's followers and by situational factors such as task complexity and time constraints. Therefore, it is important to be able to correctly diagnose important situational factors and be able to adopt the most appropriate and effective leadership style for that situation. Third, it is important to augment transactional leadership behaviors, such as providing contingent rewards, with transformational leadership behaviors (such as setting the example) in order to significantly impact the performance of followers.

The review of leadership research provided earlier also disclose a variety of hints concerning how you may improve your effectiveness. If we focus on the four major approaches—trait, behavioral, contingency, and transformational—it is possible to come up with more specific suggestions.

It is extremely difficult to change one's personality or level of intelligence, but certain traits that impact effectiveness can be changed. One of these is your level of technical expertise. The person who knows more than others about how to get a job done correctly will generally be the more effective leader. These skills are learned through formal education and through on-the-job training and experience. Yukl (1989) lists six traits (identified in the assessment center approach to measuring managerial potential) that are related to managerial success: energy level, organizing and planning skills, interpersonal skills, cognitive skills, work-related motivation, and personal control of feelings and resistance to stress. It is possible for leaders, through education, training, and experience to improve on many of these traits and become more effective.

Leaders make various assumptions about followers. McGregor (1960) called these assumptions Theory X and Theory Y. While these assumptions may or may not represent a "trait," these assumptions can be changed through education and training. Leaders who believe that followers have an inherent tendency to avoid work and therefore need to be directed, guided, or coerced (Theory X) will treat subordinates in a very different manner than leaders who believe that followers are capable of self-direction, and accept and seek responsibility (Theory Y). Leaders who believe in Theory X assumptions often create an

Theory X
Theory Y

self-fulfilling
prophecy

unfortunate side effect—the self-fulfilling prophecy. This occurs when the expectations of leaders actually bring about that behavior. In a study sponsored by the U.S. Marine Corps (Lau & Pavett, 1984), I recall interviewing an officer who assigned new recruits to challenging or menial jobs based on the information provided in the recruit's personnel folder. If the recruit's drill instructor had labeled the recruit a "dirt-bag," he was assigned to a menial job by the officer (e.g., raking sand). What would you predict the subsequent behavior of that recruit to be? Systematic studies by Eden (1992), conducted with the Israeli Defense Force, described how managerial expectations regarding subordinate performance impacted on motivation and subsequent performance.

Based on the findings from Ohio State and the University of Michigan, two major leadership components were identified—consideration and initiation of structure. Recent research has identified a number of other important leadership behaviors (information gathering and dissemination, technical problem solving, and decision making/resource allocation) It is important to recognize that leadership also involves management, and that both are important contributors to leadership effectiveness. There are some critical behavioral skills that probably cut across all situations and followers (Hughes, et al., 1993). First, the *ability to communicate* with your followers is important, ensuring that they understand what you are really saying. This involves active listening skills, two-way communication, and paying attention to nonverbal behaviors. It is also important to provide constructive, specific, descriptive, and timely feedback to followers. Followers expect feedback and, if it is not provided, your followers will not be able to improve. This requires that you "manage by walking around." Second, it is important to pay attention to *human resource management skills*. These include providing a thorough socialization process (such as a realistic preview of the jobs that followers will perform), training and coaching, providing challenging jobs, and bias-free performance appraisals that reflect actual levels of performance. Third, it is important to *motivate* your followers. Follower motivation, those behaviors that involve efforts directed at accomplishing organizational objectives, is strongly influenced by the behavior of leaders. Leaders must be able to identify what motivates followers and then attempt to satisfy these needs through contingent, positive rewards such as praise and recognition; setting goals that are specific, measurable, challenging, and accepted; and/or participative leadership which often results in higher follower commitment, involvement, and empowerment. Finally, the behavioral model suggests that *networking and political skills* contribute to effectiveness. Effective and successful leaders develop a network of contacts within and outside the organizational unit, develop favorable relationships with their superiors, and learn how the political system works in the organization. These activities enable leaders to gain upward influence in the organization, and to gain valuable resources that are necessary for getting the job done. Relevant activities include joining professional organizations and clubs, exchanging information, inviting outsiders to visit your unit, and visiting customers. To be an effective and successful leader, one must focus on both networking and the management of human resources.

The recent emphasis on transformational leadership also provides some suggestions on improving leadership effectiveness. The ability to create a vision, a view of where the organization is now and where it should be in the future, strongly contributes to leadership effectiveness. It is necessary to share that vision with followers through the development of rhetorical skills that inspire followers. It is also important to set the example by being an effective role model. Studies conducted at the service academies (e.g., Adams, et al., 1981; Curphy, 1992) and at VMI (Atwater, et al., 1996) have identified that some military leaders are more inspirational than others and have significant impacts upon effectiveness. According to Bass (1990), these skills can be developed through training.

Spitzberg (1987) estimated that over 500 colleges or universities offered some type of leadership training. In 1992, for example, the University of Richmond established the Jepson School of Leadership which offers undergraduates a degree in leadership studies. Leadership training is also an essential part of the armed forces. Recent changes in recruit characteristics, the technological complexity of weapon systems, and the emphasis of having to do more with less, make the need for effective leadership training more important than ever.

The first challenge is to discover what the military identifies as "good leadership." An ideal military leader is seen as an inspiring, dynamic, heroic, role-model; an individual who subordinates would follow up a hill into enemy fire (Atwater & Yammarino, 1993). This suggests that a focus on transactional leadership skills and bridging the gap between theory and practice is important. Secondly, it may be true that no amount of training will produce good leaders if the individual being trained lacks basic leadership potential. Thus, a focus on selection programs that identify personal traits related to leadership effectiveness continues to be important. Third, the conceptualization of leadership most consistent with the military ideal seems to be an emphasis on transformational leadership training.

At a conference held at the U.S. Naval Academy in 1987, one session was devoted to the ways that the service academies provide leadership training. Several common characteristics were identified. All of the academies emphasized formal classroom training integrated with the opportunity to lead and to receive performance feedback. The major emphasis appears to be on transactional leadership training, while transformational leadership could only be learned through observation.

At the U.S. Naval Academy, leadership begins as soon as the freshman class arrives. This period, "Plebe Indoctrination," lays the foundation for the next four years (Katz, 1989). This process, and the third-class summer cruise, develops an understanding of followership as well as leadership. In addition, all second classmen are required to take formal leadership training. Leadership training at the U.S. Air Force Academy begins with a selection program (high school standing, athletic involvement, club offices held, etc.) which is used to screen candidates for leadership potential (Gregory, 1989). Like West Point, training is developed around the concept of learning how to be a good leader by starting with how to be a good follower. Under the direction of a commissioned officer attached to each squadron (generally a captain or a major), each cadet receives a theoretical background of leadership in formal classroom settings, has the opportunity to observe correct applications and practice leadership in structured and unstructured situations and receive feedback from his or her superiors. Leadership development at West Point also includes formal classroom training and emphasizes the connection between the study and practice of leadership (Prince, 1989). Cadets are placed in leadership positions each year and evaluation and feedback are directed at improving leadership performance.

The common themes of these military leadership training programs are a focus on contingency leadership principles, followership that precedes leadership activities, leadership experiences combined with feedback, and formal, classroom training designed to provide the theoretical basis for leadership experiences. Since a variety of empirical studies have demonstrated that transformational leadership augments or supplements transactional leadership, it appears that the effectiveness of these training programs could be enhanced through the introduction of training in this area. One technique that has proven to be effective at the U.S. Naval Academy involved a process whereby followers provided anonymous, written feedback to upper classmen. The result of this feedback process was to lower the discrepancy between self-ratings and subordinate ratings of transformational leadership and to improve subsequent leadership performance (Atwater & Yammarino, 1992).

Challenges of the Future

Three major challenges confront today's military leader. The first challenge concerns the transferability of leadership across different countries and cultures; the second involves the leadership of a diverse workforce characterized by women and minorities; and the third challenge concerns how to lead in Total Quality Management organizations.

Contingency theories suggest that leadership effectiveness is affected by the leader, the follower, and by situational factors. An additional situational factor concerns cultural differences. For example, Dumaine (1992) reported that General Schwarzkopf suppressed his tendency toward impatience during the Persian Gulf War to engage in "philosophizing" with the leaders of Saudi Arabia because it was their way of decision making. This behavioral change impacted his effectiveness. Since many military leaders have been and will continue to be deployed to different countries or exposed to different cultures, it is important to consider how this factor may impact leadership effectiveness. In addition, the recent internationalization of American business, the emergence of a world economy, and the finding that American managers when assigned overseas have significantly higher failure rates than their foreign counterparts, has resulted in an increased interest in leadership across different countries and cultures.

In the American culture there is a tendency to emphasize individualism, fairness, competitiveness, assertiveness, risk-taking, informality, and the sharing of power. These cultural factors have influenced the development of leadership theory and the prescriptions taught in American leadership training programs. Do the leadership skills learned in one culture and management techniques such as goal setting, contingent rewards, or TQM work as well in other cultures? Do leadership styles vary across different countries or cultures?

In the Japanese workplace there is a tendency to emphasize collective achievement through group effort and to downplay individual effort. Decisions are made through participation and consensus and there is a stress on loyalty to the organization. In German culture there is a tendency to emphasize formality, the importance of hierarchical relationships, and a more directive leadership style.

The American manager who stresses contingent rewards and individual achievement with a Japanese workforce, who engages in informal working relationships with a German workforce, or who introduces participative leadership in a Latin American or Arab workforce may encounter considerable resistance and be ineffective. It is important to recognize the importance of cultural differences and follower expectations. Leadership skills and management techniques may or may not be generalizable, depending on the similarities between one culture and another, and effective leadership style does seem to vary across different countries or cultures. Since many American servicemen and women will be assigned overseas and will work with and supervise members of a foreign workforce, it is important to recognize these cultural differences. It is also important that military personnel receive intercultural training before and during their deployment.

The last several decades have seen the entrance of women and other minorities into all levels both in the military and civilian workplace. Diversity in the workplace is a fact of life no leader can afford to ignore. The second challenge to today's leader involves effective recruiting, training, and utilization of followers from a variety of backgrounds—gender, race, age, and education—and to value and capitalize on the unique capabilities that each of these groups brings to the workplace. Valuing diversity ensures that all individuals are given equal opportunities in the workplace and that merit, not irrelevant factors such as race or gender, is the only important factor in personnel decisions. This is a right guaranteed by Title VII of the 1964 Civil Rights Act and by other related decisions such as laws against pregnancy discrimination and sexual harassment. These changes represent a challenge and an opportunity for leaders to significantly influence the job satisfaction and job performance of women and minorities and to improve the competitive advantage of the organization. Discrimination charges in late 1996 at Texaco and charges of sexual harassment at Aberdeen Proving Ground, however, indicate that some organizations "just don't get it."

The third challenge involves leadership in a Total Quality Management (TQM) environment. Early in the 1980s, it was recognized that America was losing its edge in productivity and quality to foreign competitors. A 1978 NBC television production entitled "If Japan can, why can't we" sparked an interest in productivity and quality, and introduced W. Edward Deming to the American public. In addition to this television production, Peters and Waterman's widely read book, *In Search of Excellence,* made its appearance in 1982, and a variety of conferences were held to identify causes and solutions to the productivity crisis (e.g., King, Lau, and Sinaiko, 1983). As a result, many private sector organizations such as Ford, Motorola and Merck and, more recently, many public sector organizations, adopted a "new" approach to management and leadership focused an empowerment, information-sharing, and changes in traditional top-down management. The Army, for example, is considered by many to be one of the leaders in this change (Smith, 1994). One component of this change is Total Quality Management. TQM was officially adopted by the Navy, for example, in February 1992. TQM represents a long-term, continual commitment to quality improvement and customer service through the use of quantitative methods.

The emphasis in TQM training is on teamwork, two-way communication, participation, involvement (giving employees more responsibility and accountability for product delivery or service), and empowerment (pushing decisions down to the level where the most qualified individual makes the decision). It requires Theory Y leadership; leadership with less direction and more coaching, facilitating, and cheerleading; and delegating decision making responsibility to lower levels of authority. Although some military units may have not enthusiastically supported TQM, and its success in the public and/or private sector has yet to be conclusively demonstrated, it appears to be an approach that may be here to stay. *If* an organization has a top level commitment to TQM, this suggests that there may be an optimal leadership style. This style involves a transformational, democratic and participative approach which will, in most circumstances, fit the expectations and ability levels of today's members of the armed forces.

There is no shortage of literature dealing with the questions raised in this chapter. In fact, nearly ten thousand articles and books have been published on the topic! There are a variety of sources listed in the reference section that can provide you with a more in-depth coverage of these important leadership topics. These include the book authored by Hughes, Ginnett, and Curphy, former professors at the U.S. Air Force Academy, entitled *Leadership: Enhancing the Lessons of Experience*. The second source, entitled *Leadership in Organizations*, was authored by a variety of contributing authors from the faculty of the Department of Behavioral Sciences and Leadership at West Point. The third *Leadership in Organizations*, authored by Yukl, provides a thorough review of leadership theory and research findings. The final source, Bass and Stogdill's *Handbook of Leadership*, provides an extremely comprehensive review of the leadership literature. Some journals that often have useful articles on leadership include the *Journal of Applied Psychology, Military Psychology, Leadership Quarterly*, and the *Academy of Management Journal*.

References

Adams, J., Instone, K., Prince, & Rice, R. (1981). West Point: Critical incidents of leadership. *Armed Forces and Society, 10,* 597–611.

Atwater, L. & Yammarino, F. (1993). Personal attributes as predictors of superiors' and subordinates' perceptions of military academy leadership. *Human Relations, 46,* 141–164.

Atwater, L., & Yammarino, F. (1992). Does self-other agreement on leadership perceptions moderate the validity of leadership and performance predictions? *Personnel Psychology, 45,* 141–164.

Atwater, L., Dionne, S., Avolio, B., Camobreco, B., & Lau, A. (1996). *Leader attributes and behaviors predicting emergence of leader effectiveness.* (ARI Technical Report No. 1044). Alexandria, VA: U.S. Army Research Institute for the Behavioral and Social Sciences.

Avolio, B., Atwater, L., & Lau, A. (1993). *A multi-rater-view of transformational and transactional leadership behavior: Key predictors of Army camp performance.* Paper presented at the National Meeting of the Academy of Management, Atlanta, GA.

Bass, B. (1985). *Leadership and performance beyond expectations.* New York: Free Press.

Bass, B. (1990). *Bass and Stogdill's handbook of leadership.* New York: Free Press.

Bass, B., & Avolio, B. (1994). Shatter the glass ceiling: women may make better managers. *Human Resource Management, 33,* 549–560.

Bennis, W. (1959). Leadership theory and administrative behavior: The problem of authority. *Administrative Science Quarterly, 4,* 259–260.

Bennis, W., & Nanus, B. (1985). *Leaders: The strategies for taking charge.* New York: Harper and Row.

Burns, J. M. (1978). *Leadership.* New York: Harper.

Clover, W. (1989). Transformational leaders: Team performance, leadership ratings and first-hand impressions. In K. E. Clark & M. B. Clark (Eds.), *Measures of leadership.* West Orange, NJ: Leadership Library of America.

Curphy, G. (1992). An empirical investigation of the effects of transformational and transactional leadership on organizational climate, attrition, and performance. In K. E. Clark, M. B. Clark, & D. R. Campbell (Eds.), *Impact of leadership.* Greensboro, NC: The Center for Effective Leadership.

Department of the Army Pamphlet 600–80. (1987). *Executive leadership.* Headquarters, Department of the Army.

Dumaine, B. Management lessons from the General. *Fortune,* November, 1992.

Eden, D. (1992). Leadership and Expectations: Pygmalion Effects and other self-fulfilling prophecies in organizations. *Leadership Quarterly, 3,* 271–305.

Fiedler, F. (1967). *A theory of leadership effectiveness.* New York: McGraw-Hill.

Fiedler, F. & Chemers, M. (1982). *Improving leadership effectiveness: The Leader Match concept.* New York: Wiley.

Fleishman, E. (1973). Twenty years of consideration and structure. In E. Fleishman and J. Hunt (Eds.), *Current developments in the study of leadership.* Carbondale, Ill: Southern University Press.

Gregory, R. (1989). Leadership training at the U.S. Air Force Academy. In L. Atwater and R. Penn (Eds.), *Military leadership: Traditions and future trends.* Annapolis, MD: Action Printing.

Hersey, P., & Blanchard, K. (1984). *The management of organizational behavior.* Englewood Cliffs, NJ: Prentice Hall.

House, R. (1971). A path-goal theory of leadership effectiveness. *Administrative Science Quarterly, 16,* 321–338.

House, R., & Dessler, B. (1974). The path-goal theory of leadership: Some post-hoc and a priori tests. In J. Hunt and L. Larson (Eds.), *Contingency approaches to leadership.* Carbondale, Ill: Southern University Press.

Hughes, R., Ginnett, R. & Curphy, G. (1993). *Leadership: Enhancing the lessons of experience.* Boston, MA: Irwin.

Katz, D. (1989). Leadership education and training at the U.S. Naval Academy. In L. Atwater and R. Penn (Eds.), *Military leadership: Traditions and future trends.* Annapolis, MD: Action Printing.

King, B., Lau, A. & Sinaiko, W. (1983). *Productivity programs and research in U.S. government agencies.* Office of Naval Research. (Contract N00014–80–C–0438).

Lau, A., Atwater, L., Avolio, B., & Bass, B. (1993). *Foundations for measuring the development and emergence of leadership behavior.* (ARI Research Note No. 93–22). Alexandria, VA: U.S. Army Research Institute for the Behavioral and Social Sciences (NTIS/DTIC AD: A273 108).

Lau, A., Newman, A. & Broedling, L. (1980). The nature of managerial work in the public sector. *Public Administration Review, 40,* 513–520.

Lau, A., & Pavett, C. (1980). The nature of managerial work: A comparison of public and private sector managers. *Group and Organizational Studies, 5,* 453–466.

Lau, A., & Pavett, C. (1984). *Searching for a few good men: Identifying the outstanding performer in the U.S. Marine Corps.* Paper presented at the National Meeting of the Academy of Management, Detroit, MI.

Likert, R. (1961). *New patterns of management.* New York: McGraw Hill.

Luthans, F., Hodgetts, R. & Rosenkrantz, S. (1988). *Real Managers.* Cambridge, MA: Harper and Row.

Mael, F. & Schwartz, A. (1991). *Capturing temperament constructs with objective biodata.* (ARI Technical Report 939). Alexandria, VA: U.S. Army Research Institute for the Behavioral and Social Sciences.

McCall, M., & Segrist, C. (1980). *In pursuit of the manager's job: Building on Mintzberg.* Greensboro, NC: Center for Creative Leadership.

McCall, M., Morrison, A., & Hannan, R. (1978). *Studies of managerial work: Results and methods.* Greensboro, NC: Center for Creative Leadership.

McGregor, D. (1960). *The human side of enterprise.* New York: McGraw Hill.

Mintzberg, H. (1982). If you're not serving Bill or Barbara, then you're not serving leadership. In J. G. Hunt, U. Sekaran, and C. A. Schriesheim (Eds.), *Leadership: Beyond establishment views.* Carbondale, Ill: Southern Illinois Press.

Mintzberg, H. (1973). *The nature of managerial work.* New York: Harper and Row.

Pavett, C., & Lau, A. (1983). Managerial work: The influence of hierarchical level and functional specialty. *Academy of Management Journal, 26,* 170–177.

Pavett, C., & Lau, A. (1982). *Managerial roles, skills, and effective performance.* Paper presented at the National Meeting of the Academy of Management, New York.

Prince, H. (1989). Leadership development at the U.S. Military Academy. In L. Atwater and R. Penn (Eds.), *Military leadership: Traditions and future trends.* Annapolis, MD: Action Printing.

Rice, R., Instone, D. and Adams, J. (1984). Leader sex, leader success, and leadership process: Two field studies. *Journal of Applied Psychology, 69,* 549–570.

Rosener, J. (1990). Ways women lead. *Harvard Business Review, 68,* 119–125.

Schneider, B. (1985). Organizational behavior. In M. Rosenweig and L. Porter (Eds.), *Annual Review of Psychology,* (Vol. 36). Palo Alto, CA: Annual Reviews, Inc.

Smith, L. New Ideas from the Army (Really). *Fortune,* September 1994.

Spitzberg, I. (1987). Paths of inquiry into leadership. *Liberal Education, 73,* 24–28.

Stogdill, R. (1974). *Handbook of leadership.* New York: Free Press.

Tichy, N., & Devanna, M. (1986). *The transformational leader.* New York: Wiley.

Vroom, V. & Jago, A. (1988). *The new leadership: Managing participation in organizations.* Englewood Cliffs, NJ: Prentice Hall.

Vroom, V., & Yetton, P. (1973). *Leadership and decision making.* Pittsburgh, PA: University of Pittsburgh Press.

West Point Associates. (1988). *Leadership in organizations.* Department of Behavioral Sciences and Leadership, United States Military Academy. Garden City Park, NY: Avery.

Yammarino, F., & Bass, B. (1988). Long-term forecasting of transformational leadership and its effects among naval officers: Some preliminary findings. In K. E. Clark and M. B. Clark (Eds.), *Measures of leadership.* West Orange, NJ: Leadership Library of America.

Yukl, G. (1989). *Leadership in organizations.* Englewood Cliffs, NJ: Prentice-Hall.

Yukl, G. & Van Fleet, D. (1982). Cross-situational, multi-method research on military leader effectiveness. *Organizational Behavior and Human Performance, 30,* 87–108.

Yukl, G., Wall, S. & Lepsinger, R. (1989). Preliminary report on validation of the Managerial Practices Survey. In K. E. Clark & M. B. Clark (Eds.), *Measures of leadership.* West Orange, NJ: Leadership Library of America.

Reuters/Corbis-Bettmann

5

Teams and Teamwork in the Military

Eduardo Salas, Ph.D.
Janis A. Cannon-Bowers, Ph.D.
Stephanie Church Payne, M.S.
Kimberly A. Smith-Jentsch, Ph.D.

* Introduction * Why Is Teamwork Important in the Military? * What Is a Team? * What Comprises Teamwork? * How Can We Measure Team Performance? * Team Training * Concluding Remarks * References

Introduction

Many, if not all important or dangerous tasks in the military are performed by crews, groups, teams or units. This means that individuals must pool their resources in order to accomplish the overall mission or goal. Therefore, in order to optimize military performance and readiness it is crucial that we understand how to select, train and maintain high performing teams.

In recent years military psychologists have devoted considerable resources toward understanding the complexities and dynamics of teamwork and team functioning (Salas, Bowers, & Cannon-Bowers, 1995). In fact, military psychologists have been studying team performance for over 50 years (Dyer, 1984); hence, a large body of knowledge has accumulated that has begun to ascertain key findings on how military teams function. The purpose of this chapter is to briefly review what military psychologists have learned about teams and teamwork in the military. We do this by answering five questions: (1) why is teamwork important in the military, (2) what is a team, (3) what comprises teamwork, (4) how can we measure team performance, and (5) how do we train teams. We finish with a few remarks about the future of military teams.

Why Is Teamwork Important in the Military?

There are a number of reasons why teamwork is important in the military. The first, is that *readiness* (i.e., the ability to fight, on-demand, anywhere) requires that personnel

The views expressed herein are those of the authors and do not reflect the official position of the organization with which they belong.

synchronize, coordinate and communicate effectively during the "fog of war". Military personnel have to process enormous amounts of information in a timely, efficient and accurate manner in order to maintain tactical superiority. It is imperative that during these life-or-death situations, military teams exhibit flawless performance including expert teamwork. Consequently, the development of critical teamwork skills cannot be left to chance alone. These skills must be assessed and developed in preparation for combat, and maintained during actual combat situations.

The second reason is that recent changes in the world order have dictated a change in military strategy. That is, today's predicted battle scenarios are characterized by battle space compression, ambiguity with respect to intent, severe time pressure, no clear or visible enemy, and ever-increasing levels of information overload. Furthermore, the military of the future will likely be called upon to accomplish a wider range of missions and objectives. Taken together, these factors place unprecedented demands on human operators and teams. In order to cope with these challenges, we must have flexible and adaptable systems. Teams hold competencies essential for motivating their teammates, backing each other up, monitoring each other, self-correcting and shifting strategies when the situation demands it. Therefore, we must identify and develop these competencies to prepare warfighters for the varied, unexpected, stressful and rapidly changing environment of the future.

The third reason why teamwork is important, which is related to those already discussed, is the nature of the tasks teams are typically called upon to perform. Specifically, tasks in the military often include decision making in highly complex environments (Cannon-Bowers, Salas, & Grossman, 1991). Tactical decision making teams are faced with scenarios characterized by rapidly unfolding events, multiple plausible hypotheses, high information ambiguity, severe time pressure, sustained operations, and severe consequences for errors. Military team tasks also include working with sophisticated technologies which requires team members to integrate information from a number of systems. In sum, in order to maintain readiness adapt to new battle scenarios and cope with the nature of team tasks, military team members must learn to coordinate their actions so that they can gather, process, integrate, and communicate information in a timely and effective manner (Salas, Cannon-Bowers, & Johnston, 1997).

What Is a Team?

While most of us have an intuitive understanding of what it means to be part of a team, military psychologists in this area have sought to define the term "team" as a means to distinguish real teams from other types of work groups. In this chapter we will use the definition of a team provided by Salas, Dickinson, Converse, & Tannenbaum, 1992:

team
> "a team is defined as a distinguishable set of two or more people who interact, dynamically, interdependently, and adaptively toward a common goal/ objective/mission, who have each been assigned specific roles or functions to perform, and who have a limited life-span of membership (p. 4)."

There are at least three aspects of this definition that warrant discussion. First, the definition states that the team members act interdependently. This means that no one team member can accomplish the team's mission or goal alone. It also means that team members are dependent upon one another. That is, the output of one team member's work affects his/her teammates. This interdependence can also take several forms. For example, sequential interdependence occurs when team members contribute to the task in some sequence, where one team member must wait for his/her teammates for input before performing his/her own task. For example, in a Combat Information Center (CIC) team, the electronic warfare supervisor would identify the type of radar emitted by an aircraft and pass this information to the antiair warfare coordinator, who would then identify the type of aircraft detected. A second kind of interdependence is called pooled interdependence. In this case, team members work on completing their own individual tasks and only engage in minimal amounts of coordinated activity with their teammate(s). For example, when aircrews are working under normal conditions, the pilot may be flying the aircraft and the co-pilot may be attending to the radio and a variety of aircraft systems. However, the two must coordinate if either one becomes confused or if a problem occurs which requires explicit coordination. A final type of interdependence is known as reciprocal. Reciprocal interdependence demands the highest degree of coordination among team members. It requires that team members make mutual adjustments on a continuous basis. This type of coordination is necessary for Combat Search and Rescue (CSAR) teams. The downed pilot must work very closely with the CSAR team to provide them with his specific location. The CSAR team must work with the pilot and coordinate to first survey the area in which the downed pilot is located and second, to send in an aircraft to perform the actual rescue. In wartime or heavy combat situations, intense coordination is required to successfully retrieve downed military aviators.

Turning back to the definition of a team offered earlier, a second aspect that is important concerns "shared, valued goals." In order to be a team, members must be working towards the same end, and they must value that accomplishment. This may seem obvious, but in many cases individuals in a team have their own goals that are not necessarily compatible with the overall team's goal.

A final point that needs to be made regarding the definition of a team concerns the term "adaptively." One of the most important features of a team is that it can adapt itself to novel situations more easily and effectively than an individual. As stated earlier, this is because teams have pooled resources (for example, the talents of multiple members) that can be allocated and reallocated as required by the task. Hence, teams are particularly well suited to tasks that have rapidly changing demands or that can turn into emergency or dangerous situations quickly.

Besides the characteristics listed above, many military teams share several other characteristics that describe how they perform. These characteristics are shown in Table 1. An important question regarding team performance is how the characteristics shown in Table 1— for example, communication, coordination and task knowledge—actually contribute to a team's success. That is, what exactly is the nature of *teamwork*?

What Comprises Teamwork?

Several years ago, military psychologists working with Navy command and control teams addressed the question of "what is teamwork?" (McIntyre & Salas, 1995). In order to answer this question they observed the performance of a variety of Navy teams performing their tasks. From this observation and subsequent analysis, this group of researchers concluded that

Table 1

Defining characteristics of teams

- There are two or more team members
- Team members possess multiple sources of information
- Team members hold meaningful task interdependencies (i.e., individuals are required to work together to accomplish the task)
- There is close coordination among members (i.e., team members' actions must be synchronized to work together
- Team members have shared common, valued goals and objectives
- Team members hold specialized member roles and responsibilities (i.e., team members have different work requirements, task relevant knowledge and responsibilities)
- Team is hierarchically organized (i.e., team has a leader)
- Task demands intensive communication processes
- Team develops adaptive strategies to help respond to change (i.e., allows the team to respond differently to situations based on experience)

taskwork skills   there are actually two tracks of skills that develop in teams. First, taskwork skills are defined as those skills associated with the technical aspects of the job. These are typically what we think of when we talk about an individual's task or job skills. Taskwork skills include all of the necessary competencies (knowledge, skills, and attitudes) that a person requires to accomplish his/her task. Taskwork is that work done by an individual, behaving autonomously as opposed interdependently. For example, in an aircrew when the pilot is learning how to fly ("stick and rudder task"), he/she is getting competent on the taskwork aspect of flying.

teamwork skills   The second track of skill that emerges in teams is called "teamwork" skills (Cannon-Bowers, Tannenbaum, Salas, & Volpe, 1995). These are a separate and distinct set of competencies that are associated with being an effective team member. It includes the knowledge, skills and attitudes (KSAs) required to work effectively with others in pursuit of a common goal. These KSAs will be discussed in detail in a later section. Suffice it to say here that one cannot assume that a team will be effective simply because its members have well developed task skills. On the contrary, effective team performance seems to require attention to teamwork skills as well.

The Nature of Team Competencies

In order to define what is required to be an effective team, military psychologists have studied military teams in a variety of settings (e.g., McIntyre & Salas, 1995; Prince & Salas, 1993; Salas et al. 1997; Stout, Salas, & Carson, 1994). Military psychologists have been interested in how, when, where, and why teams behave the way they do (Dyer, 1984; Salas et al., 1992). From this work, it is now possible to compile a list of competencies—knowledge, skills and attitudes—that contribute to teamwork. These are summarized in Table 2.

Table 2

Examples of competencies required for effective teamwork

Type of Knowledge	Definition
Cue-Strategy Associations	The association of cues in the environment to appropriate coordination strategies
Task-Specific Teammate Characteristics	Task-related competencies, preferences, tendencies, strengths, and weaknesses of teammates
Shared task models	Shared models of the situation and appropriate strategies for coping with task demands
Teamwork skills	Ability to comprehend the required skills and behaviors necessary for successful team performance
Team Orientation	Process by which information relevant to task accomplishment is generated and disseminated to team members

Type of Skill	Definition
Adaptability	Process by which team members use information from task environment to adjust strategies through flexibility, compensatory behavior, and the dynamic reallocation of functions
Shared Situational Awareness	Process by which team members develop compatible models of teams' internal and task environment
Mutual Performance Monitoring	Process by which team members give, seek, and receive, task clarifying feedback
Motivating Team Members/ Team Leadership	Process by which team members direct and coordinate activities of, and motivate other team members, assess team performance, and establish a positive atmosphere
Mission Analysis	Process by which team resources, activities, and responses are organized to ensure integrated and synchronized and completion of tasks

Type of Skill	Definition
Communication changed	Process by which information is clearly and accurately ex- between two or more team members
Assertiveness	The willingness of team members to communicate ideas and observations in a manner which is persuasive to other team members

Type of Attitude	Definition
Collective Efficacy / Potency	Belief that the team can perform effectively as a unit when given some set of specific task demands
Shared Vision	Commonly held attitude regarding the direction, goals, and mission of a team
Team Cohesion group;	Total field of forces which act on members to remain in the an attraction to the team as a means of task accomplishment
Collective Orientation	Belief that team approach is better than individual one
Importance of teamwork	The attitudes that the team members have toward working as a team

Note. Adapted from Cannon-Bowers, et al., 1995 with permission.

According to Table 2, team members require certain types of knowledge to be effective. This is knowledge about the tasks, the environment and other team members. It includes knowledge of: cue/strategy associations, task-specific teammate characteristics, shared task models, teamwork skills, team orientation and others (for a full description see Cannon-Bowers et al., 1995). We elaborate on two with examples. First, Cannon-Bowers and colleagues have ascertained that in order to be effective, team members in information-rich situations must make associations between cues (i.e., target identification) in the environment and coordination strategies (i.e., what to coordinate to whom). That is, a pattern of cues in the environment tells team members what should be done in terms of coordination with teammates. It is this knowledge that allows team members to adjust their strategy quickly to radically evolving scenarios. Research has also shown that team members must also possess knowledge about the preferences, tendencies, strengths and weaknesses of their teammates. Specifically, team members adjust their behavior depending on who their teammates are. For example, when Bob is copilot, Jane behaves accordingly; however she adjusts her strategy when Joe is copilot because she knows what his particular strengths and weaknesses are. It is this type of team knowledge that allows teams to optimize their resources in accordance with situational demands.

Besides knowledge, team members must have certain team-related skills (as shown in Table 2). Team members do things to accomplish their mission. They need skills that allow them to perform actions in an effective manner. Research has shown that, among others (see Cannon-Bowers et al., 1995), team members adapt to unpredictable situations, monitor each other, provide motivational reinforcement, perform self-correction, exhibit flexibility, assertiveness and organization, use close-loop communication and resolve conflicts.

Finally, in order to be most effective, team members must hold certain task-related attitudes (see Table 2). That is, how team members feel about the task, the environment and the team influences teamwork. There is considerable research in this area. For example, cohesion—the willingness to remain in the group—has been studied in the Army (Dyer, 1984) and the Israeli Defense Forces (Soloman, Milkulincer, & Hobfoll, 1986). Also recently, Driskell and colleagues (Driskell & Salas, 1992) have shown that team members' disposition to receive and value inputs from others—i.e., collective orientation—facilitates teamwork. Similarly, the belief that the team can perform a task—i.e., collective efficacy—has also been shown to enhance teamwork (Guzzo, Yost, Campbell, & Shea, 1993).

How Can We Measure Team Performance?

A very important step in being able to train and maintain high performing teams is being able to accurately and meaningfully assess the team's performance.

When evaluating team performance, it is important to look at both the individuals who make up the team as well as the team itself. Individual knowledge, skills, and attitudes related to the work being performed ensure that each team member has the ability to do his/her own job. However, team performance is not simply the sum of its members' individual knowledge, skills, and attitudes. Team performance requires effective taskwork and teamwork skills in addition to individual expertise.

What to look for

For purposes of measuring team performance, it is useful to make the distinction between taskwork and teamwork skills as we did earlier (Morgan, Glickman, Woodward, Blaiwes, & Salas, 1986). Recall that taskwork skills are those skills that are related to the execution of the task or the mission itself (e.g., a pilot's stick and rudder skills, knowledge of equipment), and that teamwork skills are those skills that are related to functioning effectively as a team member (e.g., backup behavior, mutual performance monitoring, error correction). Therefore, to accurately diagnose team performance both types of skills need to be accessed. Why? Because, taskwork often explains the outcome of the team's performance or the end result (e.g., accuracy, latency, shoot/no shoot decisions), while teamwork describes the process (i.e., the moment-to-moment actions) that led to the outcome. In other words, measures of team process describe the performance strategies exhibited and explain how or why a particular outcome occurred (e.g., quality of team communication, adequacy of team leadership).

Although successful outcomes are the ultimate goal of team performance and team training, measurement of the processes that led to the outcome are critical for diagnosing performance problems. This is because sometimes teams arrive at correct outcomes due to luck or in spite of flawed processes. Measures of both outcomes and processes are necessary in order to identify strategies that, on average, lead to more effective outcomes. These individual and team processes then become the focus of evaluation and training.

Steps to take

There are three steps to measuring team performance. Each step has a unique purpose and involves distinct activities to be performed by the person doing the measurement (the assessor). The first step involves describing both individual and team behavior. The purpose of this step is to accurately capture and describe individual and team behavior in order to determine the basis for measurement. The assessor activities for this step include observing team behavior in order to capture moment-to-moment changes in individual and team task performance and identifying critical interactions between team members.

The second step of team performance measurement is to evaluate behavior. The purpose of this step is to evaluate the team behaviors that have already been identified in order to determine performance standards. The assessor activities for this step include evaluating the behaviors of many different teams using several different methods. This ensures that the performance standards apply to various teams which makes them both generalizable and reliable.

The third and final step of team performance is to diagnose causes of behavior. The purpose of this step is to determine what causes effective and ineffective performance. This step is crucial since it helps to identify the type of feedback or correction the team needs. The assessor activities for this final step include identifying and evaluating contributing factors, and task and mission parameters that may contribute to less effective behavior. Diagnostic information that indicates how team performance can be improved should also be identified during this step.

Developing tools

A number of activities are performed when developing team performance measurement tools for training. For example, team task analysis is conducted to identify the tasks that are performed by the individuals and the team as well as the appropriate teamwork and taskwork knowledges, skills, and attitudes (Salas & Cannon-Bowers, 1997). Specific objectives for the training or what the trainees will be expected to be able to do at the end of training are created. These objectives guide the development of tools to assess the desired cognitions, behaviors and attitudes needed.

Types of tools

A variety of tools can be used to measure team performance. The appropriate tool is dependent on what is being measured, as well as the purpose of the measurement. As mentioned earlier in this chapter, both individual and team-level efforts can be assessed and should be distinguished. In addition, outcomes and processes should also be given attention. Many types of measurement tools can be employed to capture individual and team-level processes and outcomes. Six types of measurement tools are described below.

- *Observational checklists* and *rating scales* capture moment-to-moment changes in performance strategies at both the team and individual level (Prince & Salas, 1993)
- *Content analysis* of communication patterns retrace factors contributing to coordination breakdowns in the team (Orasanu, 1990).
- *Critical incidents* demonstrate linkages between decision processes and important team outcomes (Morgan et al., 1986)
- *Automated performance recording* determine the latency of both team and individual actions; it can also be used to identify individual taskwork problems such as frequent keystroke errors (Salas et al., 1997).
- *After action reviews*, or team debriefing sessions, collect information on individuals' mental processes and decision biases that may not be obvious to the assessor (Dwyer, Oser, Salas, & Fowlkes, 1997).

These same six tools are also displayed in Figure 1 which illustrates which tool is appropriate given the desired unit of analysis.

Overall, an effective team performance measurement system has the following functions (Cannon-Bowers & Salas, 1997). (1) It identifies the team processes that are linked to key, team outcomes. (2) It distinguishes between individual and team level contributions as well as deficiencies. (3) It produces an assessment that can be used to deliver specific performance feedback that is applicable to both the team and the task. (4) It produces accurate and reliable evaluations that can be defended, when necessary. (5) It is operationally easy to implement and use (Cannon-Bowers et al., 1991).

In addition, a valid performance measurement system provides the foundation for effective team training. First, it identifies the training objectives and the content. Second, it allows for the tracking of participants' training progress. Third, it evaluates the effectiveness of different training strategies. Finally, it produces developmental feedback for trainees. In order to support the learning process, a team performance measurement system must produce

	Individual	Team
P r o c e s s	• Observational checklists • Rating scales • After action reviews	• Observational checklists • Rating scales • Current analysis • After action reviews
O u t c o m e	• Expert opinion • Automated performance recording	• Expert opinion • Automated performance recording

Figure 1: Individual and Team Process and Outcome Measurement Tools

assessments that can be used to deliver specific performance feedback. Performance measures can enhance learning if they provide behavioral information (e.g., reduce excess chatter) rather than general information (e.g., improve communication) to team members, if they are easy to use "real-time" during an exercise, and if they provide immediate feedback.

Team Training

Of central concern to those involved with military teams is how to train them. In recent years a number of research studies have been conducted with military crews. Out of this work, several training interventions emerge that appear to be effective in training teams. These are listed in Table 3 and summarized in the following sections.

Cross-training

Cross-training is a type of team training in which team members rotate positions in order to develop an understanding of the basic knowledge necessary to successfully perform the tasks, duties, and/or positions of the other team members. In addition, cross-training gives team members an overall framework of the team task and how each particular individual's task is important to it (Travillian, Volpe, Cannon-Bowers, & Salas, 1993). Military psychologists have demonstrated empirically that cross-trained teams outperform teams without such training. For example, one laboratory study that demonstrated the benefits of cross-training was conducted by Volpe, Cannon-Bowers, Salas, and Spector (1996). In this study, 80 male undergraduate students were assigned to 40 two-person teams. Half of the teams received cross-training on how their teammates' equipment operated while the other half only received training on their own functional responsibilities. All teams completed a 30 minute mission in a PC-based F-16 aircraft simulation. The objective of the mission was to "shoot down" as many

cross-training

Table 3

Military Team Training Strategies and Definitions

Strategies	Definitions	References
Cross-training	A team training technique which increases knowledge of individual team positions and the team as a whole by allowing each team member the opportunity to experience the tasks and duties associated with other roles within the team.	Travillian, Volpe, Cannon-Bowers, & Salas, 1993 Volpe, Cannon-Bowers, Salas, & Spector (1996)
Aircrew Coordination Training (ACT)	A method used to train crews in essential teamwork skills by providing team members with information about these skills, providing demonstrations of how and when these skills are used, giving individuals the opportunity to practice teamwork skills, and providing feedback to individuals on their performance.	Prince & Salas, 1993; Stout, Salas, & Fowlkes, in press
Guided team self-correction	A training strategy in which the team leader or instructor poses questions to the team regarding the problems the team encountered while completing their tasks. This guided discussion allows the team to generate feedback about their performance.	Blickensderfer, Cannon-Bowers, & Salas, 1994; Smith-Jentsch, Payne, & Johnston, 1996
Team coordination and adaptation training	This training strategy strives to improve teamwork skills during periods of high stress and/or workload by teaching teams how to maximize preplanning sessions, periods of low workload, the way information is anticipated and distributed among team members, and how to redistribute tasks to reduce workload.	Serfaty, Entin, & Volpe, 1993
Assertiveness training	A training intervention which either attempts to change attitudes towards the use of assertiveness behaviors using lectures or attempts to change actual behaviors through the use of role-plays and simulation exercises.	Ruben & Ruben, 1989; Smith-Jentsch, Salas, & Baker, 1996
Team leader training	A training technique which uses role-play exercises and simulations to teach team leaders how to be effective coaches/facilitators.	Caminiti, 1995; Tannenbaum, Smith-Jentsch, & Behson, 1997

aircraft as possible. Volpe et al. concluded that the cross-trained teams interacted more effectively with each other, as determined by a measure of overall teamwork. In addition, the cross-trained teams used more efficient communication strategies than those teams in the no-cross-training condition. Cross-training appears to be a valid intervention to improve team performance.

Aircrew Coordination Training (ACT)

The development of cockpit resource management (CRM) training was stimulated in the 1970s when National Aeronautics and Space Administration (NASA) determined that a high number of aviation accidents were occurring as a result of human error. The primary purpose for CRM training was to reduce the number of human errors committed and enhance mission effectiveness (Povenmire, Rockway, Bunecke, & Patton, 1989; Thornton, Kaempf, Zeller, & McAnulty, 1991). CRM training has been integrated into the curriculum for all pilots in the Air Force, Army, and Navy as well as commercial industries (see Weiner, Kanki, & Helmreich, 1993).

Aircrew Coordination Training (ACT) in the military is patterned after CRM training programs (Prince & Salas, 1993). Course content includes information designed to increase pilot's awareness of the need for coordination in the cockpit. The Navy developed and demonstrated a methodology that could be used by all aviation communities to create a validated, mission-oriented and skill-based training for aircrew coordination (Prince & Salas, 1993; Stout, Salas, & Fowlkes, in press). The methodology consisted of: (1) defining aircrew coordination skills among aircrews, (2) specify training objectives, (3) determining the best instructional strategy to support the objectives, (4) training instructors in how to observe and provide feedback to aircrews, (5) developing measurement and evaluation tools for feedback, and (6) providing guidelines for creating scenarios for aircrews to practice requisite skills. In general, the ACT comprised of: (1) providing aircrews with information about teamwork skills in the cockpit, (2) demonstrating the behavior via videotapes, (3) providing opportunities to practice (e.g., role plays, low fidelity flight simulators) the skills, and (4) receiving feedback on their performance. Topics like communication, situational awareness, leadership/ followership, decision-making, and mission analysis are discussed and taught. A key component of their methodology is that pilots have the opportunity to practice the skills they learn in high fidelity (very realistic) simulators and are evaluated on their ability to coordinate effectively with others.

Aircrew Coordination Training

Guided team self-correction

Encouraging feedback between team members is another technique that is often employed in team training. Salas, Cannon-Bowers, and Blickensderfer (1995) pointed out that feedback in a team environment should:

1) enable each team member to perform his/her individual task
2) demonstrate the contribution of an individual's performance to the performance of other members
3) demonstrate the contribution of an individual's performance to the performance of the team as a whole.

Team members often have all the information and expertise they need to identify and solve their own problems (Blickensderfer, Cannon-Bowers, & Salas, 1994; Smith-Jentsch, Salas, & Baker, 1997). The process whereby the team diagnoses important team processes and develops effective solutions to these problems has been referred to as team self-correction (Blickensderfer et al., 1994). It has been argued that this process could help to foster correct expectations and intentions among team members and ultimately contribute to more effective team performance.

guided team
self-correction

When team self-correction is led by the team leader or instructor, it has been referred to as guided team self-correction (Smith-Jentsch, Payne, & Johnston, 1996). Here a leader or instructor questions the team, usually in training, about specific teamwork processes. The team identifies their own problems which become targets for improvement in subsequent team training exercises. This process is guided in that the leader or instructor imposes structure on the team's feedback by asking focused, open-ended questions about specific teamwork processes that have previously been determined to be important to the performance of that team (Smith-Jentsch et al., 1996).

Guided team self-correction has been shown to improve training performance in both the laboratory (Koslowski, Gully, Smith, Brown, Mullins, & Williams, 1996) as well as onboard Naval cruisers (Smith-Jentsch et al., 1996; Tannenbaum, Smith-Jentsch, & Behson 1997). Teams who were debriefed by leaders trained to provide guided team self-correction outperformed teams debriefed by leaders in control groups.

Team coordination and adaptation training

Teams adapt to increased levels of stress by shifting their coordination strategies. Serfaty and colleagues recently developed a team training strategy aimed at improving teamwork during periods of high stress (Serfaty, Entin, & Volpe, 1993) by altering their coordination and reducing communications overheard. This strategy provides team members with specific information on how to best optimize (1) preplanning sessions, (2) idle periods (3) information dissemination, (4) information anticipation, and (5) redistribution of workload.

Team Adaptation
and Coordination
Training

Serfaty and colleagues have tested this team training strategy—called Team Adaptation and Coordination Training (TACT)—with naval command and control teams in the laboratory. Teams performed, under realistic conditions, a simulated tactical decision-making scenario. Teams trained on TACT outperformed those that did not receive the training. That is, effective teams provided more information to the team leader and exhibited more anticipatory behavior.

Assertiveness
training

Assertiveness training

Effective team decision-making requires that individual team members' unique knowledge, skills, ideas, and observations are recognized as resources that are available to solve a problem (Smith-Jentsch, Salas, & Baker, 1997). In certain environments, team members' ability and willingness to state their concerns assertively can mean the difference between life and death (e.g., nuclear power plant accident management, air crews, emergency medical teams).

In the context of team interaction, assertive communication is needed when providing feedback to other team members, stating and maintaining opinions, offering potential solutions, initiating action, offering and requesting assistance or backup when needed. Research has shown that we can not predict which team members will be assertive in a team setting on the basis of their tendencies to be assertive toward strangers or in personal relationships (i.e., friends & family) (Smith-Jentsch et al., 1996). Team member assertiveness seems to be linked specifically to the belief that it is appropriate to use this type of communication in the context of a team. However, effective team performance-related assertiveness is much more than just an attitude. Being clear and direct without putting other teammates on the defensive requires a significant degree of skill.

Industrial assertiveness training is on the rise in a wide range of settings; from hospitals to automobile factories (Ruben & Ruben, 1989). Such training generally involves a brief one-shot training seminar that takes one of two approaches. The first approach attempts to change attitudes with the presumption that behavior change will follow (Helmreich, 1984, 1987; Helmreich, Foushee, Benson, & Russini, 1986). This approach stems from the belief that non-assertiveness in the work-place is more a function of "status-typed expectations" than a lack of interpersonal skill (Driskell & Salas, 1991). Organizations which take this view typically make use of charismatic lecturers and organizational propaganda, promoting the concept of assertiveness much like a salesman (Ruben & Ruben, 1989). In contrast, other organizations take a second more skill-based approach which emphasizes the active practice of specific task-related assertive behaviors in role-plays and/or simulation exercises followed by individualized performance feedback (e.g., Prince & Salas, 1993).

Previous research has demonstrated that active practice and feedback in training is essential for improving team performance-related assertiveness. Training which consisted of lecture, and critique of behavioral models improved attitudes toward using assertiveness in a team, however did not improve performance (Smith-Jentsch, et al., 1996). This finding is consistent with the view that team performance-related assertiveness has a critical skill component.

Team leader training

Team leader training

Recently, there has been a shift in the way that team leadership is viewed. Traditionally, leaders' primary function was to monitor, control, and make decisions for the team. While team leaders still perform these functions, a greater emphasis is being placed on their role as a coach/facilitator (Caminiti, 1995). This includes drawing out team members' unique contribution to the team as well as managing "team climate", or the norms and expectations that team members have regarding their interaction. Research suggests that team members' perceptions of a team's climate are strongly influenced by leader behavior. These perceptions, in turn, drive team interaction patterns (e.g., team member assertiveness), which can facilitate or hinder team performance.

As is the case with industrial assertiveness training, team leader training is conducted in a variety of ways that range in the degree to which they employ active practice and feedback. Similar to the skills involved in team performance-related assertiveness, the skills involved in being an effective coach/facilitator are best learned through practice in role-play exercises, and team simulations.

Recent research has demonstrated that leader feedback skills could be improved by a brief (2 hr.) training seminar which included information, demonstration, practice and feedback (Tannenbaum, et al., 1997). Five Navy officers who received this type of training were compared to a control group of five officers who received training unrelated to briefing their teams. Results showed that trained leaders were more likely to critique themselves during a team debrief, as well as request feedback on their own performance from the team. Trained leaders also guided their teams in critiquing themselves by probing them about specific teamwork behaviors. In response to this improved briefing style, team members were more likely to admit their own mistakes and to offer suggestions to others. When the performance of teams who were briefed by trained and untrained leaders was compared, results indicated that the improved briefings influenced team performance as well.

Concluding Remarks

Crews, units, teams, and groups are the cornerstone of the military. It is through these collectives that the military maintains its readiness levels. Therefore, the importance of teamwork cannot be overlooked. The military must develop and maintain effective teamwork to ensure tactical superiority. Military psychologists have made numerous advances in trying to understand the complexity of this behavioral phenomenon. We have learned what is a team and what are the characteristics that define it. We have learned what comprises teamwork. It is more than the sum of individual performances. Effective teamwork requires that team members possess a set of knowledge, skill, and attitude-based competencies. We have also learned about assessing and measuring teamwork. Tools that are scientifically-based but practical have been developed. Finally, we have learned about a variety of strategies available for team training. The future of the science and practice of teamwork in the military is promising. While there are new challenges, military psychologists are well positioned to help the military create effective warfighting teams.

References

Blickensderfer, E. L., Cannon-Bowers, J. A., & Salas, E. (1994, February). Feedback and team training: Team self-correction. *Proceedings of the 2nd Annual Mid-Atlantic Human Factors Conference* (pp. 81–85). Washington, DC: Human Factors Society, Inc.

Caminiti, S. (1995, February 20). What team leaders need to know. *Fortune, 131* (3), 93–100.

Cannon-Bowers, J. A., & Salas, E. (1997). A framework for developing team performance measures in training. In M. T. Brannick, E. Salas, & C. Prince (Eds.), *Team performance assessment and measurement: Theory, methods, and applications.* Hillsdale, NJ: LEA.

Cannon-Bowers, J. A., Salas, E., & Grossman, J. D. (1991, June). *Improving tactical decision making under stress: Research directions and applied implications.* Paper presented at the International Applied Military Psychology Symposium, Stockholm, Sweden.

Cannon-Bowers, J. A., Tannenbaum, S. I., Salas, E., & Volpe, C. E. (1995). Defining competencies and establishing team training requirements. In R. A. Guzzo & E. Salas (Eds.), *Team effectiveness and decision making in organizations* (pp. 333–380). San Francisco: Jossey-Bass Publishers.

Driskell, J. E., & Salas, E. (1991). Overcoming the effects of stress on military performance: Human factors, training, and selection strategies. In R. Gal & A. D. Mangelsdorff (Eds.), *Handbook of military psychology* (pp. 183–193). Chichester, NY: John Wiley & Sons.

Driskell, J. E., & Salas, E. (1992). Collective behavior and team performance. *Human Factors, 34,* 277–288.

Dwyer, D. J., Oser, R. L., Salas, E., & Fowlkes, J. E. (1997). *Performance measurement in distributed environments: Initial results and implications for training.* Manuscript submitted for publication.

Dyer, J. (1984). Team research and team training: A state-of-the-art review. In F. A. Muckler (Ed.), *Human Factors Review* (pp. 285–323). Santa Monica, CA: Human Factors Society.

Guzzo, R. A., Yost, P. R., Campbell, R. J., & Shea, G. P. (1993). Potency in groups: Articulating a construct. *British Journal of Social Psychology, 32,* 87–106.

Helmreich, R. L. (1984). Cockpit management attitudes. *Human Factors, 26,* 583–589.

Helmreich, R. L. (1987). Theory underlying CRM training: Psychological issues in flight crew performance and crew coordination. In H. W. Orlady & H. C. Foushee (Eds.), *Cockpit resource management training: Proceedings of NASA/MAC workshop.* NASA Ames Research Center: CP-2455.

Helmreich, R. L., Foushee, H. C., Benson, R., Russini, W. (1986). Cockpit management attitudes: Exploring the attitude-performance linkage. *Aviation, Space, & Environmental Medicine, 57,* 1198–1200.

Koslowski, S. W. J., Gully, S. M., Smith, E. M., Brown, K. G., Mullins, M. E., & Williams, A. E. (1996). Goal focus and advanced organizers: Enhancing the effects of practice. In K. A. Smith-Jentsch (Chair), *When, how, and why does practice make perfect?* Symposium submitted to the Eleventh Annual Conference of the Society for Industrial and Organizational Psychology, San Diego, CA.

McIntyre, R. M., & Salas, E. (1995). Measuring and managing for team performance: Emerging principles from complex environments. In R. Guzzo & E. Salas (Eds.) *Team effectiveness and decision making in organizations* (pp. 149–203). San Francisco: Jossey-Bass.

Morgan, B. B., Jr., Glickman, A. S., Woodward, E. A., Blaiwes, A. S., & Salas, E. (1986). *Measurement of team behaviors in a Navy environment.* (Technical Report No. NTSC TR–86–014). Orlando, FL: Naval Training Systems Center.

Orasanu, J. M. (1990). *Share mental models and crew performance.* Paper presented at the 34th annual meeting of the Human Factors Society. Orlando, FL.

Povenmire, H. K., Rockway, M., Bunecke, J. L., & Patton, M. W. (1989). Cockpit resource management skills enhance combat mission performance in B-52 simulator. *Proceedings of the Fifth International Symposium on Aviation Psychology* (pp. 489–494). Columbus: The Ohio State University.

Prince, C. A., & Salas, E. (1993). Training and research for teamwork in the military aircrew. In E. L. Weiner, B. G. Kanki, & R. L. Helmreich (Eds.), *Cockpit resource management* (pp. 337–366). Orlando, FL: Academic Press.

Ruben, D. H. & Ruben, M. J. (1989). Why assertiveness training programs fail. *Small Group Behavior, 20,* 367–380.

Salas, E., Bowers, C. A., & Cannon-Bowers, J. A. (1995). Military team research: 10 years of progress. *Military Psychology, 7* (2), 55–75.

Salas, E., & Cannon-Bowers, J. A. (1997). Methods, tools, and strategies for team training. In M. A. Quinones & A. Ehrenstein (Eds.), *Training for a rapidly changing workplace: Applications of psychological research* (pp. 249–279). Washington, D. C.: APA Press.

Salas, E., Cannon-Bowers, J. A., & Blickensderfer, E. L. (1995). Team performance and training research: Emerging principles. *Journal of the Washington Academy of Sciences, 83* (2), 81–106.

Salas, E., Cannon-Bowers, J. A., & Johnston, J. H. (1997). How can you turn a team of experts into an expert team?: Emerging training strategies. In C. Zsambok & G. Klein (Eds.), *Naturalistic decision making.* Hillsdale, NJ: LEA.

Salas, E., Dickinson, T. L., Converse, S. A., & Tannenbaum, S. I. (1992): Toward an understanding of team performance and training. In R. W. Swezey & E. Salas (Eds.), *Teams: Their training and performance* (pp. 3–29). Norwood, NJ: Ablex Publishing Corporation.

Serfaty, D., Entin, E. E., & Volpe, C. (1993, October). Adaptation to stress in team decision-making and coordination. *Proceedings of the 37th Annual Human Factors and Ergonomics Society Annual Meeting, Santa Monica, CA,* 1228–1232.

Smith-Jentsch, K. A., Payne, S. C., & Johnston, J. H. (1996). Guided team self-correction: A methodology for enhancing experiential team training. In K. A. Smith-Jentsch (Chair), *When, how, and why does practice make perfect?* Paper presented at the Eleventh Annual Conference of the Society for Industrial and Organizational Psychology, San Diego, CA.

Smith-Jentsch, K. A., Salas, E., & Baker, D. P. (1997). Training team performance related assertiveness. *Personnel Psychology, 49,* 909–936.

Soloman, Z., Milkulincer, M., & Hobfoll, S. E. (1986). Effects of social support and battle intensity on loneliness and breakdown during combat. *Journal of Personality and Social Psychology, 51,* 1269–1276.

Stout, R. J., Salas, E., & Carson, R. (1994). Individual task proficiency and team process behavior: What's important for team functioning. *Military Psychology, 6,* 177–192.

Stout, R. J., Salas, E., & Fowlkes, J. E. (in press). Enhancing teamwork in complex environments through team training. *Group Dynamics.*

Tannenbaum, S., Smith-Jentsch, K. S., & Behson, S. (1997). Team leader training teams. To appear in J. A. Cannon-Bowers & E. Salas (Eds.), *Decision making under stress: Implications for training and simulation.*

Thornton, R. C., Kaempf, G. L., Zeller, J. L., & McAnulty, D. M. (1991). *An evaluation of crew coordination and performance during a simulated UH-60 helicopter mission.* Unpublished manuscript, U.S. Army Research Institute Aviation Research and Development Activity, Fort Rucker, AL.

Travillian, K. K., Volpe, C. E., Cannon-Bowers, J. A., & Salas, E. (1993). Cross-training highly interdependent teams: Effects on team process and team performance. *Proceedings of the 37th Annual Human Factors and Ergonomics Society Conference* (pp. 1243–1247). Santa Monica, CA: Human Factors Society.

Volpe, C. E., Cannon-Bowers, J. A., Salas, E., & Spector, P. (1996). The impact of cross training on team functioning. *Human Factors, 38,* 87–100.

Wiener, E. L., Kanki, B. G., & Helmreich, R. L. (Eds.). (1993). *Cockpit resource management.* San Diego, CA: Academic Press, Inc.

Official U.S. Marine Corps Photo

6

Military Performance under Adverse Conditions

Gerald P. Krueger, Ph.D.

* Introduction * Climatic and Environmental Extremes * Effects of Heat
and Cold on Performance * Effects of Altitude on Performance
* Mechanical Forces * Acoustical Noise * Acceleration * Vibration
*Motion Induced Sickness and Military Performance * Atmospheric Mix
* Toxic Gases and Fumes * Radiation Exposure and Performance
*Military Operational Considerations * Chemical Protective Clothing
* Continuous Operations * Sustained Performance and Sleep Loss
* Concluding Remarks * References

Introduction

Soldiers, sailors, marines, and airmen engage in military occupational activities in a variety of adverse working conditions. A psychologist might quickly mention the many psychological stresses of military life—the sort that contribute to the emotional make-up of those who wear a military uniform. Being away from family and friends for extended periods, or feeling command and peer pressures to perform, or experiencing fear of being injured or killed by an enemy, each can constitute adverse conditions of a sort. However, this chapter is not about those psychologically laden stresses. Rather, it focuses on general ambient environmental stressors that accompany living and working conditions in harsh and dangerous military environments; or those which come with operating large equipment and weapon systems that may present hazards to the health and safety of operators. This chapter addresses adverse working conditions which impose stress upon military personnel and affect their performance in training, operational readiness exercises, or combat.

Military forces often are deployed to harsh environments ranging from cold wintry forests, to hot humid tropics, hot dry deserts, high mountainous altitude, high aerospace altitude, and to depths under the sea. Typical military work stations consist of small crew compartments in tanks and armored personnel carriers designed to traverse rugged terrain, or in high performance jet aircraft and helicopters winging the skies, or fast attack surface boats skimming ocean waves, and submarines to negotiate the sea bottoms. These working environments present a myriad of work stressors which usually involve combinations of adverse environmental conditions as high acceleration, noise, vibration, noxious fumes, smokes, obscurants, or chemical, biological and radiological contamination on the battlefield. Compounding the usually high physical and mental workload of military equipment operators are crew compartment and battlefield stressors of intense heat, toxic fumes, and atmospheric contaminants as well as other demanding conditions involving requirements of sustained operations and accompanying sleep loss, high psychological stresses, threats of indigenous biological diseases etc.

At times these work stressors and adverse conditions can dominate and decisively affect the performance of individuals and thereby determine mission effectiveness. In a somewhat abridged version of many chapters in the *Handbook of Military Psychology* (Gal & Manglesdorff, 1991), several of these adverse working conditions are described in this chapter. The information presented is from psychologists, physiologists, human engineers, ergonomists, other behavioral scientists, and occupational medicine specialists.

Climatic And Environmental Extremes

Effects of Heat and Cold on Performance

Humans have a complex thermo-regulated biological system to maintain the body core temperature in a range near 37°C (~99°F) with a variation of just ±2°C. If the temperature at a human cell exceeds 45°C, heat coagulation of proteins takes place. If the temperature reaches freezing, ice crystals break the cell apart. The body's regulatory system must keep its outer skin layer temperatures below the 40s (°C) to protect itself from conditions that are too hot; or keep temperature well above freezing to keep from being too cold. If deep body temperature deviates ±6°C from its set value it is usually fatal (ASHRAE, 1985); but if temperature fluctuates just a few degrees, physical and mental task performance can be impaired (Kroemer et al., 1986).

metabolic heat

Due to diurnal changes in body functions throughout the day, our body temperature fluctuates slightly; but the main impact upon the thermal regulatory system results from interaction of metabolic heat generated in the body (mostly through work or exercise) and from external energy gained by being in hot surroundings, or lost to the atmosphere in a cool environment. Cellular structures, enzyme systems, and many other functions are directly affected by changes in the body's core temperature (Kroemer et al., 1986). Heat strain and heat illness are the biggest risk to soldiers working in hot environments (whether it be hot dry desert or hot wet tropics) (Burr, 1991).

In an excellent review of this topic, Kobrick and Johnson (1991) indicated physicists and physiologists can precisely measure human responses to intense heat and severe cold, but it is difficult to accurately predict job performance of individuals and groups under specific climatic conditions. Psychological performance decrements and critical changes in physio-logical states do not necessarily occur together; in fact, psychological changes often precede physiological deterioration. "Thermal stress" research involving extreme climates, and "subjective comfort" studies, dealing with generally moderate conditions, involve different combinations of temperature, humidity, wind speed, and exposure time. In practical terms, two hours of thermal stress in one study may be no worse for performance than six hours of a moderate temperature in another, making generalizations about the effects on psychological performance difficult (Kobrick & Johnson, 1991).

Kobrick and Johnson (1991) reviewed fifty years of research on the effects of heat and cold stress on psychological performance. With respect to **heat**, performance on vigilance tasks becomes impaired above 90°F (32°C) with best performance at or about 85°F (29°C)/ 63% relative humidity (RH). Manual tracking becomes somewhat degraded at or above 85°F (29°C), but differences in exposure periods complicate that picture. Kobrick and Johnson (1991) state that reaction time, sensation, and psychomotor performance each have been extensively studied in the heat, but results vary, indicating in some cases impairment, or improvement, or no change in performance. The exact psychological tasks likely to be impaired during severe climatic exposures still cannot be stated specifically. Performance varies widely among individuals. Performance of various types of cognitive tasks has been reported as impaired for exposures above 100°F (38°C), but no change and even improved performance have also been reported (Kobrick & Johnson, 1991).

vigilance tasks

manual tracking

psychomotor performance

Performance in high heat. The following list of points (on *heat*) are summarized from Kobrick and Johnson (1991) with additional comments by this author:

1. To varying degrees, heat affects performance on different types of tasks. Since heat has a cumulative blunting effect, tasks of continuous low demand with relatively low arousal value, especially if they are boring and repetitive, are affected most (e.g. vigilance, low-activity sentry or surveillance duty, routine watchkeeping, etc.). Frequently occurring interesting tasks tend to be less affected by heat.

2. Heat affects people's performance differently according to their skill level. Those who are well trained at their jobs are better able to withstand ambient heat.

3. Crucial tasks should be learned and practiced in the heat, since conditions under which they are performed are altered by heat. Heat exposure brings typical body reactions such as sweating, creating performance conditions different from those for the same task done under cool conditions (e.g. sweat running into the eyes can blur vision; controls become slippery due to sweaty hands; eyepieces and headsets become uncomfortable, painfully hot or physically unstable on the head).

4. Hot environments create perceptual problems such as mirages, visual distortions, and optical illusions due to heat shimmer and glare, and can result in reduced or inaccurate visual task performance.

5. Protective uniforms, gloves, respirator masks, helmets, hoods, survival vests, personal gear carried on belts or shoulder harnesses, and especially the wearing of chemical protective clothing (CPC) can all interact with heat to affect performance. Such personal clothing and equipment can obstruct major sensory and information-gathering zones of the body. Impermeable CPC encapsulates soldiers; perspiration trapped inside suits condenses on viewing ports of face masks, restricting vision. Gloves become slippery inside and slide against the hands, making sensory impairment worse; cloth glove liners become soaked and bunch up at finger joints. Additionally, hands and feet can swell due to heat-related tissue edema, reducing joint motility and manual dexterity.

chemical protective clothing

6. Solar heat makes metal equipment surfaces very hot to touch, difficult or painful to handle and thereby can retard performance. Armored tanks and personnel carriers become heat sinks in desert sun, lead to arm and leg burns when climbing about hulks of hot metal, and, when vehicles are "buttoned up" for chemical-biological operations in the desert, require much energy to cool crew compartments. This is a particular problem for operator performance when artificial cooling systems are inoperative or unavailable.

Performance in the cold. The ability to maintain body temperature within a narrow range is also important in the cold. Cold temperature, directly cools skin and affects the viscosity of fluids in the joints, reducing manual dexterity, manipulative ability, and sensitivity of the skin to stimulation (LeBlanc, 1956). General lowering of body temperature results in hypothermia, and local freezing of body tissues can lead to frostbite. Wind-chill effects pose a greater danger to immobile soldiers than to exercising or working soldiers. When uniforms are wetted by sweat, an immobile soldier is particularly threatened by wind chill. Prolonged exposure of feet in cold water causes immersion foot or trench foot injuries, and both are serious problems for dismounted troops operating in cold marshy areas. Dry and windy conditions at high altitude or in the desert keep drinking water requirements high even when it is cold. Dehydration increases the risk of cold injury and contributes to discomfort in the cold (Krueger, 1993).

hypothermia

immersion foot

trench foot

Not as much research has been conducted on effects of cold on psychological performance. Among Kobrick and Johnson's (1991) list of general findings for *cold*:

1. Cold exposure primarily affects psychomotor and manual dexterity aspects of task performance. These impairments are largely due to stiffening of the muscles, joints and possibly synovial joint fluids, resulting in a mechanical "locking up" of the biomechanical capabilities of the hands. Manual dexterity tasks are affected whenever the hands are cooled to a critical peripheral skin temperature of 55°F (12.7°C). Therefore the crucial factor in maintaining dexterity is hand temperature despite the temperature level of the rest of the body. Keeping the hands warm is essential for preserving operational capability.

synovial joint fluids

2. Loss of dexterity in the cold is accompanied by reduced tactile sensitivity, resulting in diminished feedback as to what the hands are doing. This may be compensated visually, but does not serve for tasks in the dark or if task components to be manipulated are hidden from sight. Sensitive tasks done bare-handed may also involve cold-soaked components, which cause further contact cooling where sensitivity is essential.

3. Local peripheral application of heat to the hands is an effective remedy for maintaining manual capability in the cold.

4. Performance strategies in the cold include breaking tasks into shorter segments, interrupted by rewarming the hands. Ergonomic redesign enlargement of critical controls, using plastic coatings on control knobs can improve manipulability, avoids contact freezing of skin to metal forces and improves manipulation while wearing gloves. Important tasks should be learned and practiced in the cold.

Ergonomic redesign

5. The properties of physical materials to be used in the cold are important. For example, hydraulic-based tracking systems of weapons may have perfect feedback

properties in temperate conditions, but in wet-cold or the arctic will develop different hysteresis levels and throw off normal tracking performance.   hysteresis

6. In severe cold, water sources and containers present important problems including freezing or bursting, contamination, and simple lack of availability of drinking water. Physical problems associated with disruptions of water supply can contribute to increased dehydration of troops, a familiar problem in arctic operations.

7. Rewarming hands to the point of sweating, followed by subsequent cooling presents another problem for manipulative performance. Mittens or gloves, wet from perspiration or other moisture, can become cold and lead to frostbite.

The U.S. Army Research Institute of Environmental Medicine at Natick, MA provides occupational preventive medicine guidance for deployment to the desert or the arctic cold. Portions of that guidance are summarized here:

Troop deployment to the desert. Soldiers who live in temperate environments face several hazards upon deployment to a harsh desert. They need to be protected from heat and sun, but also dangerously cold winds in winter. Soldiers must be as prepared to keep warm at night as they are to keep cool in the day. Hazardous blowing dust and sand cause asthma attacks, chafing and skin infection, and eye irritation and infection.

Soldiers should attain peak physical fitness and heat acclimatization* prior to deploying to a desert. Significant acclimatization can be attained in 4–5 days; whereas full acclimatization takes 7–14 days of 2–3 hours per day of exercise in the heat. Physically fit troops acclimatize more rapidly than those less fit. Military units on alert for deployment should immediately optimize the physical training program and their state of heat acclimatization (i.e. spend more time exercising in heat), (Krueger, 1993).

Countermeasures for arctic deployments. (Young, et al., 1992; Burr, 1993). Soldiers   countermeasures
must have enough clothing and shelter to keep warm and be adequately equipped for exposure to cold. Loose, layered, dry clothing should be worn, as all enhance the insulating value for the body. Layers of clothing allow easy adjustment for comfort. Soldiers should regularly change clothes, should completely dry and warm their feet, and regularly change their socks, especially wet socks, 2–3 times per day. Direct effects of the wind should be avoided. Dry, windproof clothing and natural or artificial barriers should be used as wind breaks. Covering the head keeps one warmer, and exposed skin areas should be minimized.

If shivering, soldiers should do something to warm themselves, such as put on more clothing. Use of large muscles generates internal heat, or if that does not help, they should seek assistance from others in getting warm before hypothermia takes over.

Providing warm food and drinks at night before sleep helps keep soldiers warm and boosts morale. Adequate drinking and food reduces susceptibility to cold-wet injury. Planned shortened periods of sentry duty will allow soldiers the opportunity to get warm.

Effects of Altitude on Performance

"Lowlander" troops deploying to much higher terrestrial altitudes (i.e. mountains) encounter a decrease in atmospheric pressure as they ascend to higher altitude. Although the proportion

---

Acclimatization: development of better physiological tolerance to environmental extremes through continued exposure. Implies that acclimatized soldiers should perform better on tasks under environmental stress than unacclimatized people.

of oxygen in the air is constant (20.9%), decreased atmospheric pressure results in proportional reduction of partial pressure of oxygen in the air, and lowers the rate at which oxygen diffuses into the blood, leading to hypoxemia. This reduces oxygen to the brain, working muscles, and other parts of the body, and eventually has profound effects on sensory processes, mentation, sleep, and physiological work capacity. If soldiers do not ascend to altitude slowly, gradually acclimatizing to decreased atmospheric pressure, they will experience mountain sickness which may develop into three serious medical problems (Banderet & Burse, 1991; Cymerman & Rock, 1994):

partial pressure of oxygen in the air

> *Acute mountain sickness* (AMS). Rapid ascent to higher altitude produces headache, light-headedness, loss of appetite, sleeplessness, lassitude, nausea, vomiting, and general malaise. AMS begins in susceptible individuals within a few hours to a day. The faster one ascends to very high altitude, the more likely he/she will be affected by AMS.

> *High altitude pulmonary edema* (HAPE). Upon ascent to substantially higher altitudes, lung capacities become leakier in response to hypoxia. Blood plasma, forced into the lungs during exercise, interferes with oxygen transport, rendering the victim severely hypoxemic and greatly compromising the capillaries. Unless treated by quick descent to lower altitude, or by breathing supplemental oxygen, the lungs continue to fill with blood until the victim gets too little oxygen to survive.

hypoxia

> *High altitude cerebral edema* (HACE). At very high altitudes, increased intravascular pressure in the skull, and/or increased cerebral blood flow essentially lead to brain hemorrhages. Onset of HACE is indicated by mental confusion, slurred speech, or uncoordinated gait; it can develop rapidly, and include hallucinations, ataxia, sleepiness, paralysis, coma, and death. The first sign of symptoms calls for immediate descent with oxygen therapy.

ataxia

Other environmental stressors at high terrestrial altitude include cold, wind, dryness, and solar and ultraviolet radiation. Because of loss of appetite at high altitude, soldiers do not eat as much, and body nutrition is then affected. High altitude cold and extreme physical activities dramatically increase caloric demands and may require special nutrients.

Work performance. The most important effect of high altitude on physical performance is reduction in the maximum rate at which the body can use oxygen to perform physical work (maximum aerobic capacity). Not only is the maximum work rate less at altitude, and progressively more so the higher one goes, but the endurance time for submaximal work also is reduced and the effort required is greater (Banderet & Burse, 1991).

Aerobic work capacity is not affected meaningfully up to 1500 m, but is progressively diminished at higher altitudes. Loss in aerobic capacity is a monotonically increasing function of altitude (Cymerman et al., 1989), approximately an average of 10% loss in capacity for each 1000 m increase in elevation above 1500 m. The decrement persists as long as one remains at altitude.

The decrement in submaximal endurance is less predictable, as it depends on time at altitude and rate of work in addition to motivational factors. When physical work load is reduced at altitude to match the percentage of aerobic capacity taxed at sea level, average endurance times are approximately the same. The amount of external work (loads carried, rate of ascent) that can be accomplished is much less, because the total work rate includes the added work of faster breathing plus the load imposed by any environmental clothing and climbing equipment. The reports on how much submaximal endurance increases with prolonged acclimatization to altitude are contradictory, ranging from a small change after several weeks to large improvements up to 60%.

Sensory and neurological effects. The senses are affected by altitude before cognitive and psychomotor performances (Banderet & Burse, 1991). Sensitivity to light changes, and visual acuity and color discrimination decrease at altitudes above 3000 m. Vision is the first sense affected by hypoxemia resulting in performance decrements due to the changes in visual sensitivity. Reaction times to visual stimuli increase as one ascends to higher altitudes and are greater with hypoxemia.

Decreased detectability of a visual stimulus results from a change in the perceptual sensitivity of the visual system. High altitude produces neural changes in dark adaptation; but contrast sensitivity is unaffected during gradual ascent to 7620 m. Auditory sensitivity appears unaffected.

dark adaptation

Mood states. Initial exposure to high altitude may produce euphoria and errors in judgment. Even moderate levels of altitude alter metabolism of biogenic amines, and result in changes in mood and affect. For example, after 1–4 hr at 4300 m, subjects rated themselves as less "friendly" with less "clear thinking" and were also "more sleepy" and "dizzy;" but surprisingly, also "happier," which is perhaps related to the initial euphoria of high altitude (Shukitt & Banderet, 1988). At 1600 m, increased sleepiness was detected the first day but returned to sea level values after 18 hr. Although cognitive performance does not degrade very much at comparatively low altitudes (1800 m), moods and other psychophysiological symptoms generally are present and may affect a soldier's motivation.

Personality changes. In expedition climbs to high mountains, personality characteristics change acutely. For example, at 3800 m, personality characteristics and cognitive functioning were similar to those at sea level (Nelson, 1982). At 5000 m, undesirable changes were evident, i.e. increased paranoia, obsessive-compulsiveness, depression, and hostility; cognitive functioning was also dramatically impaired. Ryn (1988) describes mental disturbances in climbers as related to the level of altitude and duration of stay at high altitude. The neurasthenic syndrome, characterized by fatigability, lack of motivation, feeling inadequate, and psychosomatic symptoms, is common at 3000—4000m. The cyclothymic syndrome which occurs at 4000—7000 m involves alternating depressed and elevated moods. Acute organic brain syndromes occur above 7000 m and result from structural or functional defects in the central nervous system. The climber's personality, emotional atmosphere associated with climbing, high degree of risk, and other biological and psychological factors are important in the etiology of such mental disturbances.

cyclothymic syndrome

etiology

Knowing the effects of varying levels of high altitude is of critical importance for many military activities, especially those carried out in conjunction with special forces operations. Banderet and Burse (1991) give an excellent review of issues surrounding the effects of high altitude on neurological functioning. With this author's additions, the following are from Banderet and Burse (1991):

apnea

1. At high altitude people report disturbed and fitful sleep where respiratory periodicity and apnea (frequent awakening) and sleep disruption occur. Deep sleep (stages 3 and 4) is decreased and periodic breathing disruptions occur. Eventually as acclimatization occurs, sleep quantity and quality begin to return to resemble that normally attained at lower altitudes.

2. Cognitive performance is more vulnerable to hypoxemia than is psychomotor performance. Complex performances are usually affected before simpler performances, and activities requiring decisions and strategies are more vulnerable than automatic processes. Performances involving visual input of shapes, patterns and contours are more vulnerable to impairments at altitude than those involving numbers, words, or characters.

3. Performance impairments at high altitude can result from increased errors, slowing of performance, or a combination of these degradations. Varied performance outcomes and strategies (such as choosing to work slower to preserve accuracy) are common since task characteristics influence the quality of performance and the number and types of errors.

Mountain altitude countermeasures. At altitude, acclimatization begins almost immediately as our brains respond immediately by initiating an increased breathing rate in an attempt to elevate oxygen concentration in the blood stream. Other physiological mechanisms automatically increase efficiency of oxygen transport from the lungs to the brain and its utilization. Most of these normal bodily responses are well-developed within 1–2 weeks.

Acclimatization, staging by low ascent to high altitude, use of supplemental oxygen and selected medications like acetazolamide (a carbonic anhydrase inhibitor), dexamethasone (anabolic steroid), and tyrosine (a precursor of the neurotransmitter norepinephrine) can all serve roles in adapting to high altitude. A balanced diet at high altitude would include: 52% carbohydrates, 33% fats, and 15% proteins, extra vitamins and up to 5 liters of water per day may be required (Cymerman & Rock, 1994).

Climbers on foot can prevent AMS, HAPE, and HACE by ascending slowly, to 3000 m, and then climbing less then 500 m per day, allowing 1 or 2 days to acclimatize at each altitude. Shuttling equipment components for the next higher camp up 300—500 m over a day or two, but returning to sleep at a lower camp at night ("climb high, sleep low") takes advantage of this gradual acclimatization method (Banderet & Burse, 1991).

Aerospace altitude. Aviation personnel assigned flying duties frequently climb to very high altitudes in pressurized aircraft (other than jet fighters) and may occasionally be exposed to low levels of hypoxia, especially when the pressurization system is not working properly. On rare occasions they might expect to experience rapid depressurization of such aircraft, and in those instances must quickly take action to obtain supplemental oxygen or suffer dire consequences of hypoxic hypoxia.

Hypoxic hypoxia is incomplete oxygenation of the blood in the lungs. Oxygen deficient blood from the lungs is returned to the heart and circulated to the body tissues. This situation exists when the partial oxygen of inspired air is lowered by exposure to high altitude or dilution of oxygen in ambient air with foreign gases. Because hypoxia is insidious and sneaks up on a person, its role in errors, bad judgments and accidents is emphasized in training both pilots and passengers on military aircraft. Aviation personnel are usually trained to recognize the symptoms of hypoxia in a hypobaric chamber.

The appearance of acute hypoxic hypoxia symptoms and their severity depend on: a) absolute altitude, b) rate of ascent, c) duration at altitude, d) ambient temperature, e) physical activity, and f) individual factors including inherent tolerance, physical fitness, emotionality and acclimatization (Crowley, 1993). The stages of hypoxia (which pertain to those riding in unpressurized aircraft, or to those climbing at high mountainous altitudes) include:

a) Indifferent stage—deterioration of night vision which becomes significant at about 5000 ft.

b) Compensatory stage—the pulse rate, the systolic blood pressure, the rate of circulation, and the cardiac output increase. Respiration increases in depth and sometimes in rate. The individual becomes drowsy, frequently makes errors in judgment, and has difficulty with simple tasks requiring mental alertness or moderate muscular coordination.

c) Disturbance stage—physiological responses can no longer compensate for the oxygen deficiency. Occasionally there are no subjective symptoms of hypoxia up to the time of unconsciousness; more often, symptoms such as fatigue are reported.

d) Critical stage—within 3 to 5 minutes, judgment and coordination deteriorate with subsequent mental confusion and dizziness, and finally severe hypoxia and incapacitation. A hypoxic individual may be able to maintain an upright posture but will probably lose consciousness while maintaining blood pressure, or may faint if a reflex fall in blood pressure occurs (Crowley, 1993).

The signs and symptoms of hypoxia include impairment of peripheral and central vision. There is a weakness and incoordination of extraocular muscles with decreased range of visual accommodation. Touch and pain are diminished or lost. Intellectual impairment is an early sign which often prevents the individual from recognizing his disability. Thinking is slow and calculations are unreliable. Memory is poor and judgment and reaction time are also affected. There may be a release of basic personality traits and emotions similar to those of alcoholic intoxication, resulting in euphoria, pugnaciousness, overconfidence or moroseness. Muscular coordination is decreased and delicate or fine muscular movements may be impossible. The skin becomes bluish in color from the presence of reduced hemoglobin in the capillaries.

The importance of the onset of hypoxia as one rapidly ascends to aviation altitudes (whether through climb-outs, or from rapid decompression of the aircraft cabin) is that upon deprivation of oxygen, there is little time-of-useful-consciousness (TUC) during which an individual is oriented, attentive and can act with both mental and physical efficiency and alertness. Although pressurization of aircraft cabins removes several of the risks of hypoxia, should such protective systems fail, the risk of decompression effects is pronounced at high altitudes. The TUC reaches a minimum value of 13 to 15 seconds at 43,000 feet when an individual is breathing air, and at 53,000 feet when breathing 100% oxygen but then suddenly loses that oxygen supply. The very short TUC during rapid decompression is a reflection of the fact that a sudden exposure of our bodies to altitude results in an outward flow of oxygen from the blood and lungs into the ambient air. Thus, as compared to an incident caused by

time-of-useful consciousness

rapid decompression, when the oxygen supply is interrupted, TUC is prolonged somewhat because our lungs are already overfilled with more pure oxygen (Crowley, 1993).

Jet fighter aircraft cockpits typically are not pressurized, and fighter pilots without sufficient oxygen supply can readily experience low levels of hypoxia as they fly in and out of altitudes above 12,000 ft. and assuredly above 15,000 ft. Having available a continuous supply of oxygen is critically important for jet fighter pilots to maintain adequate flight performance, health and safety. This is particularly true during intense aerial maneuvers typical in aerial dog fights.

For passengers and other crew members, having a fully working cabin pressurization system is critically important for health and safety. Cabin altitudes in pressurized aircraft are maintained below 2438 m (<8,000 ft.) equivalent to ensure the performance of aircrews (Ernsting, 1978). Most modern passenger aircraft are pressurized to the equivalent of 5000 ft to 8000 ft (1500 m to 2400m), with the aircraft flying up to a little over 40,000 ft (12,200 m). In the event of decompression, quick, easy access to supplemental oxygen can be life saving.

Mechanical Forces

Acoustical Noise

One might readily believe military forces thrive on generating noise. Almost anywhere one turns an ear, weapons and equipment of all sorts produce noisy environments in training or in combat. Rifle, machine gun, armor tank, cannon artillery fire, missile launches, and use of explosives all impose intense *impulse noise* on military personnel who fire such weaponry, or who are in the vicinity when firing commences. Air and ground crews, communications personnel, missile repairmen and countless others are frequently exposed to *continuous or steady state noise* associated with jet or diesel engines, tracked vehicles, generators, electronics gear, and a myriad other noise sources of a sustained and continuous nature.

impulse noise

Noises in the military work environment: a) threaten hearing loss, b) hinder voice communications, and c) constitute another stressor in the adverse environmental spectrum. The noise intensities of modern military equipment are much higher than those found in most industrial jobs. Long duration steady state noise levels in aircraft can range from 100 to 115 dB sound pressure level (SPL) and those encountered working around generators and engines as high as 110 to 120 dB. Riding as crew members, or passengers in high performance jet aircraft, helicopters, ground tracked vehicles such as tanks and armored personnel carriers exposes to continuous noises that dramatically interfere with unaided voice communications, and as a function of exposure time, present health hazards to hearing conservation.

blast overpressure wave

Impulse peak levels at the ear of a soldier firing an M-16 rifle can be as high as 160 dB, and for the crew firing the M-198 howitzer with maximum explosive charge, as high as 185 dB for a single firing (it can also be accompanied by a blast overpressure wave directed rearward from the gun tubes toward the crew members, threatening the air containing organs of the body) and even higher for shoulder launched rockets and anti-aircraft missiles. Repeated exposures to such impulse noises, as when on a firing range, or firing a machine gun etc. merely increase the influence of such noises on communication ability and hearing conservation risk. In rapid artillery fire scenarios, gun crews might fire as many as ten rounds in a rapid succession volley from their gun tube, then shortly might follow with another volley, and another.

Hearing loss. Noise induced hearing loss, and therefore damage risk criteria, are attributable to a number of factors including the frequency bandwidth characteristics and time course of the noise exposure, the distance away from the noise source, whether there are blocking materials between the noise source and the exposed ear, and other variables. Price (1986) points out the importance of higher sound frequency rifle firing in establishing damage risk criteria, whereas the lower frequency impulses of guns such as howitzers may be slightly less hazardous. Blast noises of extremely high intensity may rupture the tympanic membrane (ear drum) and result in damage to the ossicular chain of the inner ear. The threshold of eardrum rupture, although dependent on the blast spectrum, starts around 184 dB (5 psi), with 50% of eardrums failing at approximately 194 dB (Moore & Von Gierke, 1991).

Too much exposure to either impulse or steady state noise causes physiological damage to the auditory system and results in permanent loss of sensitivity over all or part of the audible frequency range. Most noise induced hearing loss is insidious (we are unaware it is happening to us) and is cumulative over time. After years of being exposed to various noise sources such as the vehicular or other noises we tend not to protect ourselves against, military personnel usually discover they are developing a hearing loss during their annual physical examination. Thousands of ex-military personnel wear hearing aids, and the most frequent form of medical compensation for U.S. military veterans continues to be for hearing loss, illustrating the importance of military noise as a health hazard. *[margin: audible frequency range]*

Equipment design standards (e.g. MIL-STD-1474) hold materiel developers accountable for designing equipment and weapon platforms that, with proper protection, can be operated safely with less hearing risk. All military services practice extensive hearing conservation programs in an attempt to preserve both the hearing and the performance of personnel. The military services have noise exposure regulations (e.g. USAF Reg 161–35) in an attempt to establish operating procedures to restrict noise exposures and encourage wearing hearing protection devices.

Speech intelligibility. Ambient acoustical noise reduces the intelligibility and comprehension of speech communications and can interfere with detection of auditory signals and warnings. These effects can be due to masking of the desired signal or communication, or to the generation of a temporary hearing threshold shift (TTS), which reduces an individual's auditory sensitivity. Interference with auditory signals or with speech communications can seriously impact the performance of military personnel (Moore & Von Gierke, 1991). *[margin: speech intelligibility; temporary hearing threshold shift]*

Ambient noise can mask speech sounds and decrease the intelligibility and thereby decrease the probability that messages will be understood. There are several techniques for indicating the predictability that speech signals will be understood in noisy environments (Kryter, 1972). A commonly accepted technique is to determine the Speech Interference Level (SIL) which involves the calculation of the average of the noise SPLs for octave bands centered at 500, 1000, 2000, and 4000 Hz. With this information it is possible to estimate what the sound pressure level of the speech signal must be at the listener's ear for it to be intelligible in that noise environment.

A number of standardized intelligibility tests, such as monosyllabic lists of sounds, are used to evaluate communication system performance (Kryter, 1972; Haslegrave, 1995). Speech intelligibility (SI) is impaired by many noisy environments. For normal hearing persons in noise environments, intelligibility of spoken words in face-to-face communication can be improved by wearing hearing protection (muffs, plugs, form fitted helmets, etc.) because the signal-to-noise (S/N) ratio at lower noise levels offers lower distortion in the auditory system. Wearing hearing protective devices, or electronically aided signal receivers and noise cancellers becomes critically important for maintaining good communication performance in many military jobs, and for hearing conservation. Additionally, in ways we do not often think about, Moore and Von Gierke (1991) indicate that for individuals whose native language is other than the one of the speaker, SI is generally affected by ambient noise.

For personnel with noise induced hearing loss (i.e. a loss of sensitivity due to past noise exposures) their ability to discriminate speech is disproportionately affected by a decrease in S/N relative to those with no measurable hearing loss (Moore & Von Gierke, 1991). For a discussion of voice input, voice interactive systems, speech output, and digital communications in noise see the chapter by Moore and Von Gierke (1991).

Performance in the presence of noise. There is a wide and diverse literature describing research on performance of cognitive and psychomotor task performance in the presence of noise. Most of this literature suggests noise contributes to a person's arousal state; and therefore at an optimum level, facilitates performance, but above certain levels the noise disrupts performance. After reviewing much of that research Moore and Von Gierke (1991) report it appears reasonable to say that:

1. With respect to monitoring or vigilance tasks, noise is most likely to impair performance when signals are difficult to detect and the task situation (e.g. the relative payoff for hits and misses when guessing the auditory signals as hearing loss patients typically do in an audiological exam in a sound booth) encourages risky decision behavior (Broadbent, 1979).
2. Performance is more likely to be affected when multiple signal sources such as other masking sounds are present and in that case noise encourages an increased tendency to sample the most probable sources of information and pay less attention to lower probability sources (Broadbent, 1971).
3. When noise has an effect on serial, sequential responding there is a tendency for speed of work (responses to sound signal targets) to increase at the expense of accuracy. This is jointly influenced by the level of noise exposure and the length of time on the task (Hartley, 1973).

Moore and Von Gierke (1991) point out that the effects of noise on performance (where information is not presented via the auditory system) are not unique, but are similar to those reported for a variety of other environmental stressors. The best course of action is to employ engineering controls and hearing protection to bring ambient noise environments within limits safe for the auditory system and to levels which mitigate adverse noise effects on cognitive or psychomotor performance.

Acceleration

acceleration Acceleration and vibration accompany operation of many military equipments ranging from ground-based vehicles, to aircraft, aerospace vehicles, and other weapon platforms. Powered automotion in helicopters, trucks, tanks, etc. expose operators to increasing acceleration with a frequency range extending up to 100 Hz, depending on the roughness of the air, road, or terrain, and the vehicle speed. Similarly, ship-at-sea motions can extend from extremely low frequencies produced by ocean waves (below 0.1 Hz) to high frequencies in high-speed surface attack ships. Military exposures can extend over several hours per day. Flying in high performance combat aircraft can increase sustained acceleration exposures up to 7 or 8 Gs during aerial combat maneuvering or the boost and re-entry phases of space flight. Exposure times range from a few seconds to several minutes (Von Gierke, McCloskey & Albery, 1991).

Effects on the body. There are three ways (mechanical, physiological and psychological) acceleration and vibration exposures adversely affect military performance; and they are not easily separated. Von Gierke, McCloskey and Albery (1991) indicate the mechanical effects of increased gravitational fields increase the weight of body parts, which become elongated or compressed along one of the three linear G vectors. For those who are not jet fighter pilots it may be helpful to remember experiences in riding an amusement park roller coaster ride which uses rapid accelerations to throw our bodies about in several competing directions to provide "more thrills." The Gz vector is the longitudinal, vertical vector from feet to head (riding steep angles upward or downward); Gx is the chest to back direction normally associated with sloping forward or backward into one's seat as a vehicle accelerates (as in rapid ascent at about a 35–60 degree angle); and Gy, the tendency to lean sideways laterally as when we ride a vehicle on a curved road or when the roller coaster severely swings our bodies from side to side. *G vectors*

Acceleration can affect the shape and function of the soft internal organs, including the heart, lungs, kidneys, liver, etc. In higher G forces, we must put forth high countering muscle forces to keep the head, torso, and limbs in desired positions. At approximately +2 Gz there is increased pressure on the buttocks, drooping of the face and noticeably increased weight of all body parts; at this level of G force it is difficult to raise oneself, and at +3 to 4 Gz it is nearly impossible. Above +3 to 4 Gz controlled motions require greater effort, accommodation and learning to offset loss of fine motor control. Speech is severely affected, yet possible up to +9 Gz if the operator is utilizing protective techniques (i.e. a G-suit). Sensory inputs can be affected, such as vision (affected through eyeball deformations) and blood loss at eye level, vestibular orientation through the semicircular canals and otoliths, and force-weight judgments in manual dexterity tasks. Acceleration protective equipment can either improve or degrade performance through mechanical interference (Von Gierke, McCloskey & Albery, 1991).

Tolerance to G. High sustained acceleration levels produce different physiological tolerance limits for force in each of the varying vector directions. The most debilitating and dangerous of the physiological effects of increased gravitational forces are those caused by changes in the hydrostatic blood pressure column between the heart and the brain. As the level of +Gz increases, blood pressure to the brain is decreased (essentially the body is accelerating upward against gravitational forces holding the blood closer to earth as in a jet pilot climbing rapidly to altitude); to almost zero brain blood pressure at about +5.5 Gz, and negative blood pressure at higher +Gz levels. Zero or negative blood pressure to the brain leads to grayout, blackout and loss of consciousness. A pilot of a high performance jet fighter aircraft may blackout at about +5.4 Gz, whereas his/her passengers may blackout at approximately +4.7 Gz pointing out the psychological influence of "control" over the situation, permitting the pilot at the controls a small amount more tolerance than his passenger.

Tolerance is greater for G loads in the transverse plane (+Gx, chest to back) than it is in the Gz plane (head to foot) because of reduced effects on the hydrostatic column. Human tolerances of as high as +16.5 Gx forward acceleration for short durations and up to +12 Gx with a duration of 3 minutes have been reported, albeit accompanied by great additional physiological effects such as respiratory difficulties, inspiratory chest pain, tracheal tugging, coughing, and sensation of weight on the thorax (Cherniack et al., 1961).

Protection against G. Any *action* or *device* that increases the pressures in the hydrostatic column between the heart and the brain during increased +Gz loads (head to foot) increases the human operator's tolerance level of G, and the time he can remain at increased G. This concept

active straining maneuver

underlies two commonly used G-protection techniques: the *active straining maneuver* and the *G-suit*. The military M-1 straining maneuver consists of pulling the head down between the shoulders (maximum shrugging), slowly and forcefully exhaling through a partially closed glottis (space between the vocal cords at the upper part of the larynx) and simultaneously tensing all skeletal muscles. By doing the M-1 straining maneuver another +1.5 Gz of tolerance can be added, raising the G-tolerance from approximately +5 Gz to +7 Gz. The straining maneuver is quite taxing and can interfere with speech, and if performed for 1–3 minutes duration, or repeatedly, pilots become severely fatigued (Von Gierke et al., 1991).

G-suit

The G-suit protection device consists of bladders inserted into a full- or partial-body coverall. The bladders inflate during the onset of acceleration, starting around 2 to 3 G, and increase as G increases to a maximum bladder pressure of 13 psi at +9 Gz. The external increase in pressure afforded by the G-suit reduces body deformation, blood pooling in the extremities, and mechanically increases internal blood pressure. A relaxed pilot wearing a G-suit has an increased G-tolerance of approximately 2 G; but when combined with the straining maneuver, G-tolerance can increase up to 3 to 5 G, allowing him to tolerate +8 Gz to +10 Gz without losing consciousness (Burton, 1974).

Rarely does acceleration affect only one factor such as speech, vision or manual dexterity. Vision, simple and choice visual reaction time, dial reading, absolute detection thresholds for foveal and peripheral vision have all been shown to be affected by G-stress. However, some of the effects can be diminished somewhat as people become more accustomed to acceleration effects. Memory is impaired at above +5 Gx levels. Complex manual control tracking is adversely affected at +6 Gz and only slightly from +1 Gx to 14 Gx (Creer, 1962). Tracking and flight control exhibits progressive impairment with increasing +Gz acceleration and somewhat less impairment with +Gx acceleration (Grether, 1971).

There is more to the story obviously. Manual dexterity, speech, vision and memory processes are all affected by high acceleration forces. For more detailed description see Von Gierke, McCloskey and Albery (1991) who state that the mechanical, physiological and

synergistic

psychological effects of acceleration on human operators all act in a synergistic, interactive way to adversely impact military performance; but they also describe numerous counter-measures designed to protect humans from these effects.

vibration

Vibration

As in the case of acceleration, vibration effects result from an interaction of mechanical, physiological, and psychological influences. There is no specific target body area or organ for low frequency whole-body vibration. If severe, whole body vibration can interfere with bodily functions and cause tissue damage in virtually all parts of the body. Oscillations of operational vehicle motions are transmitted to humans in military environments through supporting seats, platforms, floors, walls and handles. In rotary wing aircraft vibration frequencies are associated with revolution rates of rotors, gearboxes and other engine or mechanical parts.

Vibration can have similar mechanical effects on the human body as does acceleration; the differences center on the alternating forces that characterize vibration. The magnitude of vibration effects depends on the frequency range of the vibration force and its relationship to the body's dynamic response (Von Gierke, McCloskey and Albery, 1991). The most troublesome whole-body vibration frequencies are in the 2 to 12 Hz range; the abdominal-thoracic visceral region has been shown to be responsible for pain occurring in the 1–2 Gz and 2–3 Gx ranges (Hornick, 1973).

Effects on performance. The effects of vibration on performance depend on many details concerning the operator's task, body position, and experience. Vibration acts on the biodynamic and psychomotor properties of man rather than on central neural processes. Vibration impacts performance either by modifying how information is perceived, or by influencing control movements mechanically (or some combination of both). Within physiological tolerance limits, vibration primarily affects the human's ability to see, as well as the control of fine motor movements. A wide range of vibration frequencies adversely affect performance through the mechanical disturbance of fine motor control, vibration of the eyes and or visual displays and increased discomfort levels which may distract the operator from the task at hand.

Control measures. It is important in vehicle design to apply control measures at the vibration source or the pathway between the source and the operator or passenger to protect personnel from vibration effects. Thus, modern truck tractors have air-cushioned ride seats to isolate drivers from some of the over-the-road vibration. Operational controls can be used to restrict engine operation, handling mode (e.g. ground speed of vehicles or selection of terrain) and the time of personnel exposure to control vibration exposures; all of which affect overall efficiency and capability (Von Gierke, McCloskey, & Albery, 1991).

Exposure to environments exceeding ISO and ANSI vibration limits (exposure curve guidelines) carry a risk of impaired working efficiency and potential fatigue known to degrade some performance. Standards were developed based upon studies of aircraft pilots and vehicle drivers and their subjective responses to habitually operating at various vibration levels.

Motion Induced Sickness and Military Performance

Motion sickness is a transient normal response to unnatural motion stimulation that accompanies many ship, airborne and space missions. It even presents itself in training simulators, especially those with computer-generated visual imagery. Motion sickness is traceable to the vestibular labyrinth in the inner ear which acts as a sensor for acceleration. The otolith organs register linear accelerations, while the semicircular canals react to angular accelerations. Certain motion patterns, especially those involving both the vertical-linear component and the combination of angular accelerations with head movements are very provocative. The common characteristic of all conditions which cause motion sickness is varying acceleration—that is, when acceleration of the person changes with time it is likely he/she will experience some motion sickness (Rolnick & Gordon, 1991).

motion sickness

otolith organs

Motion sickness is characterized by development of pallor (mainly extreme facial paleness), cold sweating, a general feeling of discomfort, nausea, and ultimately by emesis (vomiting). Many people experience motion sickness as a devastating condition which makes them highly passive, apathetic and depressed; seasickness may also make them drowsy and sleepy.

Incidence of motion sickness. The incidence of motion sickness depends on many different factors, including individual susceptibility, the type and magnitude of the stimulus, etc. Any person with normal vestibular function can succumb to motion sickness, if the type of motion is appropriate and continues for a sufficiently long time. The small size of naval fast attack missile craft (FAC) and their high speeds render them unstable at sea and therefore likely to provoke seasickness. Rolnick and Gordon (1991) indicate that 62% of the sailors aboard the Israeli SAAR Missile Boats reported episodes of emesis, while 80% experienced nausea on their first cruise. NASA has only recently admitted that 100% of astronauts experience varying degrees of motion sickness on space missions.

The interaction between vestibular information and the visual sense is highly relevant in explaining motion sickness. The incidence of seasickness on a ship's bridge is lower than it is below deck. When on the bridge, one can see the outside world, and thus there is agreement between the information coming from the eyes and from the vestibular system. Below the deck, the individual's whole visual world is moving with him, thus providing the mistaken impression that he is not moving in relation to the outside world (Rolnick & Gordon, 1991).

As experience at sea is acquired some adaptation usually takes place. Generally, upon return to land, or a return to motionless standing on the ground, the symptomatology of seasickness and other forms of motion sickness rapidly disappears.

Effect on performance. Assessing direct effects of motion versus the motion sickness on performance is difficult, and the literature presents conflicting reports. However, it is clear, motion sickness as an uncontrolled aversive event, affects mood and motivation, and a person's readiness to carry out tasks. It can give rise to a profound helplessness reaction which is manifested through cognitive, emotional and motivational deficits. If one is vomiting, it is difficult to continue performing, no matter what the task. In many studies, the participants simply chose to stop performing when motion sickness interfered with their tasks. For a review of laboratory attempts to study motion sickness see Rolnick and Gordon (1991).

There are various approaches to reducing the salience of motion sickness in the military. Human factors engineering, preselection of personnel, desensitization training, behavior therapy, biofeedback, and pharmacological intervention are all useful countermeasures worthy of consideration for likely motion sickness conditions.

Atmospheric Mix

Toxic Gases and Fumes

Military training and combat frequently expose weapon system crewmen to mixtures of potentially toxic fumes. Armor crewmen work in confined spaces amid short bursts of highly concentrated propellant gases from their own weapons. Battlefield smokes and obscurants used to disguise movements, as well as combustion products from projectile propellants, exploding munitions, fires, and vehicle exhausts all affect the eyes, nose and throat of ground-pounding infantrymen and vehicle crews. From a health and performance perspective it is important to ensure exposure to toxic fumes is kept to a minimum. Appropriately designed crew compartment ventilation systems, filters, and protective masks should be provided.

A common air mixture threatening soldiers is combustion products which originate from sources such as burning of projectile propellants, exploding munitions, fires, vehicle exhausts, etc. (Benignus, 1991). Concentration and composition of gases vary depending upon material burned, available oxygen, adequacy and operation of ventilation systems, etc. In addition to concern for general health of personnel, there is legitimate concern for effects of toxicants upon performance of military tasks. If the central nervous system (CNS) is affected, even temporarily, operator errors or delays can reduce crew effectiveness and affect survival of combatants and those who are dependent on them. Even small decrements are of interest in performance of critical tasks.

central nervous system

Gases. Gases known to have effects on the CNS and on performance which appear in most of the contaminated atmospheres are carbon monoxide (CO), carbon dioxide (CO_2), hydrogen cyanide (HCN) and a host of other chemicals which are sensory irritants such as aldehydes, hydrochloric acid, phosgene and ammonia. In atmospheres contaminated by fire products, hypoxia, a reduced level of oxygen (O_2) also occurs. Benignus (1991) presents a cogent summary of gases and conditions that can affect performance, along with a statement of the supposed biological or physiological mechanism of action.

hydrogen cyanide

Carbon monoxide (CO). Of all the contaminants in military settings, CO is the most ubiquitous. CO is a clear, colorless, odorless gas—the byproduct of incomplete combustion from burning projectile propellants, fires in the environment and engine exhaust. CO buildup in crew compartments of fighting vehicles (e.g. tanks, armored personnel carriers, missile launch vehicles, etc.) can be exacerbated by inadequate, malfunctioning or misused ventilation equipment. Shipboard fires, especially in closed environments like submarines, can lead to high levels of CO. Breathing moderate levels of CO for a relatively short time can cause brain damage, and larger quantities can quickly kill a person.

Military personnel who smoke tobacco have elevated CO blood levels and may, therefore, be at greater risk to behavioral impairments by additional environmental exposure (Benignus, 1991). CO enters the body by inhalation and is diffused across alveolar membranes in the lungs where it is dissolved into blood and quickly becomes bound to hemoglobin (oxygen carrying red blood cells) to form carboxyhemoglobin (COHb) and thereafter competes with O_2 in the body, not permitting enough oxygen to be taken up by the lungs.

carboxyhemoglobin

Effects of CO exposure are largely due to the hypoxic effect produced by reduction of hemoglobin available to carry O_2. Benignus says dose effect studies of COHb do not give consistent results but at some level of COHb portray performance decrements for vigilance, signal detection, precision tracking performance, time discrimination, and other neuro-behavioral effects.

Carbon dioxide (CO_2) is a metabolic byproduct given off from the lungs when we breathe and it can be accumulated from respiration in tightly closed spaces such as fighting vehicles, shelters, aircraft cockpits, or submarines in which ventilation devices or respirators may malfunction or be misused. It is also found as an environmental pollutant whose sources include all combustion such as engines, heating by fuel burning and accidental or combat-related fires.

Whatever the source, the presence of carbon dioxide in inspired air produces increased cerebral vasodilation (expansion of blood vessels) resulting in increased brain blood flow, and leading to elevated cerebral tissue partial pressure of oxygen. This triggers stimulation of an increase in ventilation, and that increased breathing rate is accompanied by generalized arousal. Difficulty breathing (dyspnea) begins between 7 and 10 % CO_2 in inspired air. Below 5% CO_2 few behavioral effects are noted. Generally, respiratory effects are not observed below 4% CO_2, but there is a great deal of variability across subjects. Behaviors which appear to be adversely affected at >5% CO_2 in inspired air are: reaction time, arithmetic computation and sensory processing. Sleep is also disturbed. All of these behaviors are important in a wide range of military settings. Furthermore, the respiratory disturbance produced by CO_2 inhalation can act as a competing response, especially in fine motor control. At higher levels of exposure, panic and eventually physiological events can have severe impacts on military performance (Benignus, 1991).

Hydrogen cyanide (HCN) is produced by combustion of certain materials commonly used in construction of military equipment, and is therefore likely present in closed spaces during fires. At high levels HCN is lethal. At low levels it could produce impairments in performance, but has not been studied in humans.

Benignus (1991) says some of the gases and toxic fumes are physiologically well understood, yet their behavioral and CNS effects remain inadequately studied and therefore poorly understood. The effects of exposure to combinations of gases are even less well understood.

Radiation Exposure and Performance

Exposure to ionizing radiations may occur in nuclear warfare, during radiation accidents, or military space operations, and can significantly disrupt task performance. Although there is no experience fighting in nuclear ionizing radiation, animal data and accidents where humans were exposed to large doses of radiation, and studies of psychological response to advanced technology weapons, give some perspective on how combatants will react and perform on a nuclear battlefield.

Radiation exposure may produce deficits in learning, altered performance of cognitive and motor tasks, and hazards of combined injury (e.g. radiation and trauma). Radiation exposure may produce an early transient incapacitation that can be influenced by task complexity, radiation dose, dose rate, and radiation type. Neurophysiological correlates of radiation-induced behavioral damage are reviewed by Mickley and Bogo (1991), who present data relevant to the prediction, prevention, and treatment of radiation-induced performance deficits. They also offer information about military radiation hazards, human irradiation, psychological effects of tactical nuclear warfare, describe neurophysiological correlates of radiation-induced behavioral damage, offer animal predictive models of human radiation effects, and describe methods of behavioral radioprotection (i.e. protection from the behavioral consequences of ionizing radiation), (Mickley & Bogo, (1991).

dyspnea

Military Operational Considerations

Chemical Protective Clothing

This section does not address chemical and biological threats, clear adverse conditions of the battlefield, rather, it addresses the use of Chemical Protective Clothing (CPC) worn by military forces to protect against chemicals, biologicals and radiological contamination on a modern battlefield. Krueger and Banderet (1997) review the important physiological and psychological considerations of wearing a charcoal lined semi-permeable overgarment and a rubber butyl protective face mask to protect against contaminants. They also review the numerous practical considerations for troops wearing what amounts to an encapsulating micro-environment (the CPC) and attempting to function normally in the field.

Chemical Protective Clothing

micro-environment

Thermal burden. For most military forces the CPC ensemble presents a serious concern for the thermal burden such protective uniforms add to troops working in the fullest protective posture offered by the CPC, especially in moderately warm to hot ambient environments. The thick textile material of CPC severely limits evaporation of body sweat through the uniforms because they typically are impregnated with an absorbent charcoal lining, permitting little evaporation. Consequently, CPC hampers the natural ability of the body to thermoregulate, leading to fluid and electrolyte losses, and increased heat stress in direct relation to the severity of ambient heat conditions and the level of physical work. Conditions do not have to be unreasonably warm to present risk of heat stress. Work generates heat which must be dissipated through the protective suit. The harder and longer the work, the more body heat must be dissipated and the greater the risk of heat stress.

In high ambient temperatures, a soldier only can work encapsulated in CPC for a few hours or less. Some field tests found endurance times on the order of 1 to 2 hours in high ambient temperatures (above 85–90°F), and laboratory tests in moderate temperatures were as long as 11 hours. If appropriate work-rest cycles are adhered to, allowing from 5 to 15 minutes rest during each hour of work can significantly extend endurance in CPC.

The CPC is also stressful in other ways. Even in a cool environment, wearing CPC results in unfavorable subjective reaction to the ensemble and adverse physiological changes. See Krueger and Banderet (1997) for details.

Metabolic cost of working in CPC. The increase in metabolic workload attributable to wearing the full CPC ensemble was measured to be an increase for men of from 7% to 26% increase in oxygen uptake on 29 of 42 standard military tasks, and for women from 5% to 29% increase. The greatest increases in metabolic cost were for tasks requiring continuous mobility (e.g. load carriage), (Patton et al., 1995).

Restriction in body movement. The CPC uniform, a big bulky overgarment not unlike snowpants, includes a protective hood over the head and shoulders, has a full facial respirator mask, and overboots. This combination, generally worn over the soldier's normal utility uniform restricts certain body movements, particularly angular movements about the joints of the body. The CPC of the U.S. military restricts head flexion as much as 20° in the ventral-dorsal plane and lateral rotation of the head by as much as 40° to 50° (Bensel, 1997). Such restrictions of head movements limit a soldier's normal visual scan, and thus more pronounced head movements are required to view the environment. The gas mask is the principal offending restrictor of head movements (Bensel, 1997; Krueger & Banderet, 1997).

Other impairments attributed to CPC. Unfortunately CPC degrades manual dexterity, psychomotor coordination, and performance of many activities accomplished with the hands. The gloves distort normal tactile feel for various tools, equipment controls, keysets and grip handles. Finger dexterity can be altered as much as 30% depending upon the thickness of the gloves (Bensel, 1997). Gas masks impose breathing resistance for their wearers. The typical modern military gas mask produces a four-fold increase in the resistance to breathing. Attempting to breathe normally with the gas mask while doing significant work can lead to breathing distress, hyperventilation, shortness of breath, tremors, and claustrophobic reactions. Performance of tasks requiring high aerobic power (e.g. running) is hindered greatly by breathing resistance, since this factor becomes more critical under high workload conditions. Thus more time is required to perform such tasks when the mask is worn.

A major finding with most CPC tests is that performance of tasks across the board just take more time to complete, generally about 1.5 times as long. Frequent, repeated training of soldiers in the CPC uniforms on all tasks that are to be performed in combat is required to resolve most of the issues with the current CPC ensembles. For future developments, much human engineering redesign is called for with almost every CPC ensemble presently available. See Krueger and Banderet (1997) for extensive review.

Continuous Operations, Sustained Performance and Sleep Loss

In part because technological developments (e.g. radar, thermal imaging, infrared image intensification systems, night vision devices etc.) provide capability to operate through the night, military forces have been planning and training to operate continuously around-the-clock in continuous operations (Krueger, 1989; 1991). Such demanding work schedules cause military forces to participate in sustained work efforts, spend long hours on the job, obtain little sleep, and often work against their circadian rhythms biological clock (Krueger, 1991; Comperatore & Krueger, 1990).

continuous
operations

circadian
rhythms

The requirements to sustain work around the clock mean large numbers of troops will lose significant amounts of sleep. Those who attempt to retrain their biological systems to become night fighters will, like factory shift workers, quickly find out how difficult that is to accomplish. Additionally, after completing "the night shift" it is even more difficult to obtain adequate sleep on a daytime battlefield filled with continuous movement of people, weapons platforms and equipment. As we have seen in other parts of this chapter, the battlefield may present its own unique environmental extremes, or adverse conditions, (e.g. the deserts of Southwest Asia during the Persian Gulf War were very hot, presented much blowing sand and flies, and plenty of equipment movement and explosive noises).

night
fighters

sleep management

Thus, developing sleep management, and disciplined planning on how to obtain adequate rest and sleep becomes a critically important variable in modern combat. Sustaining 100 hour battles is likely to be the norm in future conflicts. Military forces must become familiar with the body and the brain's sleep and rest needs, our circadian rhythms, and the implications of "shift lag," fatigue, sleep loss and their effects on military performance (Krueger, 1991).

Concluding Remarks

Many of the adverse conditions of the modern battlefield described above have been studied in isolation for their effects on performance. However, often it is the combination of these factors which affect soldier performance the most. As Von Gierke, McCloskey and Albery (1991) pointed out, studies of combinations of these factors usually present interesting findings: combinations of environmental stressors affect military performance differently than when applied in isolation. Thus it is incumbent upon those who use such data to inquire in depth about how they were collected, and to piece together evidence from several sources to improve predictions of how soldiers, sailors, airmen and marines will perform in the face of so many adverse work conditions.

References

Banderet, L. E., & Burse, R. L. (1991). Effects of high terrestrial altitude on military performance, Chapter 13, pp 233–254. In: R. Gal & A. D. Mangelsdorff (Eds.) *Handbook of Military Psychology*. New York: John Wiley & Sons.

Benignus, V. A. (1991). Effects of atmospheric mix and toxic fumes on military performance, Chapter 17, pp 313–333. In: R. Gal & A. D. Mangelsdorff (Eds.) *Handbook of Military Psychology*. New York: John Wiley & Sons.

Bensel, C. K. (1997). Soldier performance and functionality: Impact of chemical protective clothing. *Military Psychology*, 9, (4) 287–300.

Broadbent, D. E. (1971). *Decision and stress*. London: Academic Press.

Broadbent, D. E. (1979). Human performance in noise. In: C. M. Harris (Ed.), *Handbook of noise control*, 2nd Ed. Pp. 17.1–17.20. New York: McGraw Hill.

Burr, R. E. (1991). *Heat illness: A handbook for medical officers*. (USARIEM Tech. Note No. 91–3). Natick, MA: U.S. Army Research Institute of Environmental Medicine. (DTIC No. AD: A238–974).

Burr, R. E. (1993). *Medical aspects of cold weather operations: A handbook for medical officers*. (USARIEM Tech. Note No. 93–4). Natick, MA: U.S. Army Research Institute of Environmental Medicine. (DTIC No. AD:A263–559.)

Burton, R. R. (1974). Man at high sustained and +Gz acceleration: A review. *Aerospace Medicine*, 10, 1115–1136.

Cherniack, N. S., Hyde, A. S., Watons, J. F., & Zechman, F.W. (1961). Some aspects of respiratory physiology during forward acceleration. *Aerospace Medicine*, 32, 113–120.

Comperatore, C. A., & Krueger, G. P. (1990). Circadian rhythm desynchronosis, jet lag, shift lag, and coping strategies. In A. J. Scott (Ed.) *Shiftwork*. Occupational Medicine State of the Art Reviews, Philadelphia, PA: Hanley and Belfus, Inc.

Creer, B. Y. (1962). Impedance of sustained acceleration on certain pilot performance capabilities. *Aerospace Medicine*, 33, 1086–1093.

Crowley, J. S. (Ed.) (1993). *United States Army aviation medicine handbook*. Fort Rucker, AL: The Society of U.S. Army Flight Surgeons.

Cymerman, A., Reeves, J. T., Sutton, J. R., Rock, P. B., Groves, B. M., Malconian, P. M., Young, P., Wagner, P. D., & Houston, C. S. (1989). Operation Everest II: Maximum oxygen uptake at extreme altitude. *Journal of Applied Physiology*, 66, 2446–2453.

Cymerman, A., & Rock, P. B. (1994). *Medical problems in high mountain environments: A handbook for medical officers* (USARIEM Tech. Note 94–2). Natick, MA: U.S. Army Research Institute of Environmental Medicine. (DTIC No. AD: A278–095).

Ernsting, J. (1978). Prevention of hypoxia: acceptable compromises. *Aviation, Space, and Environmental Medicine*, 49, 495–502.

Grether, W. F. (1971). Vibration and human performance. *Human Factors*, 13, 203–205.

Hartley, L. R. (1973). Effects of noise or prior performance on serial reaction. *Journal of Experimental Psychology*, 101, 255–261.

Haslegrave, C. M. (1995). Auditory environment and noise assessment. Chapter 17, pp 506–540. In: J. R. Wilson & E. N. Corlett (Eds.) *Evaluation of human work: A practical ergonomics methodology* (end ed.). London: Taylor & Francis.

Hornick, R. J. (1973). Vibration. Chapter 7. In: J. F. Parker & V. R. West (Eds.), *Bioastronautics data book*. Washington, DC: National Aeronautics and Space Administration, U.S. Government Printing Office.

Kobrick, J. L., & Johnson, R. F. (1991) Effects of hot and cold environments on military performance, Chapter 12, pp 216–232. In: R. Gal & A. D. Mangelsdorff (Eds.) *Handbook of Military Psychology*. New York: John Wiley & Sons.

Kroemer, K. H. E., Kroemer, H. J., & Kroemer-Elbert, K. E. (1986). *Engineering physiology: Physiologic bases of human factors/ergonomics*. Amsterdam: Elsevier.

Krueger, G. P. (1989). Sustained work, fatigue, sleep loss and performance: A review of the issues. *Work and Stress*, 3, 129–141.

Krueger, G. P. (1991). Sustained military performance in continuous operations: Combatant fatigue, rest, and sleep needs, Chapter 14, pp 255–277. In: R. Gal & A.D. Mangelsdorff (Eds.) *Handbook of Military Psychology*. New York: John Wiley & Sons.

Krueger, G. P. (1993). Environmental medicine research to sustain health and performance during military deployment: Desert, arctic, high altitude stressors. *Journal of Thermal Biology*, 18, 687–690.

Krueger, G. P., & Banderet, L. E. (1997). Effects of chemical protective clothing on military performance: A review of the issues. *Military Psychology*, 9, (4) 255–286.

Kryter, K. D. (1972). Speech communication, Chapter 5, pp 161–226. In: H. P. Van Cott & R. G. Kinkade (Eds.) *Human engineering guide to equipment design*. Washington, DC: U.S. Government Printing Office.

LeBlanc, J. S. (1956). Impairment of manual dexterity in the cold. *Journal of Applied Physiology*, 9, 62–68.

Mickley, G. A., & Bogo, V. (1991). Radiological factors and their effects on military performance, Chapter 19, pp. 365–385. In: R. Gal & A. D. Mangelsdorff (Eds.) *Handbook of Military Psychology*. New York: John Wiley & Sons.

Moore, T. J., & Von Gierke, H. E. (1991). Military performance in acoustic noise environments, Chapter 16, pp. 295–311. In: R. Gal & A. D. Mangelsdorff (Eds.) *Handbook of Military Psychology*. New York: John Wiley & Sons.

Nelson, M. (1982). Psychological testing at high altitudes. *Aviation, Space, and Environmental Medicine*, 53, 122–126.

Patton, J. F., Murphy, M., Bidwell, T., Mello, R., & Harp, M. (1995). *Metabolic cost of military physical tasks in MOPP 0 and MOPP 4*. (USARIEM Tech. Rept. No. T95-9). Natick, MA: U.S. Army Research Institute of Environmental Medicine. (DTIC No. AD: A294–059).

Price, G. R. (1986). Hazard from intense low-frequency acoustic impulses. *Journal of Acoustical Society of America*, 67, 628–633.

Rolnick, A., & Gordon, C. R. (1991). The effects of motion induced sickness on military performance, Chapter 15, pp. 279–293. In: R. Gal & A. D. Mangelsdorff (Eds.) *Handbook of Military Psychology*. New York: John Wiley & Sons.

Ryn, Z. (1988). Psychopathology in mountaineering: mental disturbances under high altitude stress. *International Journal of Sports Medicine*, 9, 163–169.

Shukitt, B. L., & Banderet, L. E. (1988). Mood states at 1600 and 4300 m terrestrial altitude. *Aviation, Space, and Environmental Medicine*, 59, 530–532.

U.S. Department of Defense. Military Standard 1474: Noise exposure limits in military equipment design. Philadelphia, PA: U.S. Government Printing Office.

Von Gierke, H. E., McCloskey, K., & Albery, W.A. (1991). Military performance in sustained acceleration and vibration environments, Chapter 18, pp. 335–364. In: R. Gal & A. D. Mangelsdorff (Eds.) *Handbook of Military Psychology*. New York: John Wiley & Sons.

Young, A. J., Roberts, D. E., Scott, D. P., Cook, J. E., Mays, M. Z., & Askew, E. W. (Eds.). (1992). *Sustaining health and performance in the cold: Environmental medicine guidance for cold-weather operations*. (USARIEM Tech. Notes 92–2 and 93–2). Natick, MA: U.S. Army Research Institute of Environmental Medicine. (DTIC Nos. AD:A254–328 and A259–926).

Photograph courtesy of Paul Bartone

7
Stress in the Military Setting

Paul Bartone, Ph.D.

* Introduction * Stress Defined * The Military Setting Defined
* Psychological Framework for Considering Stress in the Military Context
* Historical Overview * Field Psychological Research Teams in the Modern
Military * Stressors on Military Operations * Dimensions of Psychological
Stress in Military Operations * Stress in the Military Family * Stress in the
Military Community * Social Support and Personality Hardiness: Social and
Personal Stress-Resistance Resources * References

Introduction

> "The stress of war tries men as no other test that they have encountered in civilized life. Like a crucial experiment it exposes the underlying physiological and psychological mechanisms of the human being." (Grinker & Speigel, 1945, p. vii)

In his classic book "The Stress of Life," Hans Selye (1978) observes that stress is a necessary part of life, and indeed that "no one can live without experiencing some degree of stress all the time." (p. xv). Furthermore, Selye notes that individuals respond differently to stress: "The same stress which makes one person sick can be an invigorating experience for another". These twin observations provide a good starting point for the present chapter, which will consider the issue of stress in the military setting from a military psychology perspective.

While stress is indeed a necessary part of being alive, and certainly every job can be stressful, some occupations are more stressful than others. The military occupation can expose its members to a greater range of stressors, including very extreme stressors, than perhaps any other human activity. But before going further, it will help to define the terms "stress" and "military setting" as they will be used in this chapter.

Stress Defined

"Stress" is a term that means many things to many people. As R. Joy wryly (and correctly!) remarks, "stress has been so conflated that it can mean almost anything." (Joy, 1996, p. v). Even the great Hans Selye, who is largely responsible for the term "stress" entering the

stress

popular lexicon, has said "if we are to use this concept in a strictly scientific manner, it is especially important to keep in mind that stress is an abstraction; it has no independent existence." (Selye, 1978, p. 53).

A common confusion is that the term "stress" is frequently used to describe two very different kinds of phenomena: (1) stimuli in the environment (both physical and psychological) that impinge upon the organism, and (2) the physical and psychological responses of the organism itself to such forces. In considering stress in the military context, it is best to preserve the term "stress" to refer to events or forces in the environment, outside the person, as opposed to subjective, internal responses. The application to environmental stimuli is emphasized by the term "stressor" or "stressors" instead of just "stress".

In ordinary parlance, both meanings of stress are heard routinely. Thus, in using the term "stress" to describe stimuli, people say such things as "There was a lot of stress at work today", or "She has a lot of stress in her family life right now". On the other hand, people also use the term to describe their own responses to environmental stimuli. For example, we hear "I am feeling a lot of stress"; or "It's exam time, and he is stressed-out". To make matters more confusing, sometimes people seem to have both meanings in mind at the same time. In saying "this is a high-stress assignment", a soldier may refer both to the demanding circumstances or stimuli associated with the job itself, as well as the subjective feelings of "stress" she experiences in response to those demands. So while the term "stress" can legitimately apply to a stimulus or a response, it is important to be aware of the difference.

For those conducting research on the possible effects of stressful conditions on human beings, it is especially critical to maintain this distinction. If stressful conditions are merely inferred to exist based on subjective responses to stress, this can lead to wrong conclusions regarding factors that might influence responses to stress, and can even mask individual differences or important personal or situational moderators of stress. For example, if researchers use psychological distress scores to indicate environmental stressors, any individual differences in response to similar stressors will go unrecognized, since higher or lower distress scores are taken to indicate higher or lower levels of environmental stress. Likewise, defining "stressors" in terms of responses in this way precludes a search for organizational resources, such as concerned leadership or cohesion, that might serve to reduce distress levels (stress response) for those lucky enough to have such resources. Again, lower distress levels are assumed to indicate less stress exposure, instead of successful coping with high levels of stress. To avoid these kinds of errors, it is necessary to understand and maintain the conceptual distinction between "stressor" and "response to stress", regardless of the specific terms used, and to strive to measure the two separately. Further discussion of the problem of confounded measures in stress and health outcome research is available in Lazarus et al. (1985), and Costa & McCrae (1985).

The Military Setting Defined

The military setting is perhaps a bit easier to define. It includes the garrison or home-station environment, the forward-deployed environment for troops stationed at overseas locations or on ships, and the deployed environment for troops on an actual mission. The military setting subsumes the range of work environments associated with the military occupation, from office and staff work, to combat infantry and peacekeeping duty. The military setting also

stimuli

responses

cohesion

garrison

114

extends to cultural, lifestyle, and community aspects of the military occupation. Finally, it is important to remember that family members and civilian employees of the military also live and work within the military setting, and are likewise exposed to special stressors related to the military lifestyle.

Psychological Framework for Considering Stress in the Military Context

Because the military environment exposes its members to a variety of unusual stressors, it is a valuable "natural laboratory" for psychologists interested in learning about human adaptation to stress. This has developed into a substantial field of activity, spanning carefully controlled laboratory studies of physiological responses to acute physical stressors, to naturalistic field studies of morale and performance under various conditions of psychological stress. The present chapter will not attempt to summarize all aspects of this burgeoning field. Instead, the focus is on fairly recent studies of soldier and family psychological stress and adjustment during actual military operations.

In approaching the problem of stress in the military context, the psychological frame- interactionism
work of "interactionism" is a useful one. This perspective emphasizes the interaction of
situation and person variables in determining human behavior (e.g., Magnusson & Endler, 1977). Following some discussion of this orientation and a brief historical overview of stress studies in the military, results from recent studies on psychological stress in the military environment are presented and discussed. In addition to identifying the types of stressors troops and families are exposed to during operations, these studies are beginning to reveal the effects of such stressors on health and adjustment outcomes. There is a growing interest in identifying both situational and person variables that can account for why individuals respond differently to stress in the military environment.

In the field of psychology, a question of quite long-standing interest concerns how human beings adapt, or fail to adapt, to the stressors of life. While the term "stress" may be relatively new, "anxiety" is not. Endler & Edwards (1982) point out that both terms are used almost interchangeably in much of the literature. The great body of Sigmund Freud's work (see for example Rickman, 1957; Freud, 1969) can be understood as a theoretical formulation on how human beings, with the help of various defense mechanisms, adjust to the essential anxiety (or stress) of life that comes from the conflict between biological instincts and the demands of society.[1] What many today would call extreme stress, William James considered as instinctual fear. Of particular relevance to the combat environment, James points to the special power of loud noises to provoke fear and terror responses in adults (James, 1890).

Lazurus and Folkman (1984) offer the classic modern position on psychological stress and adaptation as an interaction (or "transaction") between characteristics of the situation, and characteristics of the person. For Lazarus, a critical aspect of an individual's response to a stressful event is the "cognitive appraisal" that gets made, and this appraisal is based partly on the person's sense of his/her own capabilities for dealing with the event. This perspective puts substantial weight on both situational (stress) and person (appraisal) factors in understanding behavior. The interactionist viewpoint in psychology (Magnusson & Endler, 1977) generally puts equal emphasis on situational and person variables influencing behavior, and even regards the debate about which is more important as a "pseudo-issue" (Endler, 1973). Still, under certain conditions, it is thought that person or situation influences will be more or less influential. For example, Mischel (1977; 1981) describes a variety of situations, such as a prison environment, thought to have such potent influence over behavior as to overwhelm person influences. Likewise, in highly ambiguous situations where rules of behavior are unclear, personality variables may have greater influence over behavior. The interactionist position provides a useful framework for understanding stress, and responses to stress, in the military context.

There are three classes of outcome variables that are important to consider, and may be influenced by stress: performance, social adjustment, and health. Stress can lead directly to impaired performance, can contribute to a variety of physical and mental health difficulties, and can also result in a variety of social adjustment problems such as family violence, divorce, and substance abuse. Figure 1 presents a schematic diagram showing these three types of outcome variables on the far right. "Stressor", on the left, represents forces in the environment that impinge on the person. These forces can be physical or psychological, and sometimes both at once. An incoming SCUD missile is very much a physical threat to the soldier, and can result in the physical destruction of his body. At the same time, such physical events for most people also carry the psychological threat of one's own possible death or injury, or of friends and family. But the impact of stress on outcomes is rarely direct and immediate; usually, it is processed or filtered through both organizational, social context variables, and personal variables. These are represented in the figure as cones or shields in between "stressor" and "response." Examples of social context variables that might influence how stressors get

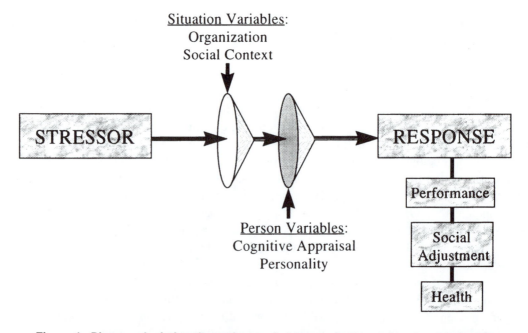

Figure 1: Diagram depicting the pathway of stressors in the environment (stimuli) to responses of the organism. Stressors are filtered through Situation Variables and Person Variables. The three principal classes of response are Performance, Social Adjustment, and Health.

processed in the military environment are unit cohesion and leadership climate (cf. Manning, 1991; Bartone & Kirkland, 1991). Person variables that could influence or moderate the stress-outcome relation include past experience, pre-existing psychopathology, and personality characteristics.

The ability to perform physical and mental tasks quickly and accurately in the military is essential. Likewise, the capacity to sustain effective performance over an extended period of time under adverse conditions ("staying power"), is also important. Performance in the military context includes both individual and group tasks and functions. Several excellent reviews of stress and performance in the military are available, including Driskell and Salas (1991), Krueger (1991), and Moore & von Gierke (1991).

Stress in the military can also contribute to a range of social adjustment problems. For example, some studies have linked higher rates of alcohol abuse to the special occupational demands of the military (cf. Long, Hewitt and Blane, 1976; Whitehead and Simpkins, 1983). The possible impact of military occupational stressors on family violence and breakup is an important area in need of further research (Prier & Gulley, 1987). Retention, or early attrition of military personnel can also be seen as a negative social or organizational outcome variable. Segal & Harris (1993) summarize data from several studies showing that retention in the U.S. Army is negatively affected by the large number of separations for missions and training.

Finally, stress can have a profound influence on physical and mental health in the military environment, whether in war or peace. The health of the force is a concern to the military organization in large part because health in turn influences performance and achievement of organizational goals.[2] Mental health in this context includes morale, which is a term of special saliency in the military used to describe both the mental fitness and motivation of individuals and groups.

Following a brief historical overview of psychology and psychiatry in the military, the remainder of this chapter will describe the kinds of stressors experienced by soldiers, and families, during modern military operations across the range of conflict. Much of this work is being done by military psychologists, and aims at understanding the nature and effects of stress in the operational environment. These studies rely heavily on questionnaire-based measures of mental health as indicators of troops' responses to stress, and assume a connection between mental health and performance. They also frequently take an interactionist perspective, in which characteristics of situations and persons are both considered to influence responses to stress.

Historical Overview

Historically, much if not most psychological research and activity within the military has been concerned with selection and placement. This concern characterized, for example, most of the work of Robert M. Yerkes and colleagues at the Division of Psychology of the Office of the Army Surgeon General during World War 1. The accomplishments of this division, which included development of the famous Army Alpha and Beta tests of general intelligence, are widely recognized as providing the impetus for the emerging field of psychological testing (Anastasi, 1968). During World War II, this tradition continued and expanded under a distinguished group of psychologists, including Robert C. Tryon, Donald W. Fisk, James G. Miller, and Henry A. Murray at the Office of Strategic Services (OSS), founded in 1942 (see the

117

final report of the OSS, 1948; Wiggins, 1973). In the same period, another group of Army social scientists, mostly sociologists and social psychologists, conducted extensive survey and attitude studies related to military personnel policies and soldier adjustment. Under the auspices of the Research Branch of the Army's Information and Education Division, this group produced an impressive collection of reports including four volumes known as the "American Soldier" studies (e.g., see Stouffer, Lumsdaine, Lumsdaine et al., 1949). Selection and placement have continued as major concerns for military psychologists since World War II (Steege & Fritscher, 1991), while ergonomics (Kaplan, 1991) and human responses to environmental extremes (Krueger, 1991) have also received considerable attention. Although studies in this latter area are conducted primarily in laboratory settings, the 1980s saw a resurgence of field studies aimed at understanding the influences on positive and negative soldier performance under various conditions (e.g., Marlowe, 1985; Siebold & Kelly, 1988; Manning & Ingraham, 1987).

Beginning with World War I and up to the present, the domain of extreme battle stress casualties has received the most attention not from psychologists, but from psychiatrists in the military (see Jones, 1995 for a comprehensive historical review of military psychiatry). As medical practitioners, military psychiatrists have focused mainly on effective treatment and return to duty of psychiatric casualties (e.g., Salmon, 1929; Jones, 1995). While individual differences in response to extreme combat stress have certainly been noted, they are generally regarded as the result of pre-existing neuroses or psychopathology (Grinker & Spiegel, 1945; Noy, 1987). In understanding what causes extreme stress reactions to combat, military psychiatrists have put the greatest importance on situational factors such as combat intensity and duration. The underlying notion is that at the extreme high end of the stress spectrum, as in very intense combat, individual differences in response to stress are minimal; overwhelming environmental stressors are overwhelming to all people alike. As military psychiatrists like to say, "Every man has his breaking point", and under the acute stress of extreme combat conditions, everyone breaks, if not at precisely the same moment. From this perspective, individual factors such as personality are deemed relatively unimportant in determining mental breakdown in combat (Marlowe, 1979; Belenky, Noy and Solomon, 1987).

An emphasis on social-situational factors is also apparent in the work of military psychologists working in this tradition. For example, Manning & Ingraham (1987) identify unit cohesion as the critical factor that moderates or buffers the impact of combat stress on performance. But in examining the possible effects of personality factors, military psychiatrists and psychologists have limited themselves mainly to psychopathological tendencies such as neuroticism and repression (Noy, 1987), while generally ignoring dimensions of normal personality such as extroversion or locus of control that might be important determiners of how soldiers respond to stress. A notable exception to this rule is found in the work of Reuven Gal (1983), who explores the influence of both personality and situational variables on Israeli soldier responses under battle. Gal reports, for example, that bravery on the battlefield is related to such individual factors as leadership ability, devotion to duty, decisiveness, and perseverance under stress, in addition to situational variables (pp. 76–79).

The next section describes a relatively new trend in the U.S. Army, toward the use of deployable "human dimensions" or psychological research teams. Following this, representative results are provided from some of these field studies conducted over the last decade with Army units on deployments. These studies all attempt to identify the major stress factors that might affect soldier health and performance on operations.

Field Psychological Research Teams in the Modern Military

"The soldier who sustains wounds in battle, the mother who worries about her soldier son, the gambler who watches the races—whether he wins or loses—the horse and the jockey he bet on: they are all under stress. The beggar who suffers from hunger and the glutton who overeats, the little shopkeeper with his constant fears of bankruptcy and the rich merchant struggling for yet another million: they are also all under stress." (Selye, 1976, p 3)

Military leaders increasingly recognize that psychological stress in military operations can have a range of serious consequences, including increased risk of death and serious injury from accidents, inattentiveness and errors of judgment, exposure (cold injuries), friendly-fire incidents, and suicide. Psychological stress can also increase the risk of soldier misconduct, alcohol abuse, and violations of the rules of engagement, as well as diminish soldier mental health, morale, and psychological readiness to perform the mission. A clear understanding of sources of psychological stress at various phases in military operations is necessary in order to counteract or reduce these stressors, and better prepare soldiers and leaders to cope with them. Leaders generally also want to identify early any special or unexpected psychological stressors, in order to take corrective action before health and performance suffers. To build the necessary knowledge, field studies with human beings are essential. Laboratory studies cannot reveal the sources of stress for deployed soldiers, nor how they cope with and adapt to these stressors. Multi-method, naturalistic studies are the best available method for gathering scientific data on the actual stressors of deployments, as well as soldier responses to these stressors.

The Persian Gulf War in 1991 provided an opportunity for Army research psychologists to study soldier health and adjustment under the stress of an actual combat deployment (Gifford, Martin & Marlowe, 1991; Bartone, 1993; Gifford et al., 1996). This work involved the deployment of field research teams into the combat theater, using interviews, observations, and surveys to identify useful lessons for future application. The activities and reports of these field psychology research teams were deemed sufficiently valuable by senior Army officials to merit continued support of the concept of psychological field research teams to study soldier stress, health and adaptation. In 1993 an agreement between the Army's Surgeon General and the Deputy Chief of Staff for Personnel established the concept of "Human Dimensions in Combat Research Teams" for deployment in future conflicts (US Army Deputy Chief of Staff for Personnel, 1993).[3]

Stressors on Military Operations

What follows is a sample of findings from psychological research efforts aimed at identifying stressors on military operations, as well as various effects of those stressors. The work described was mainly done under the auspices of the Walter Reed Army Institute of Research. Other organizations also conduct field psychological studies with military populations (e.g., Army Research Institute for the Behavioral and Social Sciences, Alexandria, Virginia; Naval

Health Research Center, San Diego, California; Office of Naval Research, Alexandria, Virginia).

Operation Just Cause (Panama, 1989)

The first modern operational application of the concept of deployable psychological or "human dimensions" research teams occurred not during, but shortly (3–6 months) after the invasion of Panama by U.S. forces in 1989. Following the operation, units that participated were surveyed and interviewed at their home stations back in the U.S. (Schneider, 1991). Reflecting the presumed importance of unit cohesion as an influence on soldier performance and adjustment, the key research question was what direct effect cohesion had on post-combat adjustment or distress (Marlowe, 1986). By Schneider's analysis, the area of greatest stress was "direct" combat exposure which included "having a buddy killed in action", "serving in a unit that fired on the enemy", "having a buddy get wounded or injured", and "engaging the enemy in a firefight". This was followed by what Schneider termed "unpredictable" combat stress, which included "being in a patrol that was ambushed", "flying in an aircraft that was shot at by the enemy", "encountering mines or booby traps", and "receiving sniper fire". Schneider termed the third most important stress factor "engaging the enemy", which included "seeing an enemy soldier killed or wounded", "firing rounds at the enemy", "seeing civilians killed or wounded", and "killing someone in action". He went on to show that (using retrospective ratings), while combat stress was by far the strongest predictor of psychological distress, unit cohesion did make an independent contribution to later adjustment. In this study, adjustment was measured by number and intensity of psychiatric symptoms reported, including depression, anxiety, and physical (somatic) complaints. The possibility that cohesion might serve to moderate the destructive effects of stress was not tested.

depression

Operations Desert Shield & Desert Storm (Persian Gulf War, 1990–1991)

The Persian Gulf War survey data collected by the Walter Reed Army Institute of Research teams corresponded to three phases of the operation: the long (6-month) period of coalition forces build-up in the Saudi desert ("Desert Shield"), the period just prior to offensive operations against Iraqi forces ("Desert Storm"), and the period following Iraq's defeat and return of U.S. forces to home stations in Germany and the U.S. (Gifford et al., 1996). The major stressors associated with anticipated combat in the periods prior to offensive operations concerned fear of enemy artillery and chemical weapons, and fear of being killed or wounded, or having a friend killed or wounded. The highest non-combat stressors during the Desert Shield phase related to isolation and family separation, lack of privacy, boredom, forced celibacy, and having to adjust behavior to a foreign culture (Marlowe et al., 1993). In the Germany-based follow-up sample, the most commonly reported combat stressors involved engaging the enemy, seeing civilians and enemy soldiers killed or wounded, and encountering mines or booby traps. (See Adler, Vaitkus & Martin, 1996 for a complete description of the Germany-based sample of Gulf War Army veterans). Using regression procedures, Bartone (1993) found that amount of combat exposure in Gulf War veterans predicted later psychological adjustment, as indicated by psychiatric symptomatology and a post hoc constructed measure of "Post Traumatic Stress" symptoms. Also, both unit cohesion and the

hardiness

personality dimension of hardiness (Kobasa, 1979) appear to moderate or buffer the ill-effects of combat exposure (Bartone, 1993). These results confirm the importance of both social (situational) and person (personality) variables in influencing how soldiers respond to combat stress.

"Post Traumatic Stress Disorder" (PTSD) was recognized as a distinct psychiatric diagnosis in 1980, in large part based upon the Vietnam War experience (Kulka et al., 1990). After the war, many American combat veterans displayed a characteristic pattern of symptoms, including (1) reexperiencing the trauma (e.g., "flashbacks" and nightmares), (2) avoidance symptoms (e.g., inability to remember certain things about the trauma, avoiding certain places or things), and (3) arousal symptoms (irritability, trouble sleeping). For a full description of PTSD and other reactions to severe or traumatic stressors, see the American Psychiatric Association (1994), *Diagnostic and Statistical Manual of Mental Disorders, Fourth Edition.*

Post Traumatic Stress Disorder

Operation Restore Hope (Somalia, 1993)

At the request of the Army's Office of the Deputy Chief of Staff for Personnel (ODCSPER), in January, 1993 a three-person "human dimensions" team deployed to Somalia with the 10th Mountain Division, for the relief/peacekeeping mission known as Operation Restore Hope. The team was multidisciplinary, consisting of a research psychologist, a social worker, and a behavioral science specialist, all on active duty. Based in Washington, DC, the team traveled to Ft. Drum, New York and then deployed to Somalia with elements of the 10th Mountain Division. They remained in Somalia until March, when the first echelons of deployed soldiers returned to the United States. A subsequent 3-person team returned to Somalia in July of 1993 for a two week visit, and included an Army psychiatrist. In both cases, data collection methods included interviews, observations, and self-report surveys.

The greatest stressor identified by the research team was uncertainty about when the mission would be complete and soldiers could go home (Gifford, Jackson & Deshazo, 1993). The other main stressors were isolation from families, ambiguity about the mission and rules of engagement, and physical discomforts. Despite these stressors, the authors report that soldiers were well-adjusted and coping effectively during the mission. One strength of this particular study is that the questionnaire was not designed in advance, but was developed after the initial field interviews and observations were complete. In this way, researchers were able to adjust the survey content to reflect the salient issues associated with this particular mission.

Operation Provide Promise (Croatia, 1993)

In October of 1992, the United States for the first time placed military units under the operational control of the United Nations, with the UN Protection Forces (UNPROFOR) in the former Yugoslavia. A Mobile Army Surgical Hospital (MASH) task force of some 300 soldiers deployed from Germany to Croatia, with the mission to provide medical care to all UNPROFOR soldiers. The U.S. portion of the overall operation was dubbed "Operation Provide Promise" reflecting the humanitarian relief focus of the mission.[4] A medical peacekeeping task force composed of Army units stationed in Germany was studied in depth over the course of their 6-month deployment (Bartone, Vaitkus & Adler, 1994; Bartone, Adler

& Vaitkus, 1996; Bartone & Adler, 1994). Army psychological research teams used multiple methods, and made repeated data-collection site visits to the unit over the course of pre-deployment training and the deployment itself. Key stressors were identified for the pre-deployment, mid-deployment, and late-deployment phases. The late-deployment data were collected shortly before the unit redeployed to their home stations in Germany.

The survey data from the mid-deployment provide a good illustration of the kinds of stressors encountered on deployments, as well as how those stressors can be related to health and morale indicators. At the mid-deployment point, the primary stressors were boredom (lack of meaningful work), isolation and lack of support from rear detachment and headquarters, separation from families and concern about family safety, uncertainty about the future, lack of recognition, sense of unfairness and lack of control over benefits and privileges (e.g., to travel), and ambiguity in the command structure.

An important question to be addressed empirically in research such as this concerns what is the relation, or possible impact of various stressors on soldier health and functioning. The most frequently reported stressor is not necessarily the one that is most destructive to health and well being. It may be the low frequency stressor that has the greatest potential impact on soldiers. By addressing this question empirically, research psychologists can help military planners distinguish between stressors that are by and large taken in stride, and those that are more serious. As an example of how such questions are explored, the following table (Table 1) shows the Pearson correlations between mid-deployment stressors, and depression, psychiatric symptoms, and morale for the medical task force in Croatia.

The correlations listed in Table 1 demonstrate (statistically significant) associations between variables. While causal links cannot be determined from simple correlations like this, it is nevertheless clear from these data that stress in the deployed environment is associated with depression, symptoms and morale. It is also apparent that certain events, though relatively low in frequency, can correlate significantly with adjustment indicators. This is the case, for example, with "Problems with unit leaders" and "Having to move my family back to the U.S."; both are fairly low frequency events for this sample, but correlate strongly with the adjustment indicators.

boredom   During the mid-deployment phase, the problem of "boredom" appeared for the first time on this mission. Boredom has been observed as a concern or stressor for troops on other operations as well. For example, Harris & Segal (1985) identified boredom as a factor that appeared to be a negative influence on morale for the first American contingent to serve in the Sinai Multinational Force and Observers. Swedish researchers have also noted boredom as a stressor among Swedish forces in UN operations in Southern Lebanon (Carlstrom, Lundin & Otto, 1990). But it was clear from our observations of the U.S. medical task force in Croatia that boredom on such missions is not a simple matter of "not enough to do." While the daily work demands were generally quite modest, there was a range of entertaining activities in which to engage. For example, regular sports competitions were held, including volleyball, basketball and soccer. Plentiful reading materials were available, as well as exercise equipment, bicycles, video movies, ping-pong and board games. On the weekends, soldiers could visit downtown Zagreb for shopping, dining, and even a movie. There were ample opportunities for entertainment and distraction. The real problem of boredom seems related to the lack of meaningful, professionally relevant work to do. Boredom appears to be an especially salient stressor in the later phases of peacekeeping and contingency operations.

Table 1

**The relation of stressors to Depression, Symptoms,
& Morale on a 6-month peacekeeping mission**

Stressors	Depression	Outcome Indicator Symptoms	Morale
Personal Health Problems	.30**	.39***	ns
Boredom	.25**	.25**	−.20*
Problems with Rear Detachment	.27**	.33***	−.23**
Concerns about Family Safety	.42***	.41***	−.22*
News reports of trouble: former Yugoslavia	.26**	.28**	−.20*
Problems with Unit Leaders	.25**	.27**	−.44***
Isolation	.34***	.35***	−.17*
Having to move my family back to U.S.	.41***	.48***	ns
Concerns about marital infidelity	.37***	.35***	ns
Delays in getting mail	.44***	.33***	ns
Trouble making phone calls	.49***	.43***	−.17*
Problems living in Europe	.41***	.43***	−.23*
Problems with co-workers	.34***	.29**	−.39***

Stressors were rated on six-point Likert scale in terms of how much trouble or concern is caused by each stressor: 0=none, 1=very low, 2=low, 3=medium, 4=high, 5=very high.

***$p < .001$

**$p < .01$

*$p < .05$

NOTE: From mid-deployment medical task force survey data, N=128. Adapted from Bartone & Adler (submitted-b). Depression measured with short CES-D (Radloff, 1977; Ross & Mirowsky, 1984). Symptoms measured with 20-item psychiatric symptoms checklist (Bartone et al., 1989). For description of morale measure, see Bartone & Adler (submitted-b).

Operation Uphold Democracy (Haiti, 1994)

In November 1994, an Army "human dimensions" research team deployed to Haiti to study soldier stress and adaptation in "Operation Uphold Democracy." Over 3,000 surveys were collected, representing about 1/3 of the deployed force (Halverson et al., 1995). Major stressors identified were poor living conditions, including concerns about sanitation, food and water, and lack of privacy; fear of disease and personal harm; family separation; heavy work schedules and lack of time off; and uncertainty about the length of deployment (Halverson et al., 1995, pp. I–ii, 14–24). These stressors were reported to be significantly related to psychological well being as measured by the Brief Symptom Inventory (Derogatis & Spencer, 1982). Unfortunately, the wording of the stress questions on the survey confounds stressors in the environment with subjective responses to these stressors.

The issue of how to measure stressors in the military environment is an important one, and one that bears some additional discussion at this point. Whatever term is applied, stress, stressor or something else completely, the research goal is to identify conditions or events, stimuli in the operational environment that can generate anxiety, tension, stress or distress for soldiers and lead to impaired functioning or health problems.

The questionnaire based measure of stress used in the Haiti study exemplifies an important measurement problem common to such studies. Soldiers are presented with a list of issues (e.g., "lack of time off", "level of sanitation in Haiti", "fear of getting a disease"), and asked to rate "how much worry or stress" these issues caused for them during the deployment, from "None at all" to "Extreme". The data obtained by this method make it impossible to distinguish event frequency, or extent of actual exposure to some potentially stressful stimulus or stimuli, from subjective responses to either real or anticipated exposure. An example will help to illustrate the problem. Imagine two soldiers in the same company, exposed to precisely the same conditions. When asked "how much worry or stress" is caused by "fear of getting a disease", the first soldier answers: "extreme", while the other answers "none at all". Though exposed to the same environmental conditions, one soldier is very worried or anxious about catching a disease, while the other is unconcerned about it. Perhaps the first soldier has a history of illness that causes him to worry more about disease in general, or perhaps he read a magazine article before deploying that heightened concerns around disease. But the way the question and response options are presented, rather than providing information about objective conditions in the operational environment, soldier responses reflect a subjective, personal assessment of their own levels of worry or stress on any given issue or potential stressor. These subjective responses are influenced by a host of factors including soldiers' idiosyncratic cognitive appraisals of threat, and their available resources for dealing with it. This also means that the stress measures and the psychological distress outcome measures are cross-contaminated (Costa & McCrae, 1985; Lazarus, DeLongis, Folkman, & Gruen, 1985), with both measuring to some degree subjective responses and states such as anxiety. Given this measurement confound, analyses that relate "stress factors" to distress levels can be risky and misleading.

Operation Vigilant Warrior (Kuwait)

In November of 1994, a three-person "human dimensions" team deployed to Kuwait to collect data with U.S. Army soldiers who had deployed as part of Operation Vigilant Warrior, an allied defense force mobilized to deter renewed Iraqi military threats toward Kuwait. The research team, composed of two research psychologists and one behavioral science specialist, arrived late in the operation as soldiers were beginning to redeploy. Approximately 600 questionnaires were administered across two brigades, and extensive interviews were conducted. The team completed its work in about 10 days, and redeployed as the soldiers themselves were returning to their home stations.

Stressors for this sample of soldiers were related to the dimensions of isolation, boredom, and lack of control or powerlessness. For example, the highest rated stress items were: "being away from my home and family", "waiting without much to do", "lack of morale, welfare and recreation equipment", "lack of time off", and "lack of personal privacy" (Bartone, Stuart and Valentine, 1994). As with the Haiti survey discussed above, the measure of stress in this survey was unfortunately confounded with subjective responses to operational stressors.

Operation Joint Endeavor (Bosnia Implementation Force–IFOR, 1996)

In 1996, the American military deployed about 20,000 troops for the international peacekeeping force known as IFOR (Implementation Force) in the former Yugoslavia. The majority of troops for this force were drawn from U.S. stations in Europe, and the anticipated length of deployment was unprecedented for missions of this type—12 months. Beginning in the pre-deployment period of November 1995, psychologists from the Walter Reed Army Institute of Research-European branch launched a longitudinal research effort to document sources of stress across phases of the Bosnia peacekeeping mission, and identify any effects on soldier psychological health and adjustment (Bartone, 1996b; Bartone, Britt & Adler, 1997). Given that the operational conditions and challenges for soldiers can vary substantially over phases of a mission, it was considered essential to take a longitudinal approach, following soldiers over all phases of the 12-month mission. A pre-deployment survey was administered to over 3,000 soldiers scheduled to deploy to Bosnia from stations in Germany, mainly in December 1995 and January 1996. After units deployed to Hungary, Croatia and Bosnia, data from the early phase of the operation (December 1995 through the April 1996) were gathered in the field using observation and interview techniques. A mid-deployment survey was administered in-theater in June 1996, and a late-deployment survey in October 1996.

Highest stressors just prior to deployment reflect concerns about family welfare and safety, insufficient time to make needed preparations, and loss of educational and job advancement opportunities. Once soldiers have deployed, a different set of stressors take the fore. For this first contingent to deploy into Bosnia in winter of 1996, the main early deployment stressors were: crowded and austere living conditions, harsh weather, high workload, mission uncertainty and concern for families. Three months into the operation, boredom, being away from home and family, and restrictions on movement and behavior emerge as major issues. For some troops, an intra-psychic conflict develops around the perceived imbalance between personal sacrifices required by the mission (loss of family, freedoms, career development), and the importance of the overall mission and one's role in it. The relative shortage of meaningful professional daily work activities leads to questions about the value of the investment, and to increased anger, frustration, guilt and depression. Because of the scope and importance of the Bosnia mission, these issues are discussed in somewhat more detail below.

A. Stressors just prior to deployment

The highest stressors in the pre-deployment phase relate to time pressure: not having enough time to complete personal business, and preparing the family for the deployment. Thirty-five percent of soldiers report these as high-stress areas. Experience shows that this is a busy time for units preparing to deploy, and soldiers and leaders work long hours in planning and preparing for the mission. This leaves less time to take care of personal and family business, placing many soldiers in a tense conflict between family and unit responsibilities. Another major stressor has to do with concern about the family: thirty-one percent of respondents report concern about the Rear Detachment taking care of their family, and 30% are troubled by the isolation of their families from friends and relatives in America. Soldiers who are about to deploy are concerned about their families being left isolated and unsupported. A large percentage of soldiers are troubled by lack of job advancement opportunities (27%), and dissatisfaction with their level of education (34%). This most likely reflects the belief that being deployed will disrupt both military and civilian education programs, which weigh heavily into promotion prospects in the all-volunteer force. A smaller percentage (22%) of soldiers report problems with unit leaders and the chain-of-command. While difficult to interpret without more data, this may well reflect the conflict between unit demands and family demands for the soldiers' time. Table 2 summarizes the pre-deployment stressors, in order of frequency endorsed by soldiers.

Table 2

Pre-Deployment stressors for American soldiers deploying to Bosnia

December 1995–January 1996

1. Completing personal business before deploying
2. Preparing my family for my deployment
3. Dissatisfied with my education
4. Concern Rear Detachment will care for family
5. Being separated from family & friends in States
6. Lack of job advancement opportunities
7. Problems with unit leaders
8. Problems with my chain of command
9. Boredom in my work

Note: Based on pre-deployment survey data (of N=3,036) of U.S. Army soldiers in Germany due to deploy to Bosnia, Croatia or Hungary, winter 1995–96 (cf. Bartone, Britt & Adler, 1996).

B. Stressors in the early-deployment period

The author had the opportunity to deploy as a psychologist with an Army "Combat Stress Control" unit for the first 3–4 months of the Bosnia mission (December-March, 1996). Army Combat Stress units are small teams of mental health specialists whose mission is to prevent stress casualties, and provide early treatment when appropriate. These teams are in a good position to develop information about the nature and extent of psychological stressors, since an important part of their mission is to identify the "stress threats" that must be addressed with prevention   prevention measures (see Stokes & Jones, 1995 for a review of Combat Stress Control programs and policies). Based on observations, interviews and consultations, the primary stressors were identified at approximately monthly intervals. These are discussed below for the period extending from just after deployment in January 1996, through about June of 1996.

The very early deployment period (January 1996) was one of intense activity and long work hours associated with the primary mission requirement of quickly moving large numbers of personnel, equipment and supplies into areas designated by the Dayton Accords, and establishing the Zones of Separation between the warring factions according to the specified dates. In general, American soldiers performed this mission with high professionalism and energy, and it can now be said that the deployment phase itself was accomplished successfully and safely. Nevertheless, it was a highly stressful period. Principal stressors for soldiers included having to leave their families just before the Christmas holiday, isolation and little contact with home once deployed, high work tempo and long hours, fatigue and lack of sleep/rest time, and crowded and Spartan living conditions, all exacerbated by some of the harshest winter weather the area had experienced in many years.

Table 3

Early-Deployment stressors for American soldiers deployed to Bosnia

January-February 1996

1. Heavy workload, long hours
2. Crowded and confined living quarters
3. Poor sanitation of latrines and living areas
4. Cold, harsh weather
5. Frequent & lengthy meetings/briefings
6. Family separation
7. Isolation (more acute for cross-attached soldiers)
8. Mission ambiguity and uncertainty
9. Poor communication, flow of information
10. "Micro-management" of junior leaders
11. Sleep deprivation
12. Lack of physical exercise
13. Little recognition

Note: Based on interviews and observations conducted during early deployment phase (cf. Bartone, 1996b).

As the operation moved into month 2 (February 1996) and living conditions began to improve, some of the very early stressors diminished and new ones surfaced. American units working on bridging operations in Croatia provide a good example. The weather started to improve, with fewer winter storms and slightly warmer temperatures. Work schedules and tempos became more regular and predictable, and more rest/sleep time was available. Living conditions at the base camps improved with the gradual provision of warm, dry living areas, hot showers, laundry and mess facilities, exchange stores, and chapel services. In addition, a growing number of American military camps were receiving special (AFN—Armed Forces Network) television and radio broadcasts, and fresh food rations were occasionally available.

In the third month of the deployment (March 1996) observations and interviews indicated soldier morale was improving as a function of steady enhancements to the quality of life, and increased predictability in work schedules. Some units were still in transit, and had not yet reached the location in theater where they would be stationed for the mission. Soldier morale in these units was generally somewhat lower, mainly due to continuing uncertainty regarding where and when they were going, and related ambiguities about the mission. Morale also appeared to vary somewhat according to location; units in more isolated locations generally experienced greater discomforts, with fewer amenities and services. Principal stress factors were related to uncertainties about the mission, lack of recognition, isolation, workload, lack of privacy, and boredom.

C. Stressors in the mid-deployment period

Survey data collected at the 6-month point indicate what the most salient stressors are for soldiers at this phase of the Bosnia mission (USAMRU-E Activities Report, July–December 1996). Family separation, or "missing my family", appears as the highest soldier concern. This is followed by "feeling confined" and "double standards". Both are issues that relate to loss of control or powerlessness. Boredom and travel restrictions also are high on the list of stressors at this mid-deployment phase of the operation. Bosnia mid-deployment stressors are listed below, arranged in order of importance for soldier respondents.[5]

Table 4

Mid-Deployment stressors for American soldiers deployed to Bosnia

June, 1996
1. Missing my family
2. Feeling confined, not able to go anywhere
3. Double standards
4. Boring & repetitive work
5. Travel restrictions
6. Problems with information flow
7. Rumor spreading
8. Feeling far away from things familiar
9. Boredom when off-duty
10. Worry that factions will continue fighting
11. Feeling cut-off from my family
12. Worry about how my family is coping

Note: Based on USAMRU-E Mid-Deployment Survey, N=1400 (cf USAMRU-E, 1996).

In order to shed additional light on the nature of psychological stressors during the Bosnia peacekeeping operation, and to further describe the role of psychologists during field operations, the following example is provided. This excerpt is taken from a field report provided by the author to a senior field commander during Operation Joint Endeavor.[6] In addition to identifying the major stressors, this report made general suggestions regarding stress "countermeasures", or leader actions likely to reduce or prevent stress-related problems around a given issue.

countermeasures

(Excerpt from Memorandum for Commander, Task Force Harmon, 13 February 1996):

> *Morale Trends.* On average, morale continues moderate to high in TF Harmon, largely as a function of continuing improvements in quality of life, and in-

creased predictability in work schedules. Morale among transient units is generally somewhat lower, mainly due to continuing uncertainly regarding where and when they are moving, and ambiguities about the mission. Morale also appears not as high among those Harmon permanent party units located outside the Harmon Life Support Area (LSA). The isolation and uncertainty are more salient factors in these outlying units, and life support services are generally not as good as those available in the LSA.

Current Stress Factors: The current stress factors closely track those reported on 31 January. The current list takes into account additional feedback from soldiers, and also rank orders the factors in approximate order of importance.

(1) Uncertainty. The major stressor for soldiers (and many leaders too) concerns uncertainty and confusion about the mission, where and when the unit might be moving next, and command policies. Unplanned unit moves are especially stressful and disturbing for soldiers. Effective countermeasures all relate to improving the flow and accuracy of information to soldiers, and minimizing changes and policy shifts. Newsletters, bulletin boards, AFN radio messages, Commanders' Calls, and leaders talking to soldiers all serve to reduce uncertainty.

(2) Lack of Recognition. Soldiers want to be recognized as professionals contributing to an important mission. Leaders can provide recognition in a variety of ways including awards, special events, and a simple "pat on the back". Leaders should remind soldiers of how their daily work activities contribute to the overall success of the Bosnia mission. Media recognition also is a powerful countermeasure to this source of stress.

(3) Isolation. TF Harmon soldiers are far away from home and family, living in a foreign, bleak environment. Many worry about the welfare of their families. Some soldiers describe weak or non-existent Rear Detachment and Family Support units. In the recent period, soldiers report increased frustration in placing morale calls through the MSE/DSN network. Countermeasures include improved mail and phone (morale calls) access, newsletters, and good family support and rear detachment activities. A leave—R&R program that permits soldiers some time with their families is important, especially given the length of the rotation. Strong unit cohesion, friends in the unit, and concerned leadership also are powerful antidotes to social isolation and family separation stress.

(4) Mission importance. Many soldiers express doubts about the importance and significance of the mission and their part in it. They need to be reminded by leaders of why this is an important mission worth the sacrifices being asked of them.

(5) Workload/OPTEMPO (Operations Tempo). Some units continue with long work days and 7 day weeks. The high OPTEMPO began prior to deployment for many. Countermeasures would include slowing the tempo, and shifting to a 6-day work week where possible. A leave—R&R program would also help by providing soldiers with rest and a change of venue.

(6) Limited MWR (Morale-Welfare-Recreation services). The Croatia/B-H environment offers little opportunity for aerobic exercise, which can be an important way to "blow off steam" as well as stay fit. Stationary exercise equipment like bike and rowing machines would help. Also, time-off is more valuable if soldiers have interesting and useful resources to utilize during their time off, such as a gym or weight room, movie theater, library, etc.

(7) Lack of privacy. Living and working with the same people in close and crowded conditions often leads to increased irritability over time. Everyone needs some privacy, or time alone. Living quarters can be configured to provide some privacy barriers. Also, some private areas for morale calls should be considered.

(8) Monotony. Daily activities are taking on a routine sameness for many units. This can lead to complacency, loss of mission focus, and depression. Counter-measures include varying the schedule on weekends, and building special events into the calendar, such as the Super Bowl Party. Holiday unit picnics and barbecues give soldiers something to look forward to. Variety in food and menus (A-rations) will also help. Additional countermeasures to consider include permitting some group travel or cultural visits to local attractions when security permits, and allowing soldiers to wear their civilian attire when off-duty and in protected areas.

Among the various stressors identified during recent deployments is one that is somewhat internal and secondary to external events. It results from a mental comparison done by the soldier in which he/she weighs out all the pros and cons of being deployed on a given mission. It can perhaps best be described as an intra-psychic conflict or dissonance, an imbalance between the cost, or personal sacrifices and discomforts that being deployed entails, and the limited perceived pay-off or benefit. For example, many soldiers perceived the value of the IFOR/Bosnia mission as minimal; a common belief was that peace established by IFOR under the Dayton Accords will not last beyond the withdrawal of IFOR forces. Soldiers also typically see little of value coming to them personally from this deployment. Their jobs are often boring and lacking in meaning, the extra pay they receive is modest, and recognition is scant. Soldiers compute a kind of mental balance-sheet or ledger, weighing out the relative costs and benefits of his/her presence on this deployment. This can lead to an intra-psychic conflict for some that may be psychologically damaging. At a minimum, deployed soldiers must tolerate an extended and emotionally difficult separation from loved ones. When this sacrifice is not offset or counter-balanced by meaningful daily work activities and an associated belief in the significance of the mission, growing frustration, bitterness and depression can result.

In partial support of this hypothesis, anger and depression were found to be the most common problems among soldiers presenting to the Combat Stress Control service in Croatia during the early deployment period (Bartone, 1996b). In a majority of cases, there were profound worries about the welfare of spouse and family members at home, coupled with a sense in the soldier that his/her presence on this mission was insignificant. Anecdotally, similar issues were apparent with many soldiers interviewed informally, along with feelings of guilt over leaving their families. Additional research is needed to identify the prevalence and significance of this intra-psychic conflict.

Dimensions of psychological stress in military operations

In addition to identifying specific stressors in military operations, it is important to ask what are the underlying, more general issues that might summarize the range of stressors observed. Can the specific stressors be classed into more general categories that make sense, providing a better understanding of soldier responses? In pursuit of this goal, Bartone & Adler (submitted-a) suggest five general areas of psychological stress that are salient on military operations.

These five dimensions capture the more detailed specific stressors quite well: Isolation, Ambiguity, Powerlessness, Boredom, and Danger/Threat (Table 5). This model is useful for summarizing the range of stressors identified in military operations, with perhaps special relevance for modern, non-combat or peacekeeping operations.

The kinds of psychological stressors described in recent U.S. military deployments, and summarized in the above model, have been noted in a variety of other military and peacekeeping operations. The dimensions of boredom and isolation are familiar ones; they continue to be important issues in the U.S. Sinai deployment (Siebold, 1996), and are also significant in the experience of Swedish and Norwegian forces in Lebanon (Lundin & Otto, 1989; Weisaeth, Mehlum & Mortensen, 1996). The powerlessness or helplessness factor has also been noted as a significant stressor in several peacekeeping operations, including Lebanon and Somalia (Weisaeth & Sund, 1982; Litz, 1996). Powerlessness can be a function of highly restrictive rules of engagement, which constrain soldiers from responding in many situations, as well as from exposure to extreme suffering of people in the area of operations, suffering which soldiers are relatively helpless to do anything about. A sense of powerlessness can also result from travel restrictions, difficulty communicating in a foreign culture and language, loss of privacy and control over living arrangements, and perceived unfairness in the allocation of valued supplies and resources ("double-standards") which the soldier is helpless to rectify. A recent study of Canadian forces in the Former Yugoslavia reported that the highest stressors were "double-standards" or unfair application of the rules, and powerlessness to change the situation (Farley, 1995).

Table 5

Dimensions of Psychological Stress on Military Operations

ISOLATION:	Deployed to physically remote locations Encountering obstacles to communication Units are newly-configured, low cohesion Individuals are cross-attached from other units
AMBIGUITY:	Mission not clear or well-defined Command structure is ambiguous Role and identity confusion, ambiguity
POWERLESSNESS:	Rules-of-engagement are restrictive Constraints on movement, action Exposure to suffering of local people Foreign culture & language Lack of privacy. . . little control over living arrangements Relative deprivation—"Double Standards"
BOREDOM:	Repetitive, monotonous routines & schedules Lack of meaningful work Over-reliance on "busy work"
THREAT/DANGER:	Danger of death, injury, threat to life or limb Mines, snipers, disease Exposure to death of others

Note: Adapted from Bartone & Adler (submitted-a)

Several studies have noted role ambiguity as a key stressor for combat-trained soldiers engaged in peacekeeping operations. (Segal & Segal, 1993; Miller & Moskos, 1995). Soldiers trained to fight sometimes have trouble adjusting to the role of peacekeeper, which often requires control and restraint. Ambiguity and uncertainty regarding rules of engagement and the purpose of the mission also have been observed elsewhere, such as in Somalia (Litz, 1996; Gifford et al., 1993), and in medical humanitarian aid operations in the former Soviet Union (Britt & Adler, submitted). The research described above on U.S. forces deployed to Croatia and Bosnia calls attention to yet another source of ambiguity on such missions: that associated with an unclear command structure in multinational and UN operations. The French experience in the Former Yugoslavia identifies ambiguity about roles and missions as a stressor, as well as confusion about the chain of command under UN operations, and enforced passivity or powerlessness to act to make things better (Raphel, 1995; Doutheau et al., 1994).

The risk of injury and death (threat/danger dimension), as well as exposure of peace-keeping troops to death and violence among local peoples or warring factions, of course varies across different operations. In recent U.S. experience, the Somalia mission represented the greatest physical dangers to troops on a daily basis. But even apparently safe and peaceful operations always carry some danger, if only from accidents and possible terrorist strikes. U.S. operations in Lebanon were peaceful until a terrorist truck bomb killed 240 Marines in 1983. More recently, the terrorist bombing of U.S. military housing in Saudi Arabia (Khobar Towers, Dhahran), which killed 19 American Air Force personnel and injured hundreds more, provides another reminder of the constant threat of terrorism faced by deployed peacekeeping and contingency forces. It is possible that the very unpredictability of such threats on peacekeeping operations, and the sharp contrast with regular and generally safe daily routines, increase the risk for post-traumatic stress symptoms or disorders afterwards. As we will see in the next section, uncertainty is also a major concern for families of deployed soldiers.

Stress in the Military Family

In the military operations described above, "separation from family" and "concern about my family's welfare" were identified as key stressors for soldiers. In the all-volunteer U.S. military, the trend is for an increasing proportion of active duty personnel to be married with children (Goldman, 1976; Department of Defense, 1988; Segal, 1989). This demographic fact of military life means that family members are a large part of the military community. The military family must adapt to a series of demands related to the military career. These demands include having to move frequently, live in foreign countries and cultures, periodic and sometimes unpredictable separations from the family, and the risk of injury or death (McCubbin & Marsden, 1979; Segal, 1986; Hunter, 1982).

One of the more difficult demands on families, and one that is increasing in frequency in recent years, is the experience of forced separation (Glisson et al., 1980; Van Vranken et al., 1984, Nice, 1980). Family separation is a common occurrence in the military. A 1985 Department of Defense survey found that 72% of enlisted personnel and 79% of officers with spouses and/or children, had been separated from their families during the preceding year, while 30% of enlisted people and 18% of officers had spent 5 months or more away from their families (Segal, 1989). These trends have continued in the years following the break-up of the Soviet Union. U.S. military forces have shrunk in size, while the pace and number of deployments have increased. Good discussions of family stressors in the military context are available elsewhere (e.g., Segal & Harris, 1993). Norwood, Fullerton & Hagen (1996) provide an overview of stress and the military family, with special attention to Navy family support efforts during the Persian Gulf War. Segal & Segal (1993) summarize what was learned about family stress and adjustment from early U.S. participation in the Sinai peacekeeping operation known as the Multinational Force and Observers (MFO). Since 1982, the U.S. has provided a new Army peacekeeping contingent to this operation every 6-months.

In addition to an excellent review of the literature on family separation and coping in the military, Vuozzo (1990) reports research results showing substantial individual differences in how spouses of Sinai-deployed soldiers responded to the separation. Spouses who relied on emotion-focused, avoidance coping strategies also reported significantly more stress-related symptoms. More recently, the Army Research Institute conducted a comprehensive evaluation of adjustment and performance of Army Reserve Component soldiers serving in the Sinai MFO (Phelps & Farr, 1996), and also examined stressors and adjustment of families. Jensen & Shaw (1996) discuss the impact of war and military-related stress on a relatively neglected group— children (see also Raviv & Klingman, 1983, for a discussion of the impact of severe stress, including war and terrorism, on children in Israel).

With the deployment of soldiers to the former Yugoslavia for peacekeeping duty in December, 1995 the U.S. military initiated a one-year separation for soldiers and their families currently living in Europe. Previously, one-year "hardship" tours were restricted to U.S. based soldiers deployed to Korea. Since the beginning of Operation Joint Endeavor, families living overseas face an increase in the frequency as well as duration of deployments. A recent study of stress, health and adjustment among 1,700 spouses of soldiers deployed from Germany and Italy for Operation Joint Endeavor in Bosnia (Bartone, J., 1997) identified the following major stressors (Table 6):

Table 6

Stressors for spouses of American soldiers deployed to Bosnia

May–June, 1996

1. Length of deployment
2. Concern about soldier's safety
3. News about the situation in Bosnia
4. Getting timely information
5. Getting accurate information
6. Separation from family & friends in US
7. Uncertainty about what the mission is
8. Reaching my soldier by phone
9. Running the household by myself
10. Problems with my children
11. Health problems of family members
12. Getting mail to my soldier

The unusually long duration of the deployment (12 months) was the most frequently endorsed (82%) stressor item on the survey. Analysis of open-ended comments from the survey revealed the intensity of strain and disruption to families caused by the long deployment. The 12-month deployment was the top issue raised by spouses in the survey comments section. The majority of spouses commenting on this issue remarked about the negative impact the lengthy separation was having on their marriages and their families. For many, the year-long deployment followed additional separations due to training and field exercises. The majority of spouses who addressed the issue of length of deployment indicated concern not for themselves (i.e., "I'm an adult, I understand that this is my husband's job"), but for the harm being done to their children. A common view was that non-combat missions such as this do not justify keeping the family apart for so long. Many expressed the intention to leave the Army, as a direct result of multiple and lengthy family separations. For those planning to remain in the Army, the desire for clear and consistent deployment and return dates was frequently mentioned.

Half of the spouses surveyed reported problems getting accurate and timely information from the soldier's unit. In environments where accurate information is lacking, rumors abound. Without reliable information about the missions their soldiers are on, spouses seek information from any source. In such a context, even rumors become a valuable source of "information". Unfortunately, rumors frequently provide misinformation, which leads to distress and an erosion of trust in unit leaders. Research with spouses of soldiers deployed to the Sinai shows that unit newsletters and mission updates once or twice a month can help sustain spouse psychological well-being (Bell, Schumm, Segal & Rice, 1996), perhaps by counteracting misinformation and rumors that inaccurate media stories may foster. Keeping spouses informed about the mission helps alleviate anxieties and fears for the safety and well-being of deployed soldiers.

Even during peace operations, the possibility of sudden death through accidents or terrorism is always present. In the next section, research conducted after a major Army air disaster exemplifies the importance of both situational and individual factors in influencing how military people respond to extreme stressors.

Stress in the military community: The case of Army Survivor Assistance Officers (SAOs)

Much of the preceding discussion has focused on stressors associated with military deployments. Stress can also occur outside of military deployments, in training, in transporting troops to and from deployments and training exercises, and in military communities. A dramatic example is provided by the crash of a chartered Army jet in 1985. The plane was carrying 248 soldiers of the 101st Airborne Division back home to their families after a 6-month peacekeeping deployment in the Sinai. Following a brief stop in Gander, Newfoundland (Canada) to refuel, the plane crashed in icy and snowy conditions. A fire burned for over 24-hours following the crash, and there were no survivors. In the aftermath, a longitudinal study was conducted to identify psychological stress and its effects among "Survivor Assistance Officers" or SAOs, Army officers assigned to assist family members of the dead (Bartone et al., 1989). Analysis of two-year follow-up data confirmed that SAO duty was highly stressful, and was often associated with health symptoms and psychological distress. Those who remained

Survivor
Assistance
Officers

healthy over the period of study, as opposed to those developing symptoms, possessed two distinguishing characteristics: social support at work, and personality hardiness (Bartone, 1989).

Social Support and Personality Hardiness:
Social and Personal Stress-Resistance Resources

Cross-sectional (using a single time-point) and longitudinal data showed that social support, especially from supervisors at work, was an important factor related to good health under stress. These analyses also showed that SAOs with the personality style known as "hardiness" (Kobasa, 1979) stayed healthy despite high stress levels. Army SAOs with these "stress-resistance resources" did experience some stress-related symptoms, but significantly fewer than others. Also, they recovered from the stressful experience more rapidly than their unprotected counterparts.

SAOs who reported their bosses or commanders were supportive, understanding, and helpful regarding their SAO responsibilities showed fewer stress symptoms over the long term than others. Although support from friends and family was also helpful, "boss support" far outweighed the value of supports from other sources. What might account for this? In a study of personal and group responses to four community disasters, Killian (1952) provides an insightful analysis that may explain why support at work is so important during times of crisis. People in society have multiple group memberships and loyalties, which under ordinary circumstances do not conflict with each other. We are loyal to our families, work groups, and larger communities, and share our time and resources among them. But in disaster, these different loyalties can clash directly, as with the fireman who deserts his post to safeguard his family when a tornado lands in town. Killian recounts a series of such incidents in which individuals experience acute conflict among competing roles that ordinarily coexist peacefully.

This is similar to the conflict Army SAOs can experience between their sense of duty to the bereaved family, continuing obligations at their work unit, and responsibility toward their own families. Social support at work might be so important because it *reduces* the conflict associated with competing group loyalties and demands during crisis periods. Such conflicts, unrecognized or ignored under ordinary conditions, can be severe and damaging following disaster or catastrophe. Leaders, supervisors and co-workers should thus be sensitive and supportive regarding competing group loyalties following catastrophic loss. Especially with dedicated, hard-working people, it may be extremely disruptive to appeal to a sense of duty to the group (e.g., organization, unit) as a means of convincing the soldier/officer to disregard other loyalties and obligations (e.g., to his dead buddies, his family). Supportive relationships at work are not likely to emerge following disaster if the framework is not built beforehand. Leaders who emphasize trust, respect, and communication besides performance and dedication are indirectly preparing their soldiers to cope more effectively with traumatic loss, if it strikes their unit.

Survivor Assistance Officers who were high in the personality dimension of hardiness also showed fewer stress-related symptoms, and recovered more quickly over time.[7] Personality hardiness is a cognitive style that appears to influence how stressful circumstances are processed and integrated into a person's life experience (Kobasa, 1979; Maddi and Kobasa, 1984). Persons high in hardiness have a strong sense of commitment to life, believe they can control events around them, and are interested and challenged by new things and obstacles. Whether or not hardiness is a fixed trait that develops early in life and never changes, or is amenable to change over the life course, remains an open question. Without attempting to answer this question, there are still valuable treatment and prevention lessons in the research findings on hardiness and stress.

For military organizations, the hardiness construct offers a useful framework for structuring work situations likely to increase stress-resiliency. These can be considered as preventive measures for reducing the negative effects of catastrophic stress when it occurs. Like the hardy individual organism, the hardy unit organism is one that has a strong commitment to the work of the unit (our mission), a sense of control over its own destiny, and enjoys challenges. Commitment in this sense requires meaningful activity, a conviction that the unit has an important purpose for existing. To a great extent, soldiers probably learn this commitment through a modeling process (Bandura & Walters, 1963), by observing it in their leaders. Leaders who believe in the importance of unit goals communicate this belief to their juniors, primarily by example. Soldiers in the "hardy unit" have a sense of control and ownership over the unit and its activities. The next higher level of command in the organization can empower or control subordinate units to varying degrees, either allowing them to exercise substantial control and influence over their activities or not.

Challenge relates mainly to training in the military context. Military organizations have ample opportunities to cultivate the sense of challenge, since there are so many new tasks and skills to master. But training should be creative and fun as well as rigorous; monotony is the enemy of challenge. Units that routinely foster commitment, control, and challenge through leadership, structure, and activities are likely to be more resilient and healthy under severe stress. A similar perspective on challenge and boredom is offered by Csikszentmihalyi (1975). He suggests that the key to fostering a sense of engagement or "challenge" is to provide situations or tasks that require slightly more of the individual than his/her present skill level permits. When task challenge slightly exceeds ability, the individual is moved to work harder and usually experiences success and gratification. If the gap between task challenge and available skills is too large, anxiety results. On the other hand, if skills exceed what is required by the task or situation, the result is boredom.

In discussing stress in the military context, an important area of concern for psychologists involves the treatment of stress-related problems. While a full consideration of treatment issues is beyond the purview of this chapter, the material presented so far leads to several interesting implications regarding early intervention strategies to prevent long-term stress related problems in the military environment.

Several researchers have suggested that stress, and traumatic stress in particular, is most psychologically damaging when the exposed person, for whatever reasons, has difficulty facing his thoughts and feelings regarding the trauma. He then fails to process and interpret the experiences surrounding the traumatic event into the rest of his life experience or cognitive schemata (e.g., Ursano and Fullerton, 1990; Pennebaker and Beall, 1986; Horowitz, 1976). From the theoretical perspective of hardiness, this failure would be highly disruptive to the sense of commitment and overall purpose and meaning in life. Effective therapies would thus provide a supportive context in which the traumatized person could begin the cognitive (and perhaps emotional) processing of experiences surrounding the trauma. Recovery corresponds to the integration of the trauma into one's general mental catalog of experiences. Such treatment is likely to be more effective as it emphasizes positive aspects of the traumatic experience. Army Survivor Assistance Officers who displayed the most healthy recovery course were also those who reported positive features to their stressful duty. The therapeutic value of retrospective attributions of positive meaning has also been found in U.S. Air Force POWs (Ursano, Boydstun and Wheatley, 1981), and in Jewish survivors of Nazi concentration camps (Antonovsky, 1979).

Two related treatment methods that emphasize cognitive processing of traumatic experiences deserve mention. Both involve the recounting of thoughts and feelings following exposure to trauma or disaster, and have proven effective in reducing distress and symptoms in stressed individuals. Arik Shalev has adapted the group debriefing method of S.L.A. Marshall (1956) to the treatment of traumatized patients (Shalev, 1994). In one recent study (Shalev et al., 1993), this method was applied successfully to a group of Israelis whose bus was attacked by a terrorist. The focus is on orally recounting events and reactions rather than feelings. A somewhat different approach involves written rather than oral descriptions of stressful encounters. In a series of studies, Pennebaker and colleagues have shown that writing about traumatic experiences reduces related symptoms for several months following treatment (e.g., Pennebaker, Colder, and Sharp, 1990). He postulates a physiological mechanism wherein writing about traumatic experiences helps individuals to organize them mentally, and releases previously operating inhibitory processes. Many of the Army SAOs commented spontaneously on the beneficial effects of having written about their experiences in the surveys they completed. The implication is it may be therapeutic for traumatized persons to have an opportunity to put their experiences into words. Whether orally or in written form, this activity apparently facilitates self-understanding and the integration of the trauma into a broader life perspective.

Most of the studies discussed or referenced in this chapter have focused on immediate or short-term effects of stress in the military context. But responses to traumatic stress can be delayed, sometimes for many years. While "Post-traumatic stress disorder", or PTSD, has received the greatest attention in this regard (Kulka, Schlenger, Fairbank et al.), other negative outcomes are also possible (Ursano, McCaughey & Fullerton, 1994). Stress-related physical health problems can appear many years after traumatic exposure (e.g., Elder, Shanahan & Clipp, 1997). Such delayed responses to stress are also of great concern for military psychologists, as scientists and practitioners dedicated not just to organizational goals, but to the short and long term health and welfare of military people.

Psychology as a discipline aims at understanding human behavior, and the forces that influence behavior. Deciphering why individuals respond differently under similar circumstances is a critical part of this endeavor. Some of the more dramatic, interesting, and puzzling human behaviors and individual differences appear in response to extreme stressors. The military context presents many such extreme stressors to troops and families alike, making it a valuable "natural laboratory" for students of human behavior. If we wish to understand not just average tendencies, but also individual responses to stress, then personal as well as situational factors must be considered. Psychological approaches that are attuned to both sets of influences hold the promise for a fuller understanding of stress in the military context.

Acknowledgments

I am very grateful to Jocelyn V. Bartone for her many critical contributions to this manuscript. The views presented herein are those of the author, and do not necessarily reflect those of the Department of Defense or the Department of the Army.

References

Adler, A., Vaitkus, M. & Martin, J. (1996). Combat exposure and Post-traumatic stress symptomatology among U.S. soldiers deployed to the Gulf War. *Military Psychology, 8*, 1–14.

American Psychiatric Association (1994). *Diagnostic and Statistical Manual of Mental Disorders, IV Edition*. Washington, DC: American Psychiatric Association Press.

Anastasi, A. (1968). *Psychological Testing*. Toronto: Macmillan.

Antonovsky, A. (1979). *Health, Stress and Coping*. San Francisco: Jossey-Bass.

Bandura, A. & Walters, R. (1963). *Social learning and personality development*. New York: Holt, Rinehart and Winston, 1963.

Bartone, J. V. (January, 1997). Operation Joint Endeavor spouse survey. Paper presented at USAREUR & 7th Army Family Support System Conference, Wiesbaden, Germany.

Bartone, P. T. (October, 1989). Long-term sequelae for family helpers following an Army air disaster. Presented at the annual meeting of the Society for Traumatic Stress Studies, San Francisco, CA.

Bartone, P. T. (June, 1993). Psychosocial predictors of soldier adjustment to combat stress. Paper presented at the Third European Conference on Traumatic stress, Bergen, Norway.

Bartone, P. T. (1996a). Family notification and survivor assistance: Thinking the unthinkable. In Ursano, R. J. & Norwood, A. E. (Eds.), *Emotional Aftermath of the Persian Gulf War: Veterans, Families, Communities, and Nations*. Washington, DC: American Psychiatric Press.

Bartone, P. T. (May, 1996b). American IFOR experience: Psychological stressors in the early deployment period. *Proceedings of the 32nd International Applied Military Psychology Symposium* (pp. 87–97), Brussels, Belgium.

Bartone, P. T. (August, 1996c). Hardiness and resiliency in U.S. peacekeeping soldiers. Annual Convention of the American Psychological Association, Toronto, Canada.

Bartone, P. T. & Adler, A. B. (October, 1994). A model for soldier psychological adaptation in peacekeeping operations. *Proceedings of the 36th Annual Conference of the International Military Testing Association* (pp. 33–40), Rotterdam, Netherlands.

Bartone, P. T. & Adler, A. (submitted-a). Dimensions of psychological stress in peacekeeping operations.

Bartone, P. T. & Adler, A. (submitted-b). Cohesion over time in a peacekeeping medical task force.

Bartone, P. T., Adler, A. B. & Vaitkus, M. A. (1996). Social psychological issues in the adaptation of US army forces to peacekeeping and contingency missions. In G. Meyer (Ed.), *Friedensengel im Kampfunzig? Zu Theorie und Praxis militarischer UN-Einsatze.* Opladen: Westdeutscher Verlag.

Bartone, P. T., Britt, T. W. & Adler, A. B. (1996). Stress, health and adaptation in Operation Joint Endeavor: Pre-deployment survey findings. Heidelberg, Germany: U.S. Army Medical Research Unit-Europe, Walter Reed Army Institute of Research.

Bartone, P. T., Britt, T. W. & Adler, A. B. (May, 1997). Stress and health in Bosnia peacekeeping operations. Presented at the Annual Meeting of the American Psychological Society, Washington, DC.

Bartone, P. T. & Gifford, R. K. (1995). Doing human dimensions research: Lessons from recent military operations. Presented at the Annual Convention of the American Psychological Association, New York.

Bartone, P. T. & Kirkland, F. R. (1991). Optimal leadership in small Army units. In R. Gal & A.D. Mangelsdorff (Eds.), *Handbook of Military Psychology.* New York: Wiley & Sons.

Bartone, P. T., Stuart, J., Valentine (November, 1994). Operation Vigilant Warrior human dimensions issues: Human dimensions field research team preliminary report. Heidelberg, Germany: U.S. Army Medical Research Unit-Europe, Walter Reed Army Institute of Research.

Bartone, P. T., Ursano, R. J., Wright, K. M & Ingraham, L. H. (1989). The impact of a military air disaster on the health of assistance workers: A prospective study. *J. of Nervous and Mental Disease, 177,* 317–328.

Bartone, P. T., Vaitkus, M. A. & Adler, A. B. (August, 1994). *Psychological issues peacekeeping contingency operations.* Poster presented at the American Psychological Association Annual Convention, Los Angeles, California.

Bell, D. B., Schumm, W. R., Segal, M. W. & Rice, R. E. (1996). The family support system for the MFO. In Phelps, R. H. & Farr, B. J. (Eds.) (1996). *Reserve component soldiers as peacekeepers.* Alexandria, Virginia: U.S. Army Research Institute for the Behavioral & Social Sciences.

Bell, D. B., Teitelbaum, J. M. & Schumm, W. R. (1996). Keeping the home fires burning: Family Support Issues. *Military Review, 76,* 80–84.

Belenky, G. L., Noy, S. & Solomon, Z. (1987). Battle stress, morale, cohesion, combat effectiveness, heroism, and psychiatric casualties: The Israeli experience. In G. L. Belenky (Ed.), *Contemporary Studies in Combat Psychiatry* (pp. 11–20). Westport, Connecticut: Greenwood.

Biville, Y. & Laffitan, D. (1996). Observations on French forces and adaptation in Bosnia. Presented at the 32nd International Applied Military Psychology Symposium, Brussels.

Britt, T. W. & Adler, A. B. (submitted). Stress, work and health during a humanitarian assistance mission to Kazakstan.

Carlstrom, A., Lundin, T., & Otto, U. (1990). Mental adjustment of Swedish UN soldiers in South Lebanon in 1988. *Stress Medicine, 6,* 305–310.

Costa, P. T. & McCrae, R. R. (1985). Hypochondriasis, neuroticism and aging: When are somatic complaints unfounded? *American Psychologist, 40,* 19–28.

Csikszentmihalyi, M. (1975). *Beyond boredom and anxiety.* San Francisco, Jossey-Bass.

Department of Defense (1988). *Defense '88 Almanac, September/October.* Alexandria, VA: American Forces Information Service.

Derogatis, L. R. & Spencer, P. M. (1982). *The Brief Symptom Inventory (BSI): Administration, Scoring and Procedures Manual.* Baltimore, MD: Clinical Psychometric Research.

DeSwart, H. W. (July, 1995). Contributions of (Dutch) military psychology to UN peacekeeping operations. *NATO/Partnership for Peace Workshop on Psychological Support for Peacekeeping Operations,* NATO Headquarters, Brussels, Belgium.

Doutheau, C., Lebigot, F., Moraud, C., Crocq, L., Fabre, L. M., & Favre, J. D. (1994). Stress factors and psychopathological reactions of UN missions in the French Army. *International Review of the Armed Forces Medical Services, 1/2/3,* 36–38.

Driskell, J. E. & Salas, E. (1991). Overcoming the effects of stress on military performance: Human factors, training, and selection strategies. In R. Gal & A. D. Mangelsdorff (Eds.), *Handbook of Military Psychology* (183–193). New York: Wiley & Sons.

Elder, G. H., Shanahan, M. J. & Clipp, E. C. (1997). Linking combat and physical health: The legacy of World War II in Men's lives. *American Journal of Psychiatry. 154,* 330–336.

Endler, N. S. (1973). The person versus the situation—A pseudo issue? *Journal of Personality, 41,* 287–303.

Endler, N. S. & Edwards, J. (1982). Stress and personality. In L. Goldberger & S. Breznitz (Eds.), *Handbook of stress: Theoretical and clinical aspects.* London, Free Press.

Eysenck, M. W. (1989). Personality, stress, arousal, and cognitive processes in stress transactions. In R.W. J. Neufeld (Ed.), *Advances in the investigation of psychological stress.* New York: Wiley & Sons.

Farley, K. M. (1995). Stress in Military Operations: Working Paper 95–2. Willowdale, Ontario: Canadian Forces Personnel Applied Research Unit.

Freud, S. (1969). *A General Introduction to Psychoanalysis.* Translated by J. Riviere. New York: Pocket Books (first published 1924, Boni & Liveright).

Gal, R. (1983). Courage under stress. In S. Breznitz (Ed.), *Stress in Israel* (pp. 65–91). New York: Van Nostrand Reinhold.

Gifford, R. K. (October, 1993). *Using field research teams to study stress and well being of deployed soldiers.* Paper presented at the Biennial Conference of the Inter-University Seminar on Armed Forces and Society, Baltimore, MD.

Gifford, R. K., Jackson, J. N. & DeShazo, K. B. (November, 1993). Field research in Somalia during Operations Restore Hope and Continue Hope. Paper presented at the 35th Annual Conference of the Military Testing Association, Williamsburg, Virginia.

Gifford, R. K., Martin, J. A. & Marlowe, D. H. (October, 1991). *Operation Desert Shield: Adaption of soldiers during the early phases of deployment to Saudi Arabia.* Paper presented at the Biennial Conference of the Inter-University Seminar on Armed Forces and Society, Baltimore, Maryland.

Gifford, R. K., Martin, J. A., Marlowe, D. H., Wright, K. M & Bartone, P. T. (1996). Unit cohesion during the Persian Gulf War. In Martin, J. A., Sparacino, L. R. & Belenky, G. *The Gulf War and Mental Health.* Westport, CT: Praeger.

Glisson, C. A., Melton, S. C. & Roggow, L. (1980). The effect of separation on marital satisfaction, depression, and self-esteem. *J. of Social Service Research, 4,* 61–76.

Goldman, N. L. (1976). Trends in family patterns of U.S. military personnel during the 20th Century. In N. L. Goldman & D. R. Segal (Eds.), *The Social Psychology of Military Service.* Beverly Hills: Sage.

Grinker, R. R. & Spiegel, J. P. (1945). *Men Under Stress.* Philadelphia: Blakiston.

Halverson, R., Bliese, P., Moore, R. & Castro, C. (May, 1995). Psychological well-being and physical health of soldiers deployed for Operation Uphold Democracy: A summary of human dimensions research in Haiti. Unpublished manuscript. Walter Reed Army Institute of Research, Washington, DC. Defense Technical Information Center Report ADA 298125.

Harris, J. J. & Segal, D. R. (1985). Observations from the Sinai: The boredom factor. *Armed Forces and Society, 11,* 235–48.

Holmes, D. S. & McCaul, K. D. (1989). Laboratory research on defense mechanisms. In R. W. J. Neufeld (Ed.), *Advances in the investigation of psychological stress.* New York: John Wiley & Sons.

Horowitz, M. J. (1976). *Stress response syndromes.* Northvale, NJ: Jason Aronson.

Hunter, E. J. (1982). *Families Under the Flag: A Review of Military Family Literature.* New York: Praeger.

James, W. (1890). *Principles of Psychology* (2 Vols). New York: Holt, Rinehart and Winston.

Jensen, P. S. & Shaw, J. A. (1996). The effects of war and parental deployment upon children and adolescents. In R. J. Ursano & A. E. Norwood (Eds.), *Emotional aftermath of the Persian Gulf War: Veterans, families, communities and nations* (pp. 83–109). Washington, DC: American Psychiatric Association Press.

Johansson, E. (1997). *In a blue beret: Four Swedish UN battalions in Bosnia.* Stockholm, Sweden: National Defense College, Department of Leadership.

Jones, F. D. (1995). Psychiatric lessons of war. In Jones, F. D., Sparacino, L. R., Wilcox, V. L., Rothberg, J. M. & Stokes, J. W. (Eds.), *War psychiatry. (Textbook of military medicine, Part 1. Warfare, weaponry and the casualty).* Washington, DC: Office of the Surgeon General, U.S. Army.

Joy, R. T. (1996). Foreword. In Martin, J. A., Sparacino, L. R. & Belenky, G. (1996), *The Gulf War and Mental Health*. Westport, CT: Praeger.

Kaplan, M. (1991). Cultural ergonomics: An evolving focus for military human factors. In R. Gal & A. D. Mangelsdorff (Eds.), *Handbook of Military Psychology*. (pp. 156–167). New York: Wiley & Sons.

Killian, L. M. (1952). The significance of multiple group membership in disaster. *American J. of Sociology. 57,* 309–314.

Kobasa, S. C. (1979). Stressful life events, personality, and health: An inquiry into hardiness. *J. of Personality and Social Psychology. 37,* 1–11.

Krueger, G. P. (1991). Sustained military performance in continuous operations: Combatant fatigue, rest and sleep needs. In R. Gal & A. D. Mangelsdorff (Eds.), *Handbook of Military Psychology* (pp. 255–277). New York: Wiley & Sons.

Kulka, R. A., Schlenger, W. E., Fairbank, J. A., Hough, R. L., Jordan, B. K., Marmar, C. R., & Weiss, D. S. (1990). *Trauma and the Vietnam War generation.* New York: Brunner/Mazel.

Lazarus, R. S., DeLongis, A., Folkman, S. & Gruen, R. (1985). Stress and adaptation outcomes: The problem of confounded measures. *American Psychologist, 40,* 770–779.

Lazarus, R. S. & Folkman, S. (1984). *Stress, appraisal and coping.* New York: Springer.

Lenz, E. J. & Roberts, B. J. (1991). Consultation in a military setting. In R. Gal & A. D. Mangelsdorff (Eds.), *Handbook of Military Psychology*. New York: Wiley & Sons.

Litz, B. (1996). The psychological demands of peacekeeping for military personnel. *NCP Clinical Quarterly, Winter, 6,* 1–8.

Long, J. R., Hewitt, L. E. & Blane, H. T. (1976). Alcohol abuse in the armed services: A review. *Military Medicine, 141,* 844–850.

Lundin, T. & Otto, U. (1989). Stress reactions among Swedish health care personnel in UNIFIL 1982–84. *Stress Medicine 5,* 237–246.

Maddi, S. R. & Kobasa, S. C. (1984). *The Hardy Executive: Health Under Stress.* Homewood, Illinois: Dow-Jones Irwin.

Magnusson D. & Endler, N. S. (1977). Interactional psychology: Present status and future prospects. In D. Magnusson & N. S. Endler (Eds.), *Personality at the crossroads: Current issues in interactional psychology.* Hillsdale, NJ: Erlbaum.

Manning, F. J. (1991). Morale, cohesion, and esprit de corps. In R. Gal & A. D. Mangelsdorff (Eds.), *Handbook of Military Psychology* (pp. 453–470). New York, Wiley & Sons.

Manning, F. J. & Ingraham, L. H. (1987). An investigation into the value of unit cohesion in peacetime. In G. L. Belenky (Ed.), *Contemporary Studies in Combat Psychiatry* (pp. 47–68). Westport, CT: Greenwood Press.

Marlowe, D. H. (1979). Cohesion, anticipated breakdown, and endurance in battle: Considerations for severe and high intensity combat. Unpublished monograph, Department of Military Psychiatry, Walter Reed Army Institute of Research, Washington, DC.

Marlowe, D. H. (Ed.) (1985). New Manning System field evaluation (Technical Report no. 1). Washington, DC: Walter Reed Army Institute of Research.

Marlowe, D. H. (1986). The human dimension of battle and combat breakdown. In R.A. Gabriel (Ed.), *Military psychiatry: A comparative perspective*. Westport, CT: Greenwood Press.

Marlowe, D. H., Wright, K. M., Gifford, R. K., Martin, J. A., Bartone, P. T. & Hoover, C. (1993). *Operation Desert Shield/Storm Codebook*. Washington, DC: Department of Military Psychiatry, Walter Reed Army Institute of Research.

Marshall, S. L. A. (1956). *Porkchop Hill*. New York: Jove Books.

McCubbin, H. I. & Marsden, M. A. (1979). The military family and the changing military profession. In F. D. Margiotta (Ed.), *Changing world of the American military*. Boulder, Colorado: Westview Press.

Memorandum dated 13 February, 1996, for Commander, Task Force Harmon, Subject: "254th MED DET (Combat Stress Control) Activities Update" (author P. Bartone).

Miller, L. L. & Moskos, C. (1995). Humanitarians or warriors? Race, gender and combat status in Operation Restore Hope. *Armed Forces & Society. 21*, 615–637.

Mischel, W. (1977). The interaction of person and situation. In D. Magnusson & N. S. Endler (Eds.), *Personality at the crossroads: Current issues in interactional psychology*. Hillsdale, NJ: Erlbaum.

Mischel, W. (1981). Introduction to Personality. New York: Holt, Rinehart & Winston.

Moore, T. J. & Von Gierke, H. E. (1991). Military performance in acoustic noise environments. In R. Gal & A. D. Mangelsdorff (Eds.), *Handbook of Military Psychology* (pp. 295–311). New York: Wiley & Sons.

Nice, S. D. (1980). *The course of depressive affect in Navy wives during family separation (Report number 80–24)*. San Diego, CA: Naval Health Research Center.

Norwood, A. E., Fullerton, C. S. & Hagen, K. P. (1996). Those left behind: Military families. In R. J. Ursano & A. E. Norwood (Eds.), *Emotional aftermath of the Persian Gulf War: Veterans, families, communities and nations*. Washington, DC: American Psychiatric Assoc. Press.

Noy, S. (1987). Stress and personality as factors in the causation and prognosis of combat reaction. In G. L. Belenky (Ed.), *Contemporary Studies in Combat Psychiatry*. Westport, Connecticut: Greenwood Press.

Office of Strategic Services Assessment Staff. (1948). *Assessment of Men*. New York: Rinehart.

Pennebaker, J. W. & Beall, S. K. (1986). Confronting a traumatic event: Toward an understanding of inhibition and disease. *J. of Abnormal Psychology. 95*, 274–281.

Pennebaker, J. W., Colder, M. & Sharp, L. K. (1990). Accelerating the coping process. *J. of Personality and Social Psychology, 58*, 528–537.

Phelps, R. H. & Farr, B. J. (Eds.) (1996). *Reserve component soldiers as peacekeepers*. Alexandria, Virginia: U.S. Army Research Institute for the Behavioral & Social Sciences.

Prier, R. E. & Gulley, M. I. (1987). A comparison of rates of child abuse in U.S. Army families stationed in Europe and in the United States. *Military Medicine, 152,* 437–440.

Radloff, L. (1977). The CES-D Scale: A Self-Report Depression Scale for Research in the General Population. *Applied Psychological Measurement, 1,* 385–401.

Raphel, C. (1995). Current aspects of stress on foreign theaters of operations. *International Review of the Armed Forces Medical Services, 1/2/3,* 128–130.

Raviv, A. & Klingman, A. (1983). Children under stress. In S. Breznitz (Ed.), *Stress in Israel.* New York: Van Nostrand Reinhold.

Rickman, J. (Ed.) (1957). *A general selection from the works of Sigmund Freud.* New York: Doubleday.

Ross, C. E. & Mirowsky, J. (1984). Components of Depressed Mood in Married Men and Women: The Center for Epidemiological Studies' Depression Scale. *American Journal of Epidemiology, 11,* 997–1004.

Salmon, T. W. (1929). The care and treatment of mental diseases and war neurosis ("shell shock") in the British Army. In Bailey, P., Williams, F. E., Komora, P. A., Salmon, T. W. & Fenton, N. (Eds.), *Neuropsychiatry. Vol. 10, The Medical Department of the United States Army in the World War* (pp. 497–523). Washington, DC: Office of the Surgeon General, U.S. Army.

Schneider, R. J. (May, 1991). *Operation Just Cause Survey Analyses: Phase III, Final Report.* Fort Detrick, MD: US Army Medical Research and Development Command, PR # DD3RMM–0333–8002–RPC.

Segal, M. W. (1986). The military and the family as greedy institutions. *Armed Forces and Society, 13,* 9–38.

Segal, M. W. (1989). The nature of work and family linkages: A theoretical perspective. In G. L. Bowen & D. K. Orthner (Eds.), *The organization family: Work and family linkages in the U.S. military.* New York: Praeger.

Segal, M. W. & Harris, J. J. (1993). *What we know about Army families (Special Report #21).* Alexandria, VA: U.S. Army Research Institute for the Behavioral and Social Sciences.

Segal, D. R. & Segal, M. W. (1993). *Peacekeepers and their wives: American participation in the Multinational Force and Observers.* Westport, Connecticut, Greenwood.

Selye, H. (1978). *The stress of life.* New York: McGraw-Hill (first published 1956).

Shalev, A., Schreiber, S. & Galai, T. (1993). Early psychological responses to traumatic injury. *J. Of Traumatic Stress, 6,* 441–450.

Shalev, A. (1994). Debriefing following traumatic exposure. In Ursano, R. J., McCaughey, B. G. & Fullerton, C. S. (Eds.), *Individual and community responses to trauma and disaster* (pp. 201–219). Cambridge, England: Cambridge University Press.

Siebold, G. L. (1996). Small unit dynamics: Leadership, cohesion, motivation, and morale. In Phelps, R. H. & Farr, B. J. (Eds.). *Reserve component soldiers as peacekeepers.* Alexandria, Virginia: U.S. Army Research Institute for the Behavioral & Social Sciences.

Siebold, G. L. & Kelly, D. R. (1988). The impact of cohesion on platoon performance at the Joint Readiness Training Center (Technical Report 812). Alexandria, VA: US Army Research Institute for the Behavioral Sciences. (DTIC No. ADA 202926).

Solomon, Z. (1993). *Combat stress reaction: The enduring toll of war.* New York: Plenum Press.

Steege, F. W. & Fritscher, W. (1991). Psychological assessment and military personnel management. In R. Gal & A. D. Mangelsdorff (Eds.), *Handbook of Military Psychology* (pp. 7–36). New York: Wiley & Sons.

Stokes, J. W. & Jones, F. D. (1995). Combat stress control in joint operations. In Jones, F. D., Sparacino, L. R., Wilcox, V. L., Rothberg, J. M. & Stokes, J. W. (Eds.), *War psychiatry. (Textbook of military medicine, Part 1. Warfare, weaponry and the casualty).* Washington, DC: Office of the Surgeon General, U.S. Army.

Stouffer, S. A., Lumsdaine, A. A., Lumsdaine, M. H., Williams, R. M., Smith, M. B., Janis, I. L., Star, S. A., & Cottrell, L. S. (1949). *The American soldier: combat and its aftermath,* Vol II. Princeton, NJ: Princeton University Press.

Ursano, R. J. and Fullerton, C. S. (1990). Cognitive and behavioral responses to trauma. *J of Applied Social Psychology, 20,* 1766–1775.

Ursano, R. J., Boydstun, J. A. & Wheatley, R. D. (1981). Psychiatric illness in U.S. Air Force Viet Nam prisoners of war: A five-year follow-up. *American J. of Psychiatry, 138,* 310–314.

Ursano, R. J., McCaughey, B. G. & Fullerton, C. S. (Eds.) (1994). *Individual and community responses to trauma and disaster.* Cambridge, England: Cambridge University Press.

U.S. Army Deputy Chief of Staff for Personnel (1993). Memorandum of Agreement on Establishment of "Human Dimensions in Combat" Research Unit.

U.S. Army Medical Research Unit-Europe (USAMRU-E) (1996). *Semi-Annual Activities Report, July-December 1996.* Heidelberg, Germany: U.S. Army Medical Research Unit-Europe, Walter Reed Army Institute of Research.

Van Vranken, E. W., Jellen, L. K., Knudson, K. H., Marlowe, D. H. & Segal, M. W. (1984). *The impact of deployment separation on Army families (Report WRAIR NP-84-6).* Washington, DC: Walter Reed Army Institute of Research.

Vuozzo, J. S. (1990). The relationship between consistency in focus of coping and symptoms among wives experiencing geographic marital separation and reunion. Unpublished Master's Thesis, College Park, MD: Department of Sociology, University of Maryland.

Weisaeth, L. & Sund, A. (1982). Psychiatric problems in UNIFIL and the UN-Soldier's Stress Syndrome. *International Review of the Army, Navy, and the Air Force Medical Services,* 109–116.

Weisaeth, L., Mehlum, L. & Mortensen, M.S. (1996). Peacekeeper stress: New and different? *NCP Clinical Quarterly, Winter, 6,* 12–15.

Whitehead, P. C. & Simpkins, J. (1983). Occupational factors in alcoholism. In B. Kissin & H. Begleiters (Eds.), *The pathogenesis of alcoholism: Psychosocial factors* (pp. 405–496). New York: Plenum Press.

Wiggins, J. S. (1973). *Personality and Prediction: Principles of Personality Assessment.* Reading, Massachusetts: Addison-Wesley.

Endnotes

[1]Holmes and McCaul (1989) provide a useful update on the defense mechanisms of repression, projection, and suppression as possible coping strategies for reducing stress.

[2]The Israeli experience is well documented by Z. Solomon (1993), who shows that exposure to combat stress can have physical as well as mental health consequences.

[3]The U.S. is not the only country expanding its emphasis on field research teams in military psychology. A number of countries involved in United Nations and multinational operations are also substantially increasing their field psychology efforts (e.g., Netherlands: DeSwart, 1995; Sweden: Johansson, 1996; France: Biville & Laffitan, 1996; Doutheau et al., 1994).

[4]An important part of the U.S. Provide Promise mission was the air-drop of humanitarian assistance (food, medical supplies).

[5]This research was embedded in a larger Bosnia mid-deployment research effort that included investigators from the Walter Reed Army Institute of Research, Washington, DC, and the Army Research Institute for the Behavioral and Social Sciences, Alexandria, Virginia. Only results from the USAMRU-E mid-deployment survey are referenced here.

[6]An important role for psychologists during a deployment is to advise and consult with leaders on a variety of "human dimensions" issues, including the identification of stressors and prevention of stress-related problems. For more on the consultative role, see Gifford, 1993; Bartone & Gifford, 1995).

[7]A study of Army casualty assistance workers in the Persian Gulf War also found personality hardiness and social support at work distinguished healthy from unhealthy workers performing this highly stressful duty (Bartone, 1996a). Also, hardiness was found to moderate the ill-effects of stress for soldiers on three separate peacekeeping operations (Bartone, 1996c).

Library of Congress

8

The Psychology of Information Warfare

Richard Bloom, Ph.D., ABPP

* Introduction * Conceptual Development * Examples * Information Warfare
as an Applied Social Science * Information Warfare Typology * Effectiveness
* Process * Education and Training * Initiator Effects * Future Trends
* Conclusion * References

This chapter is intended to compel students to seek a career in information warfare (IW). The chapter contains (1) an introduction highlighting some of IW's complexities; (2) a conceptual development; (3) examples; (4) attempts to answer questions about IW as an applied social science; and (5) comments about IW's future.

Introduction

Let's begin with some terms. What is information? It comprises anything from which people derive meaning. And people derive meaning from anything, for even nothing is something and even no meaning can be something's or someone's meaning. (Some authorities and lay people believe that meaning itself is the information, not that from which meaning is derived. Others believe both source of meaning and meaning are information. And there are still other intellectual positions on this subject—cf. Aristotle (1941/c. 340 B.C.), Kant (1958/1781), and Nietzsche (1954/1873).) Information warfare (IW) is the collection, protection, transmission, and modification of information to help achieve political objectives. A political objective denotes the control of a finite resource in a world of infinite need for that resource. How is the political objective achieved? IW can be applied directly to information or to anything affecting it, e.g., how it's created or stored. Ultimately, IW achieves the political objective through influencing human behavior. Some would posit that this occurs through the mind. I'm not sure what a mind is . . . but more of that shortly.

information

information
warfare

political objective

mind

Except for the last admission of ignorance, sounds pretty simple. But readers beware. IW can mask huge complexity. Let me remove the mask. The term itself—IW. Much of what I will label as IW has at other times been called perception management; psychological operations; psychological warfare; deception; command, control, communications, and intelligence countermeasures (C3ICM); propaganda; active measures; terrorism; political persuasion; influence operations; sabotage; reflexive control; dirty tricks; hacking;

machiavellianism

neurolinguistic programming; cyberwarfare; and mere wisdom, ingenuity, cunning, or machiavellianism. One may become chastened or bemused to read treatises from United States (US) military service schools which label the same activity with any one of a number of these terms depending on the fad of the moment (cf. Hill's bibliography, 1995; Information warfare, 1995.) The term IW—along with the rest of the usual suspects—bears the burden of denotations and connotations—some shared by all people, some by some people, some by only a single individual. In fact, the very decision to choose one term over another may elicit philosophical, political, social, cultural, and personal consequences among others for the author and for those who choose to believe or disbelieve the author. Of course, the same applied for scholastics inveighing on how many angels could dance on the head of a pin, scientists facing the Inquisition, and women undergoing the Salem witch trials.

Another thing about terms and definitions. Whatever an author may choose, seemingly relevant phenomena are included and shut out (cf. Allen, 1993). The author's task can then become easier or more difficult. With an existential twist of Aristotle's Golden Mean or Goldilocks' Three Bears, if the choice is just right, attacks may come from all sides, including one's own multiple selves.

Choosing examples to illustrate IW can be dangerous—figuratively and literally. The author may wish to be insightful and enlightening. However, some readers or associates of readers may be sincerely surprised and distressed that a specific event, e.g., the founding mythology of a nation or cult, is labeled as an IW example. Others are merely distressed. At this point, the author may need a fast getaway car as well as impeccable research and writing skills. Our martyrs are not terrorists but freedom fighters, and you, dear author, will soon be referred to in the past tense. Can you say Salman Rushdie?

And then there's the matter of citations. Citations are the intellectual accountant's bottom line of epistemological rectitude. How do you know what you know? By the wheelbarrow full of references which you dump at the end of the chapter—or at the bottom of each page, depending on format. But how relevant are the citations? How compatible with the chapter's assumptions, conclusions, screeds? Has the author even read the citations? Much like controversies about theories of emotion, does one cry and become sad or become sad and then cry? Does one write a chapter and look for research—or, literally, merely citations—to support the chapter, or read and evaluate research before, during, and even after writing? Unlike controversies over emotional theory, there seems to be an ethical prescription underlying the use of citations. Most researchers publicly advocate reading research before writing. But do they do as they say, say what they do, or talk the talk but not walk the walk? And the reader? The author can only guess what the reader's standards for believing might be and must somehow believe that these standards are, well, believable.

And what are the author's motives—conscious and otherwise? Publicly or privately held, advocated, admitted, confessed, and otherwise? To further one's career? To grind some professional axes? (One may be grinding the real thing, if one intends to activate the surprised and distressed mentioned above against oneself or others.) To reinforce or stimulate needs for affiliation or complexes of malignant narcissism. Given that the author is writing for someone, even if just for one of one's own many selves, between and within the lines there must be much of a personal nature. Hopefully this may induce less of a predilection for plagiarism, violating proprietary concerns, sovereign resources, transcultural ethics, or putative morals.

IW also may activate messy issues from the philosophy of social science. What is the optimal level of analysis to describe IW? Biochemical or electrophysiological phenomena within a neuron? Hormonal interactions throughout the body? Unconscious intrapsychic consequences? (I assume there's an unconscious. Since I'm unaware of my own, I must be unconscious of my unconscious—quod erat demonstrandum.) Interpersonal, interorganizational, or larger and larger concatenations of variables? And why are studies seeking to show cause— that IW technique A helped achieve political objective B—based on establishing the correlation between the two? Further studies seeking to show that IW influences human behavior via an effect on the mind must first show what a mind is. In the social sciences, unfortunately, much of what seems most interesting, e.g., the mind, are only hypothetical constructs. That is, they seem to help us understand the world, obtain meaning from it, through the constructed hypothesis that they exist in some absolute sense. But they may not. They may be no more than reflections of consensual word usages about particular observations and situations. The same is the case for other hypothetical constructs often associated with IW—perceptions, attitudes, instincts. (As the reader may later surmise, influencing the presence, absence, connotation, and denotation of hypothetical constructs may itself be a technique or proximal objective of IW. So may be writing and publishing an IW book chapter.) Suffice it to say—borrowing from Herman Melville and then Henry Murray (Robinson, 1992)—that the deeper we dive in these directions, the shakier our very foundations of reality become. So, let's stop here.

hypothetical constructs

Conceptual Development

Enough of allusion and illusion. What follows is a likely story of how and why IW developed in human intercourse. As the reader may be beginning to appreciate, a story may be as close as one can get.

In the beginning—irrespective of beliefs in creationism or evolutionary theories—all of human life was and has continued to be political. So many needs, so little resources. (So many resources, so little needs makes a good story, but not a likely one.) Resources may comprise anything. Food, water, shelter, weapons, money, people, and fantasies are just some examples. There are different lists of needs in psychological, philosophical, and religious texts. Some examples include hunger, thirst, sex, esteem, achievement, masochism, and spirituality. (Perversely—assuming I'm trying to encourage normative beliefs in two separate terms, resources and needs—resources can be needs and needs resources, just as in behaviorist theory reinforcers can affect the probability of behavioral occurrence but behaviors themselves can be reinforcers (Welsh et al, 1992). This should create a conceptual threat but also an operational opportunity for the IW practitioner.

Now the kicker. Humans have the unfortunate characteristic of living within a disparity between need and resource. Even when one need is satisfied, another cries out. Even when a need is satisfied, it doesn't remain so. And it seems as if new needs or previously unrecognized needs continue to crop up. As well, resources that used to satisfy a need, now only partially satisfy, or have no effect, or even increase the need. In a truly Sisyphean task, all humans can even realistically hope to sporadically achieve is to manage the overall disparity between needs and resources at some fairly tolerable level. How much ability and will one has through time to arrive at this far from lofty peak of achievement—a peak remained on only

Sisyphean task

power

anthropomorphic

gynopomorphic

hypostatized
constructions

momentarily—defines one's power. This definition of power applies to individuals, dyads, groups, tribes, clans, organizations, nations, societies, cultures—all human entities and accompanying anthropomorphic—pace, gynopomorphic—and hypostatized constructions.

The question then becomes how to obtain and maintain power. Because of the political nature of the human condition and the ever-changing nature of resources and needs, any particular technique for obtaining and maintaining power—i.e., a power technique of which IW is only one—will only work some of the time. (This is why Johnny One Note often has a bad time of it.) Moreover, there are internal and external limits to one's quest for power. The more obvious internal limits being one's own capacities of will and ability, the external being one's competitors—potentially all other humans and all or some of other species—and the physical properties of one's environment.

Obtaining and maintaining power can best be effected by casting a wide net for power techniques. Instead of depending solely on oneself, employing IW allows one to directly or indirectly influence others to help oneself. And IW is attractive not only because it garners the witting or unwitting help of others in surmounting limitations. For humans are social beings, a characteristic most ironic given one thus has a need which can be sated only by at least some of one's competitors for this and other needs (Aristotle 1941/c. 328 B.C.). (If there is a Holy Spirit, a holy sense of humor is part of the package.) So IW can kill two birds with one stone, even if the stone is not the previously alluded to Sisyphean boulder.

A further point. Because of the political nature of the human condition, the ever-changing nature of resources and needs, the continual flux of disparities between needs and resources, the varying success of power techniques, and the social need for those who are competitors for what we need, conflict is the currency of power and human relations. Peace, crisis, and war as commonly defined by laypeople, pundits, and alleged experts alike, are but three intervals on the continuum of conflict (cf. Croffi-Revilla, 1996). All variants of human activity are interrelated through this continuum. Contemplating this statement, the reader may begin to understand why vitriolic comments from leaders of the People's Republic of China (PRC) towards efforts of the US Government (USG) to spread democracy, free markets and cultural products—let alone to restrict exports of military assets—are logical and adaptive, not the raving rhetoric of paranoid schizophrenics. The same applies to the Iranian government's efforts restricting, if not banning, satellite dishes among its citizens (Sciolino, 1997). This last statement is not an apologia for oligarchs, but baldly identifies the ineluctable IW features of intercourse among nation-states.

With this observation about the PRC and stressing what has been already inferred by the reader—that IW has been employed and analyzed throughout recorded history (Liu Hsiang, 1996/c. 20; Ibn Khaldun, 1950/c. 1379; Shamasastry, 1909)—my story about IW is about to reach the era of contemporary history. Let's begin with a current look at why IW is still employed. (The reader can consider this a supplement to the material on how IW developed as a power technique.)

First, IW often costs less than a conventional war of attrition or even a small military battle to achieve a political objective. A photograph of mutilated women and children may increase international pressure to call off an attack cheaper than having to defend against the attack. Second, IW can help compensate when other military means are not available (See the discussion on the referent Palestinian in EXAMPLES below.) Third, IW can lead to future use of an adversary's resources. Massive aerial bombing, unless used as an IW technique, precludes exploiting the value of the people and material which are destroyed. Fourth, all actions and nonactions, have psychological significance. IW merely recognizes and potentially exploits this significance to achieve political objectives. In essence IW is ongoing like it or not, regardless of conscious intent. Fifth, in representative democracies and other political entities with a rule of law, IW may generate domestic support easier than other military means. This statement carries with it the assumption that enough people who matter believe that conflict is the currency of human relations and that influencing other people without injuring or killing them is less of a legal and moral transgression. If people believe significant conflict is only sporadic phenomenon, military destruction may be more readily considered as a special case. Of course, this begs the question of IW that achieves political objectives through death and destruction. (Interestingly, some people may denounce on moral grounds all IW as insidious "mind control" and applaud trench warfare as somehow morally pure. Yet, these same people may also manipulate, selectively communicate, or dissemble to help obtain and maintain their own power in their own, frankly insignificant and often sordid lives—the welfare of the collective be damned.) Sixth, IW can affect psychological processes intrinsic to concepts of security—discriminating friend from foe, supporting deterrence, delineating balance of power. Seventh, IW can be implemented and be effective without the target ever being aware. (Even in a covert and clandestine assassination someone usually knows or suspects that someone else is unnaturally dead.) Not all information from which one derives meaning is a product of IW, save to the clinically paranoid and, perhaps, some authors of IW texts. Because information is continuously impinging on the target, only the most compulsive could even have the response set, let alone the ability to separate the wheat from the chaff.

Now the reader should be ready to consider other material on the current practice of IW starting with its "stuff." Some of the anything, i.e., information, from which people derive meaning (see INTRODUCTION) and which IW uses includes words, phrases, signs, symbols, patterns, pictures, designs, and behaviors—the last including refraining from behavior as well. These examples are non all-conclusive, are partially overlapping, and reflect controversies in the arts and sciences about language, meaning, and representation. These difficulties do not stop successful IW practitioners be they US forces developing pamphlets encouraging Iraqi soldiers to surrender during the Persian Gulf War (Freund, 1991) or Indonesian military authorities affecting international perceptions of counterinsurgency operations in East Timor (Dyer, 1996).

IW affects military phenomena in several ways. It may utilize military personnel, weapons, or other resources as part of a specific technique. It may affect the will and ability of military personnel or the integrity of material. (Even without the use of or planned effects on military assets, the achieving of a political objective will have some consequence for these assets.) And most importantly, IW often influences behaviors of military decisionmakers and civilian controllers of the military. (Sometimes in the excitement of developing and implementing plans, practitioners ignore at their peril IW's sine qua non—the behavioral consequences leading to achieving a political objective.)

IW may comprise combinations of truthful and deceptive meaning. (Note that truthful and deceptive information both can engender truthful and deceptive meaning.) In other cases, e.g., direct action on information apparata through Iranian confiscation and repair of US

Embassy transmission assets during the embassy takeover in Tehran (Kelley, 1994), through US forces destroying Panamanian C3I sites in Operation JUST CAUSE (Hockstaeder, 1990), or through the random penetration or deactivation of military personal computers (Chandler,

nonsequitur 1989), the meaning conveyed by an IW technique may be a nonsequitur.

The best IW techniques often are conceived through observation of what happens in the everyday world, viz., the human ecology. For example, the proximal cause for the 1996 military revolt in Eastern Zaire may have been the announcement by a provincial official that the Banyamulenge were to be expelled (McKinley, 1996b). An IW practitioner may wait for an analogous situation to convey—truthfully or not— similar information, i.e., an impending expulsion of people, to achieve a similar result,—here the political objective of a revolt. Another example. One of the few legal ways for paramilitary or military forces of a host country to enter certain areas of a foreign embassy without formal authorization is in case of a fast-breaking emergency which demands action before notification of appropriate authorities (Kristoff, 1996b). The IW practitioner may create such a situation to help achieve a political objective.

One reason observing the everyday world for IW concepts often is valuable is that some people employ IW without conscious awareness. This may be tantamount to speaking prose one's whole life without realizing it. However, psychological conflict may lead to relabelling behavior, viz., involvement in IW activities, unconsciously judged as too threatening to be viewed in terms of what it really is. (The same psychological phenomenon affects some people who engage in torture and murder.)

Promising IW techniques also may be conceived through observing other species as well. (Most observers seem to adhere to rudimentary sociobiological tenets and ascribe a political environment to these species. This is helpful, but not necessary.) IW variants of camouflage, cover, and tactical deception have most benefitted from studying ethology (See Bloom, 1996a.) (As far as I know, the wild kingdom has not received any royalties.)

But merely observation? Of what help is theory? After all, one cannot observe without at least some implicit theory. Perhaps, that's all that is needed. Yet on a more basic level, theoretical constructions of IW seem to blind more than enhance vision because of their metatheoretical foundations. (Since some readers may find the rest of this theoretical paragraph tough going, the relative uselessness of theory may be just as well.) Researchers

empiricism often laud theory based on empiricism as reality-bound when it may be merely fact-bound— i.e., assumed real through social consensus (see Brinton, 1965); the mechanistic approach which applies a linear, hydraulic, procrustean constraint on that which is observed; the

materialist materialist approach which often renders robust mental phenomena as moot to explanatory pursuits; the rationalist approach which suffers from being no more than fevered ravings.

rationalist Perhaps all approaches reflect the casuistry of psychologists (Nietzsche, 1954/1888). Ultimately, a theory of IW is no more than an implicit exposition of a researcher's phenomenology, or perhaps a national character. (Witness such Soviet concepts as activity or reflective control (Strizinec, 1985) or Sudanese animistic possession (Holt, 1961).

hermeneutics Contextualism, hermeneutics, deconstructionism, and other approaches may someday be

deconstructionism relics towards an end of ideology masquerading as a scientific stance.

In any case theoretical constructions may be helpful—if carefully applied. The two most common constructions based on psychological research can suggest why what we observe seems to be happening. They can suggest causal hypotheses and identify what needs to be operative in specific IW techniques for specific situations. One psychological construction comprises combinations of information source, information message, and target characteristics, and evolved from propaganda research in World War II (Hovland et al, 1949). The other, the elaboration likelihood model (ELM), comprises combinations of the same characteristics dependent on the import of the derived meaning for the target (Petty & Cacioppo, 1986). (ELM has been taught to some IW practitioners in Western democracies as of the late 1980s (personal observation.)) The most recent integrations of these two constructions (cf. Mutz et al, 1996) show some promise, but still are founded largely on the concept of attitude with its same seemingly unresolvable difficulties (cf. Eagly, 1992). Another theoretical construction, not based on psychological research, is the descriptive work of Libicki (1995.) However, any descriptive approach has certain intrinsic problems. (See *Typology* below under IW AS AN APPLIED SOCIAL SCIENCE.)

A final word about the IW target. To repeat yet again, this refers to the human entity whose behavior will achieve the political objective which is the whole point of an IW technique. People murdered through terrorism or confused through altered text into making an entire computer system inoperative are not targets. Neither are information apparata destroyed or modified. They are way stations whose change in status is deserved or not, depending on the observer's ideology.

> EXAMPLES. Before reading the following IW examples, note that I am not sure that any or all of them occurred, occurred in the manner described, or are related to IW. However, there are data suggesting a strong possibility. To be more definitive is somewhat difficult. Why is this?
>
> Throughout human history, detailed written planning documents and elaborations of assassinations, propaganda, sabotage, disinformation, and other examples of IW often have not been prepared, or if they have been, often are not saved by their planners and implementors. But even if such documentation were found or similar data produced—as with memoirs, court testimony, torture-induced divulgings and the like—, estimates of their reliability and validity must be tempered with the expressed and hidden motives of their authors.
>
> Note also that examples of IW can be created which have in all probability not yet occurred, but may in the future. Whether terrorist employment of nuclear weapons or a nation-state's deactivating vital telecommunications infrastructure of a putative ally during peacetime, the constraint is one of technology limitation and, perhaps, will and inclination, not of imagination or abstract reasoning ability. It is probably the case that an interaction through time among science, technology, globalization, and human psychology will make the hypothetical real. On the other hand, evolutionary psychologists might posit that it may someday be adaptive not to engage in IW. This would predicate a huge change in the human psyche, one that will be very unlikely in the lifespan of this chapter's readers.
>
> And as previously mentioned (See INTRODUCTION), citing an event as an example of IW may have consequences beyond the written page. This is so even if the event never really occurred and even if it is true in all respects. The reader needs to at least ponder whether the author of an IW chapter might write

somewhat differently, if the only negative consequences to consider were the derision or discounting from one's professional colleagues.

In any event, what follows can be viewed as stories illustrative of the sorts of things occurring in IW.

Political Objective. To secure a homeland for non-Jewish Arabs in the British mandate of Palestine.

Discussion. (This is heavily modified from a book chapter by the author (Bloom, 1991.)) Through air hijackings, murders, speeches, political demonstrations, social service work, diplomatic meetings, and money transfers, the Palestinian Liberation Organization (PLO) has been extremely successful in psychologically priming many people throughout the for the moral necessity of achieving its political objective. The PLO's greatest success up to this writing, has been the creation of the referent, Palestinian, with specific denotations and connotations suggesting a non-Jewish, non-Jordanian, non-Israeli people without a homeland, Palestine. The referent has induced a change in meaning of the related term, Palestine, as well. That for centuries the territory labeled as Palestine has changed in size, shape, and name, and has been occupied by many racial, ethnic, political, and religious groups, has become moot or ignored in international debate about the validity of competing claims. The referent, Palestinian, has become analogously embraced and employed by "informed commentators," formally sanctioned international observers, and even foes of the PLO.

Through the referent, Palestinian, much power has already been gained. The referent impels an image of an aggrieved, wronged people who are owed something. It impels the perceived need by many international authorities and observers for negotiations, sanctions, and votes. It obviates the need for an army and facilitates concurrent justification of terrorism and requests for political and material aid. Given that the very perception of individuals as a group, let alone a people, can depend on ever-varying real and imagined characteristics, the continuous maintenance of the *Palestinian* people is an exemplary consequence of IW.

As of this writing, IW has actually led the PLO's public adversaries, the Israeli government, to allow a new political entity, the Palestinian National Authority (PNA) to exist, to control security personnel, and to be armed—with Israeli assets. It has led to the sharing of security-related intelligence. It has led to a public Israeli government position that, yes, there will be some sort of political entity with territory called Palestine for those represented by the Palestinian National Authority (PNA)—largely constituted by PLO representatives. It has led to the point where the PNA can titrate political violence directly through its own assets or indirectly through those of others, e.g., Hamas, Hezbollah, to effect political pressure on the Israeli government—pressure which affects the quantity and quality of Israeli military intelligence, operations, logistics, research and development, and various policy and planning functions. And as the distance seems to incontrovertibly shorten to the ultimate political objective, Israeli military troops and citizens continue to be murdered.

(This brief analysis is not meant to pass judgment on policies towards a Palestinian issue. Instead, it highlights an approach taken with some modifications by many so-called movements for liberation. Of special note are the low-frequency, high impact combination of violence; the perseverative demands for recognition as a "people;" the protestations of seeking freedom and democracy when the reality will likely be anything but; the often forced maintenance of suffering among one's followers, e.g., refugee camps in squalid conditions; the careful modulation and use of international aid as a political weapon; and the real possibility for success against a vastly superior military foe. In this regard IW is the prototype rejoinder to the few remaining adherents of a nuclear weapons linchpin to contemporary security, be it mutually assured destruction, massive retaliation, sufficiency, or the latest concoction of security intellectuals.)

Political Objective. To obtain political and material support for a political elite, decrease the support of its adversaries, or counter efforts of the elite.

Discussion. (The following has been edited from Leventhal (1996) with his permission.) Out of the former Yugoslavia have come a number of new countries, e.g., Bosnia. In attempts to obtain political and material aid in fighting its adversaries, there are credible reports that Bosnian government forces have engaged in deliberate sniping attacks against its own civilians and were responsible for mortar attacks on the Sarajevo market that were blamed on Serb forces.

[Addendum to Leventhal (1996): Anti-government protestors in Serbia have used a radio station, B-92, to get the word out about their cause. When the Serbian government shut down the station, protesters then used the Internet to get their message across not merely throughout Serbia but the world. In fact within hours of the closure, government officials in Europe, international humanitarian agencies, and journalists were being provided information by the opposition via the Internet. (And in another blow to Serbian attempts to control the opposition, the USG announced it would broadcast B-92 programming, after the station was shut down by the Serbian government. The Serbian authorities promptly allowed the station to reopen.) As of this writing, B-92 authorities are in the process of making a deal with an Amsterdam-based access service to record radio programming digitally and broadcast it over the Internet 24 hours per day. And if the Serbian government attempts to cut Internet lines, there are plans to fax thousands of eggs to flood government fax machines (Hedges, 1996.) At present many people with access to the Internet can obtain the official opposition's point of view on the world wide web (Myers, 1996; personal observation.)]

Moving now to Africa, note that the Libyan government has funded a film, The Maltese Double Cross, which falsely advocates that a renegade Central Intelligence Agency group in league with Syria, Iran, and Palestinian terrorists organized the bombing of Pan Am 103. The film has been broadcast by television networks in the United Kingdom, France, Germany, and Australia and is designed to help weaken sanctions against the Libyan government—sanctions which make it more difficult to obtain and maintain military and other security assets—for harboring the alleged perpetrators of the bombing and preventing legal adjudication of the case.

During the 1993 UN intervention in Somalia, General Mohammed Farah Aideed repeatedly placed women and children at the head of columns of demonstrators with gunmen behind them. This was to ensure that Somali women and children would be among the casualties after Aideed's gunmen provoked an attack. According to US Embassy reporting, some Somalis who allegedly died in fighting between Aideed forces and those of the UN would get up and walk away after news cameras had stopped filming. Also, in early September 1993, Nigerian peacekeeping forces were due to replace Italian forces within a UN sector. The night before, an Aideed representative warned nearby residents with a bullhorn that black US soldiers "disguised" as Nigerians would invade the area the next morning. He urged residents to attack them, which they subsequently did.

Turning now to the Mideast, note that during Operation DESERT STORM the Iraqis placed aircraft on ancient ziggurats to encourage damage to archaeological sites. In one case, they deliberately damaged a mosque in Basra to make it appear as if bombing by the US and its allies had inflicted the damage. The Iraqis also portrayed damage to civilian sites dating from the Iran-Iraq war as if it were caused by US and allied bombing. Iraqi authorities also claimed that a bombed biological warfare site was a "baby milk" factory.

In yet another example the Iraqis also used a mid-level official in their foreign ministry as a supposed random passerby at a site where the Cable News Network was filming damage to civilian areas. She appeared, casually dressed, and spoke to the camera in fluent English, about the "criminal" bombing of Iraq and past American injustices against the "Red Indians." She also appeared on French television, speaking fluent French.

Political Objective. To weaken the security policies and assets of competitor nation-states.

Discussion. (See Leventhal, 1996.) *Intelligence Newsletter* was a known conduit for Soviet IW operations and seemingly continues in this vein for the Russians. It reveals a great deal more information about the activities of Western intelligence services than the Russian ones, promotes a heavy reliance on open-source collection, advocates restricting clandestine intelligence collection and covert action, and argues that the public should have the greatest possible access to the latest cryptological technology. If implemented, these recommendations would result in a weakening of Western military and civilian intelligence services and others which are to some degree subject to the rule of law.

The Soviets and then the Russians have also directly and indirectly promulgated through interviews, newspaper articles, and the like that preventing nuclear weapons proliferation would be significantly hindered by, first, the disestablishment of the Soviet Union, and later by the enlarging of the North Atlantic Treaty Organization (NATO.) The idea here is that the "scare" of nuclear weapons proliferation might persuade international leaders and population segments to support—even help reconstitute—a Russian/Soviet empire and reject NATO expansion into Central and Eastern Europe.

Political Objective. To achieve political power through military victory.

Discussion. I have previously reported (Bloom, 1991) about the Ugandan prophetess who convinced her followers that they could be victorious in battle against a far superior foe by rubbing a special oil on their bodies which would protect them against bullets (Ugandan rebel prophetess, 1987.) More recently, before an artillery attack by Rwandan troops against Zaire, six Zairian soldiers danced naked across the border from Rwanda separating Goma from Gisenyi.

They chanted and sang—seemingly believing that they had been granted immunity from death by a priest who anointed them with water (McKinley, 1996a).

Political Objective. To maintain some missile capacity for political intimidation of adjacent and proximal nation-states.

Discussion. As of this writing the Iraqi government may be hiding more operational missiles than UN monitors have suspected (Crossette, 1996b). Although required to destroy all means of mass destruction in the aftermath of the Persian Gulf War, the Iraqis have removed turbo pumps—a key missile component which Iraq does not manufacture—from some missiles before the latter were destroyed and buried. Also a shipment of Russian-made gyroscopes—another key missile component—was apparently being smuggled into Iraq when it was intercepted by Jordanian authorities. UN monitors have found other gyroscopes hidden at the bottom of the Tigris Canal in Baghdad in another crude, deceptive effort to maintain the capability of political intimidation in the Mideast.

Political Objective. To decrease support for a nation-state one is occupying.

Discussion. The government of the People's Republic of China has warned the Walt Disney Company that the release of a new film about the Dalai Lama could cripple Disney's efforts to expand into China (Weinraub, 1996.) As of this writing the movie, Kundun, is still scheduled for release. Chinese authorities seemingly fear a groundswell of pressure against their control of Tibet and thus a security threat to their political authority. Even if Disney does not release the film in China, international audiences—perhaps merely exemplars of radical chic, perhaps not—may still be capable of moral outrage and engage in boycotts and other activities supporting anti-Chinese sanctions. Moreover, through bootleg cassettes and digitized transmission, some Tibetans and Chinese are likely to see the movie—a threat analogous to the passing of audiocassettes containing the Ayatollah Khomeini's exhortations leading up to the fall of the Shah of Iran in 1979.

Political Objective. To better preempt and counter adversaries through information superiority. (The meaning of this term is sill being developed. Here it denotes superior IW. (See INTRODUCTION.))

Discussion. (The following is taken from Madsen (1993).) A computer virus—apparently written by a Palestinian student—first appeared at Hebrew University in Jerusalem. It spread to computers of the Israeli intelligence agency, the Mossad. It was then transmitted to subnetworks comprising the US Defense Data Network.

As another example, Western German computer hackers employed by the Soviet KGB were able to penetrate sensitive computer systems at Lawrence Berkeley Laboratory and the USG contractor MITRE. From these computers the hackers were able to access the US Advanced Research Projects Agency Network/Military Network, then 50 military computers at the Pentagon, various defense contractors, the Los Alamos Nuclear Weapons Laboratory, the Argonne National Laboratory, the US Air Force Space Systems Division at El Segundo, CA, and other military bases throughout the US, West Germany, and

Japan. Data accessed included US Army plans for nuclear, chemical, and biological warfare in central Europe.

Madsen also includes computer penetration examples allegedly perpetrated by the intelligence agencies of a number of other countries involving but not limited to the North Atlantic Treaty Organization, South Africa, the Mideast, and Asia. Is it any wonder that the military services have been focusing such attention on the IW threat in the past few years (Author, a; Author, b; Warfare in an information age)?

IW as an Applied Social Science

IW Typology. I have already commented on problems in identifying and classifying IW in CONCEPTUAL DEVELOPMENT and EXAMPLES above, but will add the following. Even the most competent and well-meaning IW practitioner may have strong disagreements about what IW is and whether a specific event is IW. This is analogous to the forensic circus of disagreeing psychological experts found in legal proceedings and may involve the same underlying phenomena. Perhaps the reader is left with the dilemma of pornography. As with Justice Potter Stewart one knows it when one sees it—or hears it, touches it, smells it, or tastes. Second, since IW can be covert or clandestine, identification and classification may become even more difficult. Third, there is a data base problem. Although interviews with alleged experts are a source of relevant information, so are websites of international newspapers describing events throughout the world. However, these sites may contain outstanding candidates for IW examples but not contain the term IW or those of its many synonyms. Without (compulsively?) reading through a welter of data, one loses valuable information.

Effectiveness. How does one know that IW works? Frankly, one doesn't. One only can surmise and develop stories. (See CONCEPTUAL DEVELOPMENT and EXAMPLES above.) So much impinges on a target that it is highly unlikely that significant causal factors can be teased out of the chaos. And IW experts, too, are continually bombarded by many stimuli. Who's to say there isn't an IW technique targeting an IW expert so that another IW technique is perceived as causing yet another real-world event? (I apologize to Nietzsche for the occasional use of the term real-world. Readers might want to consult his discussion on the nerve stimulus and the percept (Nietzsche, 1954/1873.) Who's to say the IW expert is not the victim of various attribution errors resulting in misperception of cause and effect, cause and intent, or causal locus? Who's to say the IW expert is not reporting on matters which cannot possibly be known (Sprangers et al, 1987)?

Unfortunately, assessing efficacy too often becomes a political vehicle within and outside of IW organizations. If desired consequences occur after an IW technique has been implemented, the technique is credited for the success. If the consequences occur before the technique was implemented, this merely shows how effective the technique is. (Will be? would have been?) Of course, if the consequences seem dire, the technique still worked, because events would have been worse without it. When not only the fate of one's country but one's career is at stake, cool, dispassionate analyses of efficacy are rare. Other more sophisticated issues such as (a) how long an effect may last; (b) paradoxical, spontaneous, differential or delayed effects; (c) moderating and mediating variables impacting on effects; and (d) base rates for claimed effects even more so. The same problems beset the psychotherapy outcome literature in both civilian and military environments (cf. Kazdin & Bass, 1989).

Until there is ample reason to think otherwise, an IW technique may increase, decrease, maintain, or have no effect on a desired or alleged consequence. And any and all of these consequences may appear through time.

Process. Assuming IW works, how does it work? What are its active, robust components? Usually, if a specific technique is perceived as effective, whatever rationale has been advanced during planning for why the technique was expected to be effective often will be perceived as responsible for the technique's perceived effectiveness. And where does this expectation come from? From biases, hopes, dreams, ideologies, anecdotes, something one saw in a movie or on television, occasionally through some systematically and intelligently derived knowledge.

When a technique is perceived as ineffective, a phenomenon that will usually be associated with the perceivers not having planned or implemented it, the initial rationale for hoped for efficacy may be advanced to explain its ineffectiveness. (There is also an entire process of group and organizational dynamics among the participants in an IW project. This process won't be described here but may be thought of as based on common principles from social and organizational psychology.) Of course, IW may be implemented without a thought to how it works. This example of 'mindless' behavior is often described as 'results-oriented' by those who believe they are not mindless and those who don't mind whether they are mindful or mindless.

On the rare occasions when individuals associated with IW perceive it to have failed, the initial rationale postulated for desired effectiveness will still be embraced. Instead, other environmental and human factors may be advanced to 'explain away' the ineffectiveness of an otherwise effective technique. This style of process and causal attribution renders the invalidation of a technique or its alleged robust components highly improbable. As in perceiving effectiveness itself, individuals often are more concerned with their own power needs—including managing their self-esteem—than that of the political entity which they represent. (It follows that people who have often little disparity between personal and national needs should be sought as IW participants. This conclusion—which can be easily rejected by noting the many current military conflicts fueled by people who choose to or cannot even differentiate between the personal and national—illustrates the weakness of depending solely on rationalism and logic as venues to knowledge.)

One important process phenomenon, however, seems to involve the expectations of a target or intermediary victims. It is often easier to elicit behavior from victims and targets if that behavior is somehow consonant with expectations. For example anti-USG IW themes are easier to believe and act on because of actual misbehavior engaged in by the USG—nuclear irradiation of unsuspecting subjects (Weiner, 1994), subjects' unknowing ingestion of or exposure to LSD-25 (Bagby, 1992; Weinstein, 1990), and attempts to murder foreign leaders, e.g., Fidel Castro (Schlesinger, 1994). The actual misbehavior reinforces expectation of more of the same, so there is a practical as well as moral, ethical, and legal reasons to avoid such behavior. (One does not have to develop and advocate a critical race or nation theory to explicate this phenomenon. (See Rosen, 1996).)

Education and Training. Is there a body of knowledge, a set of procedures, experiences, or personal characteristics which facilitate effective development, planning, implementing, and evaluating of IW? There is no Holy Grail which has not been contaminated with fantasy. Perhaps only luck or instinct prevails. This seems to be the case for clinical psychology—military or civilian—and for 'street' insight. For how does one capture the essence of a master therapist? A successful prostitute? Is there a difference?

Without a solid empirical or other evidentiary base on IW typology, effectiveness, and process, education and training efforts may provide only groundless reassurance to IW neophytes or grounds for positive and negative reinforcement, omission training, and punishment at the hands of those who politically control IW assets. Either way there is no suitable criterion with which valid education and training modules can be built.

But couldn't one merely emulate the life of an IW expert? Let's try Machiavelli—often cited as a master of IW in texts and lectures. He actually experienced a very checkered career. More of his life was spent trying to rise in the political power hierarchy of his time than actually being positioned where he wanted to be (Machiavelli, 1940/c. 1513). He suffered many reversals and seems to have risen no higher than as secretary to the top-ranking military council in Florence. He even was tortured after Florence had fallen to the Medicis—after which he wrote his great works, e.g., *The Prince*. Perseverance, perhaps, or dogmatism. But an expert? And even if he were an expert, what would we choose to emulate? The torture? His bathroom ritual? Probably that which we assumed was face valid for IW expertise. And, if so, why bother even pretending to develop systematic knowledge? (Those who argue that one should instead look for correspondences among many IW experts still run into the same problems with an error factor compounded many times over.)

So with trepidation and without defense, I'll list the following for the consideration of the budding IW practitioner.

- some sense of what life is all about from a combination of reading and living life
- some sense of alienation allowing one to pull back from direct experience enough to analyze or infer from it
- some sense of creative conceit that one can indeed influence others
- some sense of an ideal towards which one is trying to move—even if that ideal is no ideal
- bureaucratic skills to properly situate and protect oneself within one's organization
- some other specialized skill—e.g., information technology, psychology, marketing, or cultural studies come immediately to mind
- within a military organization, skills for the niches from which one contributes to IW—planning, analysis, administration, briefing, operations, research and development, consultation

Initiator Effects. Does the role of IW practitioner affect the person filling the role? Does the practitioner overestimate the degree to which people may intentionally and effectively influence each other? Or believe IW to be more benign than many disinterested observers? Or become nihilists, alienated, antisocial, or remain untouched totally by the profession? But would such consequences be the result of the IW role or a reflection of the sort of people who self-select into the field? In fact, following CONCEPTUAL DEVELOPMENT above, do not all individuals have IW careers and differ only in selecting political spheres of operation ranging from that of national security to the breaking of hearts? The overarching conceptual difficulty is that the situations chosen by individuals also affect those individuals.

Future Trends

Subtle and Obvious IW. Most practitioners of IW have focused on face valid operations, e.g. little content disparity between conveyed information and the desired meaning of the intermediary victims or target. I once posited (Bloom, 1991) that more attention should be directed towards information that conceivably has little face validity with a desired meaning but is nevertheless associated with the latter's presence and, more importantly, with improving the consistency between meaning and resulting behavior. I posited that lessons could be learned related to the subtle-obvious item distinction in psychological tests (Dubinsky & Gample, 1985) and the empirically-keyed approach in psychometrics. I no longer think this last to be the case given current personality assessment research (cf. Boone, 1995.) However, the subtle-obvious distinction still merits attention, for the threat of subtle meaning—if impelling target behavior—would significantly complicate the life of those who must preempt and counter IW.

Subliminal IW. As I posited before (Bloom, 1991), more research will be done in this area. The appropriate design should be multivariate encompassing all sensory modalities. One main problem is not so much demonstrating the effects of subliminal information as establishing predictability. Another is identifying the intrapsychic processes, if any, moderating or modifying the relationship between information and resulting target behavior. (This becomes less of a problem if one believes one has mindless targets or one is mindless of them.)

Secondary problems include the technical constraints of IW techniques as well as cost-benefit analyses of subliminal versus supraliminal approaches to IW (Bornstein & Pittman, 1992). Advances in information technology may facilitate resolving these secondary problems.

The Power of the Powerless. With increasing globalization, long-accepted, security-related concepts are changing. Because in more and more ways earth's inhabitants are all on the same "ship," impending catastrophe in one part of the world can more easily have significant effect in other parts. Thus the Iraqi government can cause economic pain in its own people as a way to loosen pressure to maintain United Nations sanctions or to literally encourage begging in some international diplomatic quarters for Iraq to be allowed to sell oil to meet humanitarian needs. The fact that the sanctions are also a pretext for Iraq not to pay its debts to Russia, France, and other countries also increases pressure on the very countries which initially placed the sanctions on Iraq (cf. Crossette, 1996a). A related issue has been the choice to allow or not to allow ethnic cleansing—a choice confronting humanitarian organizations worldwide. For example in one Zaire, has it been more humane to ethnically cleanse people and minister to their physical needs or leave them to die (Gourevitch, 1996)? Another example is the political leverage of the North Korean government through the threat of its potential destabilization for Asian security if Pyongyang's economy worsens or its internal political controls weaken (Kristoff, 1996a). Consequences of this have been US and South Korean announcements of probable food aid to North Korea, the commencement of building a nuclear reactor capability for it, and the likelihood of further aid and support based on a fig leaf of publicly professed regret by North Korea for a covert military intelligence operation gone awry (Kristoff, 1996c).

Computers, Software, Semiconductors and Integrated Circuits, and Telecommunications. Cyberwarfare, a variant of IW, may see significant growth as information technology continues as linchpin of military and civilian activities supporting national and international security (Arquilla & Renfeldt, 1992; Lohr, 1995). This will include the destruction, modification, and penetration of hardware and software, as well as the utilization of classified and unclassified data bases for purposes as varied as military planning, the transferring of funds, and the besmirching and blackmail of military and political leaders. (See Devost et al, 1996; Golden, 1996; Molander, 1996; Sakkas, 1991.) A greater role of non-state actors is to be expected.

But there is much more to this than the application of military and other assets to damage or modify hardware and software. International law (cf. Anthes, 1995) and economic and communications policies may be the quintessential cyberwarfare technique and military practitioners of IW need to be represented on appropriate interagency committees. (For example until November 1996 encryption software had been classified by the USG as a weapon under the Arms Export Control Act.) It is for this reason among others that ongoing executive, legislative, judicial conflicts in many countries on the export of encryption software are so important. The recent decision of a U.S. District Court in San Francisco that USG attempts to control the export of encryption technology are an unconstitutional restriction on freedom of speech may give solace to the enemies of freedom who may exploit the finding to their advantage (Markoff, 1996). (An analogy is the use of free elections in Algeria by Muslim fundamentalists in the attempt to legally take political power and ensure there are no further free elections (Ibrahim, 1991).

Economic policies on sanctions, tariffs, duties, and subsidies and other nontariff barriers to trade in information technology including military to civilian spin-offs and civilian to military spin-ons will also be more a part of an overall IW effort. The Singapore accord known as the International Technology Agreement is but one example of an ongoing process with IW import (Schiesel, 1996a). The specific security implications of this tentative agreement to ease tariffs on a broad range of information technology products by the year 2000 have not yet been coherently identified. (One source of significant resistance to this accord has been that many governments view their telephone networks as strategic assets and have used high tariffs to protect local suppliers. IW has already been used in attempts to influence this sort of resistance as well as cultural attitudes which often underlie emotional resistance to nontariff barriers.)

Other IW-relevant international economic and communications policies involve copyright law in cyberspace which may affect accessibility of information and engender security-related sanctions and conflict (Schiesel, 1996b) and the ongoing computer chip negotiations between the US and Japan. (Some executives in the US semiconductor industry are even suspicious that Japan's government representatives significantly and intentionally overstated the share of its market held by foreign companies during negotiations (Pollack, 1996). Cross-border satellite broadcasts present not only opportunities and threats for IW practitioners, but also vehicles to induce anti-social behavior among otherwise law-abiding citizens and political conflict among nation-states (DePalma, 1996).

Another trend will be the ever increasing broadening of what is subsumed under C3I and targeting a competitor's C3I nodes—regardless of whether conflict is hot or cold, regardless of whether the node is fully developed, still in the planning stage, officially controlled by commercial or government authorities. IW targeting C3I will include all intelligence and counterintelligence efforts—not just technical intelligence but human as well.

Increasing efforts to increase the speed of computers will also be exploited by IW practitioners. For example, the forming of a consortium by Japan's major electronics companies to combine microprocessors on the same computer chip as memory may be one vehicle to increase speed significantly (Japanese form microprocessor consortium, 1996). If successful, IW will exploit the results analogous to prior exploitations of greater fidelity and bandwidth by radio warriors, of the variety, mobility, and stealthiness of broadcast platforms for audio and video transmissions, and the variety of materials for posters and pamphlets (cf. Warlaumont, 1988; Wasburn, 1992).

(Recent US IW projects such as Radio Free Asia exemplify, however, that regardless of technical superiority, adhering to basic principles, e. g., understanding the operational and political context in which IW is implemented, remains paramount (See Snyder, 1997; Tyler, 1996).

Advances in virtual reality technology (VRT) may even further increase the effects of visual and auditory information on behavior, and when integrated with auditory, olfactory, tactile, and even gustatory modalities may help elicit behavior profoundly difficult to modify. (However, there may be untoward and unpredictable effects (Bloom, 1997). The huge outpouring of aid which Mohammed Amin accomplished with photographs for the Ethiopian famine during the 1980s will be topped by some budding VRT expert for good or for evil (cf. Mohammed Amin, 1996).

Accessible data bases, e.g., via websites, may help and hurt both IW practitioners, their intermediate victims, and their targets (cf. The CIA starts to use the Internet, 1994). For example, the US National Technical Information Service continues to add documents related to terrorism threats and countermeasures to its data base (McKenna, 1996). These include reports on aviation security, chemical and biological warfare agents, and the countries sponsoring terrorist activities. Adversaries can exploit this information against each other and must even wonder whether the very addition of new documents to a data base is an IW technique itself. As well, proprietary keyword and data search procedures will have more IW significance—affecting accessibility and relevance of information. Business news data bases afford IW practitioners more lead time to plan variants of cyberwarfare even before their targets take delivery of what therefore may already be exploited assets—unless this has been preplanned by the targets as a deception operation. (See Two companies to share $237 million order from Army, 1996.) Also, political descriptive and analytic data bases, as well as live and taped broadcasts, can provide opportunity both for terrorists and counterterrorist during an ongoing operation. (See 3 ex-hostages in Peru speak kindly of captors, 1996; Two guards reported glued to soap opera, 1996.)

Will continued advances in computers and telecommunications foster freedom or its opposite through IW? Although the preponderance of "informed observers" seem to posit the former and sing the praises of free, instantaneously transmitted information creating a democratic global village, a very strong case can be made for the latter. (See Rodan's (1996) article on information technology and political control in Singapore.) It is also thought-provoking that convicted terrorist supporters (Foderaro, 1996), political front groups for terrorist organizations and the organizations themselves (Sims, 1996), the UN, many embassies, and some USG security agencies, e.g., the CIA, the Federal Bureau of Investigation, and the US Air Force, already have their own websites (personal observation). (As with any feature of IW, these present not only opportunities for web site developers, but their adversaries as well. (See Schiesel, 1996c.)

The Essence of Security. The concept of security will continue to subsume and integrate military, economic, and value-oriented components of political entities—be they nation-states or the yet undefined. The implication is that military IW experts will be needlessly constraining themselves by analyzing, targeting, and seeking to influence only traditionally military personnel and assets. The inclusion in this chapter of military and so-called non-military examples is intentional. The closest coordination among all IW personnel will be crucial to being appropriately proactive and reactive in the security arena of the 21st century. Therefore, IW researchers will more intensively mine relevant literature outside the narrow confines of any one academic discipline. So, too, more government, military, university, and commercial centers will attract such researchers, especially as description, inferences, and development of influence techniques become more highly funded.

In a world of increasing ethnic conflict and anarchy, IW also will more likely feature attempts to encourage militias, paramilitary units, and renegade military organizations to accept the Geneva conventions during interludes of political and criminal violence. Many of these entities are more likely to discount, ignore, be ignorant of, or distrust the principle of selflessly providing care to all sides (Bonner, 1996). If IW is not successful, there may be increasing difficulty in implementing the humanitarian agendas of relief and aid organizations. The premeditated murder of 6 Red Cross workers in their sleep in Chechnya exemplify the problem. Already the Red Cross has modified one of its dearest principles and has occasionally accepted military protection.

And it may become more difficult to develop some IW measures—antiterrorism and counterterrorism. The varieties of motivators to engage in terrorism and the interrelatedness of political phenomena seem to guarantee a more elusive foe with more threatening means—a postmodernist terrorism, if you will (cf. Laqueur, 1996). Without reinforcing values compatible with informing about "suspicious" behavior, even the most sophisticated approaches to head off terrorism will not be sufficient (cf. Bohlen, 1996). Yet the downside of constricting freedom renders this option problematic in a society valuing democracy and a rule of law.

epistemology *The Perishability of Epistemology.* The effectiveness of IW techniques will continue to vary through time, as does the reliability and validity of information analyzed to develop techniques. This is because knowledge itself may be transformed through sociohistorical (cf, Ibn Khaldun, 1950/c. 1379) and evolutionary psychological forces (Buss, 1995.) Thus the study of IW will continue to be ongoing. Information believed to be accurate must be reassessed. There will be no IW "cookbook" to take off the shelf.

IW and Economics. Economic phenomena will become more frequently involved in IW. An example following Friedman (1996)—perhaps somewhat in jest—relates to Saddam Hussein. Friedman posits that Saddam and his representatives may have been playing the oil futures market to increase Iraqi financial assets during a time of international sanctions. To develop almost a "sure thing," Saddam hints that Iraq will or won't comply with UN demands in the aftermath of the Persian Gulf War. When he suggests compliance, he sends oil prices falling, noncompliance, prices rising.

IW and Power. IW will be perceived as ever more important by all who seek power. This is supported by Huntington's (1996) contention that global politics and the use of military assets is being reconfigured along cultural lines—i.e., ethical and moral values, socio-cultural and perceptual preferences and styles. It is also supported by the ever more pervasiveness of television and other multimedia depictions of alternate and altered realities (cf. Gerbner, 1986) and help explain the struggle to exploit these realities during terrorist operations (Fujimori angered, 1997). Here IW practitioners are confronted not only with ongoing influences on how we think, feel, and are motivated, but the challenge of harnessing the sources of these influences to achieve political objectives within the confines of specific political contexts.

Given the above, IW experts may more often head military organizations—much like pilots, infantry experts, or submariners do now. Formally credentialed behavioral scientists—at least those who don't display their professional degrees as body armor protecting intense feelings of inferiority—might even have greater opportunity. (Aware of previous examples, e.g, Henrik Verwoerd, a psychologist and founder of apartheid in South Africa (Lapping, 1987), or Radovan Karadzic, a psychiatrist and internationally indicted war criminal from Bosnia (Tribunal plans to file warrants, 1996), the reader is left to ponder what this bodes for the human condition—or, for that matter, internecine, interprofessional rivalries.)

Conclusion

The IW challenge for the 21st century is to adapt to new technologies and new political, sociocultural, and psychological phenomena with the same timeless concepts (cf. Barnett & Lord, 1989; Scott, 1995). I write these last words on Christmas eve 1996. With the timeless wish for peace on earth, good will, love. A paean to a future without war—information or otherwise. But all for naught.

References

Allen, B. (1993). On the definition of propaganda. In R. Marlin, (Ed.) *Propaganda and the ethics of rhetoric*. Ottawa: The Canadian Journal of Rhetorical Studies. pp. 1–12.

Anthes, G. (June 5, 1995). New laws sought for info warfare. *Computerworld, 29,* 55.

Aristotle. (1941/c. 340 B.C.) On interpretation. In R. McKeon (Ed.), *The basic works of Aristotle*. NY: Random House. pp. 43–44.

Aristotle. (1941/c. 328 B.C.) Politics. In R. McKeon (Ed.), *The basic works of Aristotle*. NY: Random House. pp. 1127–1146.

Arquilla, J., & Renfeldt, D. (1992). *Cyberwar is coming*. Santa Monica: Rand (P–7791).

Author. (Undated, a). *Air Force information applications in the 21st century*. New World Vistas: Air and Space Power for the 21st Century. Washington, D.C.: Department of the Air Force.

Author. (Undated, b). *Cornerstones of information warfare*. Washington, D.C.: Department of the Air Force.

Bagby, R. M. (1992). A review of *Psychiatry and the CIA: Victims of mind control* by H. M. Weinstein. *Contemporary Psychology, 37*, 152.

Barnett, F. R., & Lord, C. (Eds.) (1989). *Political warfare and psychological operations: Rethinking the U.S. approach.* Washington, D.C.: National Defense University.

Bloom, R. W. (1991). Propaganda and active measures. In R. Gal & A. D. Mangelsdorff (Eds). *Handbook of military psychology.* London: John Wiley.

Bloom, R. W. (November 22/29, 1996a). Editorial: An ethological approach to information warfare. *International Bulletin of Political Psychology, 4,* 7–10.

Bloom, R. W. (December 6, 1996b). Commentary concerning T. Leventhal's *The Need for a United States Government Capacity to Analyze and Counter Foreign Perception Management Operations: Part I. International Bulletin of Political Psychology, 5,* 4. (http://www.pr,erau.edu/~security)

Bloom, R. W. (January 22, 1997). *Psychiatric applications of virtual reality technology.* Paper presented at the Medicine Meets Virtual Reality: 5 International Conference. San Diego, CA.

Bohlen, C. (October 23, 1996). Killing of Mafioso's wife by informant raises debate. http://www.nytimes.com. *The New York Times*

Bonner, R. (December 22, 1996). Killing of 6 workers causes soul searching at Red Cross. *The New York Times.* p. 9.

Boone, D. (1995). Differential validity of the MMPI-2 subtle and obvious scales with psychiatric patients: Scale 2. *Journal of Clinical Psychology, 51,* 526–531.

Bornstein, R. F., & Pittman, T. S. (Eds.) (1992). *Perception without awareness: Cognitive, clinical, and social perspectives.* NY: Guilford Press.

Brinton, C. (1965). *The anatomy of revolution.* NY: Vintage Books. pp. 3–26.

Buss, D. M. (1995). Evolutionary psychology: A new paradigm for psychological science. *Psychological Inquiry. 6,* 1–30.

Chandler, D. (October 30, 1989). Despite safeguards, viruses are spreading. *Boston Globe.* p. 29.

Croffi-Revilla, C. (1996). Origins and evolution of war and politics. *International Studies Quarterly. 40,* 1–22.

Crossette, B. (October 29, 1996a). UNICEF says thousand of Iraqi children are dying. *The New York Times.* http://www.nytimes.com.

Crossette, B. (December 19, 1996b) U.N. says Iraq may be hiding more missiles than suspected. *The New York Times.* http://www.nytimes.com.

DePalma, A. (December 30, 1996). Space, the TV frontier now: Cross-border satellite beams get mixed reception. *The New York Times.* pp: C1;2

Devost, M. G., Houghton, B. K., & Pollard, N. A. (1996). *Information terrorism.* http://www.geocities.com...olHill/2468/itpaper.html

Dubinsky, S., & Gample, D. (1985). A literature review of subtle-obvious items on the MMPI. *Journal of Personality Assessment, 49,* 62–68.

Dyer, G. (October 24, 1996). Waiting for 1989 in East Timor. *Chicago Tribune*. p. 31.

Eagly, A. H. (1992). Uneven progress: Social psychology and the study of attitudes. *Journal of Personality and Social Psychology, 63*, 693–710.

Forderaro, L. W. (December 21, 1996). Crisis draws attention to jailed American's Internet site. *The New York Times*. http://www.nytimes.com.

Fujimori angered at press visit to besieged house. (January 2, 1997). *The New York Times*. p. A5

Freund, C. P. (January 27, 1991). The war on your mind: Psychological warfare from Genghis Khan to POWs to CNN. *Washington Post*. p. C3.

Friedman, T. (October 13, 1996). Follow the money. *The New York Times*. p. 13.

Gerbner, G., Gross, L., Morgan, M., & Signorelli, N. (1986). *Living with television: The dynamics of the cultivation process*. In J. Bryant & D. Zillman (Eds.) Perspectives on media effects. Hillsdale, NJ: Erlbaum. (pp. 17–40).

Golden, T. (October 21, 1996). Tale of CIA and drugs has life of its own. *The New York Times*. http://www.nytimes.com/y...ational/crack-story.html

Gourevitch, P. (September 9,1996). Neighborhood bully: How genocide revived President Mobutu. *The New Yorker*. pp. 52–57.

Hedges, C. (December 8, 1996). Serbs' answer to tyranny? Get on the Web. *The New York Times*. pp. 1; 8.

Hill, C. (1995*). Bibliography for information warfare*. Newport, RI: Naval War College Library.

Hockstaeder, L. (October 6, 1990). In Panama, civilian deaths remain an issue. *Washington Post*. p. A23.

Holt, P. M. (1961). *A modern history of the Sudan*. NY: Grove Press.

Hovland, C., Lumsdain, A., & Sheffield, F. (1949). *Experiments in mass communication*. Princeton: Princeton University Press.

Huntington, S. (1996). *The clash of civilizations and the remaking of world order*. NY: Simon & Schuster.

Ibn Khaldun. (1950/c. 1379). In C. Issawi. (Trans. & Ed.) *An Arab philosophy of history: Selections from the Prolegomena of Ibn Khaldun of Tunis*. London: Murray.

Ibrahim, Y. M. (December 19, 1991). Algeria and Tunisia intensify anti-fundamentalist efforts. *The New York Times*. p. A10.

Information warfare: Legal, regulatory, policy and organizational considerations for assurance. (July 4, 1995*). Research report for the Chief, Information Warfare Division (J6K). The Joint Staff*. Science Applications International Corporation (SAIC). Contract No. MDA903–93–D–0019.

Japanese form microprocessor consortium. (December 19, 1996). *The New York Times*. http://www.nytimes.com.

Kant, I. (1958/1783). *Critique of pure reason*. NY: The Modern Library. pp. 25–27.

Kazdin, A., & Bass, D. (1987). Power to detect differences between alternative treatments in comparative psychotherapy outcome research. *Journal of Consulting and Clinical Psychology*, 57, 138–147.

Kelley, J. (January 4, 1994). Propaganda pours from embassy's shell. *USA Today*. p. A2.

Kristoff, N. (August 25, 1996a). U.S. legislator sees food crisis in North Korea. *The New York Times*. p.6.

Kristoff, N. (December 20, 1996b). Tokyo pushes for conciliation and caution. *The New York Times*. http://www.nytimes.com.

Kristoff, N. (December 30, 1996c). 'Deep regret' sent by North Koreans. *The New York Times*. pp: A1; A5.

Lapping, B. (1987). *Apartheid, a history*. NY: Braziller.

Laqueur, W. (1996). Postmodern terrorism. *Foreign Affairs*, 75, 24–36.

Leventhal, T. (1996). The need for a United States Government capacity to analyze and counter foreign perception management operations. In R. Bloom, (Ed.) *International Bulletin of Political Psychology*, 5, 1–5; 6, 1–7.

Libicki, M. C. (August, 1995). *What is information warfare?* ACIS Paper 3. Institute for National Strategic Studies. Washington, D.C.: National Defense University.

Liu Hsiang. (1996/c. 20). In J. I. Crump (Trans.), *Chan-Kuo Ts'e*. Ann Arbor: Center for Chinese Studies.

Lohr, S. (September 30, 1996). A new battlefield: Rethinking warfare in the computer age. *The New York Times* (http://www.nytimes.com/1.../cyber/week/0930war.html) pp. 1–6.

Machiavelli, N. (1940/c. 1513). *The Prince*. NY: Modern Library.

Madsen, W. (1993). Intelligence agency threats to computer security. *International Journal of Intelligence and Counterintelligence*. 6, 413–488.

Markoff, J. (December 19, 1996). Judge rules against curbs on export of encryption software. *The New York Times*. http://www.nytimes.com.

McKenna, J. T. (November 25, 1996). AERObyte. *Aviation Week and Space Technology*. p. 65.

McKinley, J. C. (November 27, 1996b). Old revolutionary is a new power to be reckoned with in Central Africa. *The New York Times*. http.//www.nytimes.com.

McKinley, J. C. (November 2, 1996a). On Zairian-Rwandan border, strife erupts into open war. *The New York Times*. http://www.nytimes.com.

Mohammed Amin. (November 26, 1996). *The New York Times*. http://www.nytimes.com.

Molander, R. C. (1996). *Strategic information warfare: A new face of war*. Santa Monica: Rand (LC95–53673).

Mutz, D. C., Sniderman, P. M., & Brody, R. A. (1996). Political persuasion: The birth of a field of study. In Authors (Eds.) *Political persuasion and attitude change*. Ann Arbor: The University of Michigan Press. pp. 1–14.

Myers, S. L. (December 25, 1996). U.S. accuses Milosevic of inciting protesters. *The New York Times*. http://www.nytimes.com.

Nietzsche, F. (1954/1873). On truth and lie in an extra-moral sense. In W. Kaufmann (Ed.), *The portable Nietzsche*. NY: Viking Press. pp. 42–47.

Nietzsche, F. (1954/1888). Twilight of the idols. In W. Kaufmann (Ed.), *The portable Nietzsche*. NY: Viking Press. p. 523.

Petty, R., & Cacioppo, J. (1986). *Communication and persuasion: Central and peripheral routes to attitude change*. NY: Springer-Verlag.

Pollack, A. (December 21, 1996). Japan says it overstated a key figure in microchip talks. *The New York Times*. http://www.nytimes.com.

Robinson, F. G. (1992). *Love's story told: A life of Henry A. Murray*. Cambridge, MA: Harvard University Press.

Rodan, G. (1996). *Information technology and political control in Singapore*. http://www.nmjc.org/jpri/wp26.html.

Rosen, J. (December 6, 1996). The bloods and the crits. *The New Republic*. pp. 27–42.

Sakkas, P. E. (1991). Espionage and sabotage in the computer world. *International Journal of Intelligence and Counterintelligence*. 5, 155–202.

Schiesel, S. (December 21, 1996a). Long road for global technology trade. *The New York Times*. http://www.nytimes.com

Schiesel, S. (December 21, 1996b). Global agreement reached to widen copyright law. *The New York Times*. http://www.nytimes.com..

Schiesel, S. (December 31, 1996c). Air Force computer invaded as hackers forge a web page. *The New York Times*. p. A9.

Schlesinger, A. M., Jr. (February 18, 1994). Plots to kill Castro preceded Bay of Pigs. *The New York Times*. p. A26.

Sciolino, E. (January 1, 1997). Reporter's notebook: Iran welcomes Western tourists—Almost. *The New York Times*. http://www.nytimes.com.

Scott, W. B. (March 13, 1995). Information warfare demands new approach. *Aviation Week & Space Technology*, pp. 85–88.

Shamasastry, R. (1909). *Kautiliya's Arthasastra*. Mysore: Wesleyan Mission Press.

Sims, C. (December 30, 1996). Growing optimism for peaceful end to crisis in Peru. *The New York Times*. pp: A1; A4.

Snyder, A. (January 1, 1997). Use satellite TV for U.S. broadcasts to China. *The New York Times*. http://www.nytimes.com.

Sprangers, M., Van der Brink, W., Van Heerden, J., & Hoogstraten, J. (1987). A constructive replication of White's alleged refutation of Nisbett and Wilson and of Bem: Limitations on verbal reports of internal events. *Journal of Experimental Social Psychology*, 23, 302–310.

Strizinec, M. (1985). Principal trends in contemporary psychology of thinking in the USSR. *Studia Psychologica*, 27, 93–105.

The CIA starts to use the Internet to gather, disseminate, and swap non-classified information among offices worldwide. (September 5, 1994). *Federal Computer Week*, 8.

3 ex-hostages in Peru speak kindly of captors. (December 24, 1996). *The New York Times.* p. A6.

Tribunal plans to file warrants. (May 24, 1996). *The New York Times.* p. A6.

Two companies to share $237 million order from Army. (December 27, 1996). *The New York Times.* p. C3.

Two guards reported glued to soap opera. (December 30, 1996). *The New York Times.* p. A4

Tyler, P. E. (December 27, 1996). U.S. radio aims at 'Tyranny' in China, but few can hear it. *The New York Times.* p. A1; A4.

Ugandan rebel prophetess says holy mission is to oust Museveni (October 24, 1987, AM cycle). *The Reuters Library Report.*

Warfare in an information age. (1996). *Parameters. 26,* 81–140. http://carlisle-www.army. mil/usawg.

Warlaumont, H. G. (1988). Strategies in international radio wars: A comparative approach. *Journal of Broadcasting and Electronic Media, 32,* 43–59.

Wasburn, P. C. (1992). *Broadcasting propaganda: International radio broadcasting and the construction of reality.* Westport, CN: Praeger.

Weiner, T. (January 5, 1994). CIA seeks documents from its radiation tests. *The New York Times.* p. A11.

Weinraub, B. (December 9, 1996). Hollywood feels chill of Chinese warning to Disney. *The New York Times.* pp. B1; B6.

Weinstein, H. M. (1990). *Psychiatry and the CIA: Victims of mind control.* Washington DC: American Psychiatric Press.

Welsh, D. H., Bernstein, D. J., & Luthan, F. (1992). Application of the Premack principle of reinforcement to the quality performance of service employees. *Journal of Organizational and Behavioral Management, 13,* 9–32.

9

Cultural Diversity and Gender Issues

Mickey Dansby, Ph.D.[1]

* Introduction * Background * Racial/Ethnic Issues * Personnel * Discipline
* Diversity Training * Human Relations * Gender Issues * Sexual
Harassment * Other Gender Issues * Future Diversity Issues * References

In their book, *All That We Can Be*, military sociologists Charles Moskos and John Sibley Butler (1996) compare the prevailing negative paradigms for racial integration in America to one institution where African-Americans have been uniquely successful:

> It is an organization unmatched in its level of racial integration. It is an institution unmatched in its broad record of black achievement. It is a world in which the Afro-American heritage is part and parcel of the institutional culture. It is the only place in American life where whites are routinely bossed around by blacks. The institution is the U.S. Army. (p. 2)

Moskos and Butler discuss the remarkable success (with caveats) of racial integration in the Army and suggest it may be instructive for the rest of American society. Because of its historical precedent (MacGregor, 1981) and its contemporary significance, race must be a key issue in any discussion of diversity and military psychology.

But race is not the only major concern for psychologists interested in diversity within the military. Incidents like the Navy's Tailhook scandal and the sexual assaults at Aberdeen Proving Grounds, Maryland, and Fort Leonard Wood, Missouri, remind us that gender issues are also of paramount importance. Therefore, in this chapter we concentrate on these two key diversity topics from a military psychology perspective. We close the chapter with some brief thoughts on the future of diversity issues in the military. But before we discuss the two major content areas, let us establish a context for understanding how the military deals with diversity and gender issues.

Background

President Truman's Executive Order 9981 (July 26, 1948), which established racial equality as a policy and called for desegregation of the military services, was a watershed event for diversity within the military. Born out of questions relating to military efficiency (MacGregor, 1981) and Mr. Truman's concern about treatment of black soldiers returning from World

175

War II (Nalty, 1986), this unprecedented policy faced stiff opposition (MacGregor, 1981). Even so, it established a tradition that would result in the successes described by Moskos and Butler (1996) and led to a long line of commissions and studies on how to implement such a broad social change (MacGregor, 1981). Equality of treatment existed on paper and in policy, but effective integration did not come easily.

Despite the revolutionary and immediate impact of Executive Order 9981, it was not until some time later that the military came to grips with the important question of training service personnel (coming from all walks of American life and all regions of the nation) on how to deal with racial issues. As Hope (1979, p. 1) points out, "Like most institutions, the Army did not 'get serious' about the problem of racism until there were riots and 'fraggings' (fire bombing of top-ranking officers and enlisted personnel)." In early 1971, the military services openly recognized the racial unrest and seriously considered what might be done to combat such divisiveness (Hope, 1979). Almost immediately, a broad strategy of education and training was implemented to meet the challenge.

Hope (1979), Day (1983), and Dansby and Landis (1996) chronicle the military's response to the racial unrest of the late 1960s/early 1970s. Perhaps the keystone in the services' strategy was the establishment of the Defense Race Relations Institute (DRRI), later redesignated the Defense Equal Opportunity Management Institute (DEOMI). Unique in American society, DRRI/DEOMI has served a distinctive purpose in preparing trainers and advisors to help the services deal with diversity issues. Dansby and Landis (1996) note five key elements in the military's approach to equal opportunity (EO) and diversity:

behavioral
change

(a) a focus on *behavioral change* and *compliance* with stated policy;

(b) an emphasis on EO and intercultural understanding as *military readiness issues*;

(c) an understanding that *equal opportunity is a commander's responsibility* and that the *DEOMI graduate's function is to advise and assist the commander* in carrying out this responsibility;

(d) a belief that *education and training* can bring about the desired behavioral changes; and

(e) reliance on *affirmative action plans* as a method for ensuring equity and diversity.

(pp. 206–207)

Several of these elements lend themselves to psychological inquiry (e.g., behavioral change, compliance, training), and psychologists have frequently been called upon to explore diversity issues in the military. Hope (1979), Day (1983), and Thomas (1988) summarize much of the race relations research that was generated in the 1970s and 1980s—perhaps the "golden period" for research on these issues in the services. A useful reference to set the context for this period is Binkin and Eitelberg's (1982) excellent overview of black participation and the unprecedented rise in black representation in the military (especially the Army). Thomas' (1988) collection of readings and reports from this era is most helpful in demonstrating the scope and depth of research conducted at the U.S. Army Research Institute for the Behavioral and Social Sciences (ARI). Though many studies were conducted by "think tanks" (like the Rand Corporation, the Brookings Institution, and the Human Resources Research Organization), the work at ARI appears to be the most cogent and best organized effort to apply a behavioral science perspective to race relations questions in the military. The

ARI's race relations research diminished after this early work, perhaps due to a redirection toward other issues or to a belief that the major questions had been resolved.[2]

For many reasons, interest in gender issues developed after that for race. Participation of women in the military was restricted by law or policy for many years, and women are still prohibited by Congress from serving in a number of direct combat military specialties. The Women's Armed Services Act of 1948, while giving women a permanent place in the military, had placed many limitations on women's service: women could constitute no more than 2% of a branch's personnel; only 10% of the women in a service could be officers; women could not serve as general or flag officers; each service was limited to one woman at the rank of O-6 (colonel or Navy captain), and that woman's rank was temporary—she held it only so long as she was head of the women's corps for that service. In 1967, Congress passed a law repealing these restrictions (though women's participation in direct combat was still prohibited). (See Holm, 1992, and Dansby & Landis, 1996, for a discussion of these issues.) Representation of women in the services increased very rapidly after 1967, and by the end of 1996, women comprised over 13% of active duty service personnel (13.4% of officers and 13.2% of enlisted members; DEOMI, 1996). The military has been a particularly popular career choice for minority women (over 40% of the women in the services are from minority groups; DEOMI, 1996), perhaps due to the many benefits of military service (e.g., gender equity in pay, retirement, and other benefits).

Gender topics were added as a significant part of DEOMI's curriculum only in the late 1970s and early 1980s, and little psychological research had been conducted on gender issues in the military until about the same time. However, gender research rapidly overtook racial/ ethnic questions as a topic of interest, and many key studies on gender issues have been conducted by several military research agencies, including the Defense Manpower Data Center (DMDC), Navy Personnel Research and Development Center (NPRDC), and DEOMI, or by individual researchers.

The initial sluggish pursuit of research on gender issues in the military may have been due to several factors, including: the smaller representation of women in the services, the early concentration of women in relatively few (usually gender-stereotyped) jobs (e.g., nursing), a belief by research sponsors that race was a more pressing concern, or the lack of "crisis" events (e.g., race riots) to focus attention on gender issues. However, questions related to integration of women into the service academies, expansion of opportunities for women in the services, and possible problems in gender relations (e.g., sexual harassment, the source of several later "crisis events") led to the rapid expansion of research on gender issues.

The military's public support for diversity concerns is codified in the Department of Defense (DoD) Human Goals. This document, signed by the Secretary of Defense, Deputy Secretary of Defense, service secretaries, and military chiefs of the services, lists a number of objectives. The basic premise of the Human Goals is, "Our nation was founded on the principle that the individual has infinite dignity and worth. The Department of Defense, which exists to keep the Nation secure and at peace, must always be guided by this principle." The DoD aspires, "To make military and civilian service in the Department of Defense a model of equal opportunity for all . . ." and "To create an environment that values diversity and fosters mutual respect and cooperation among all persons..." Military psychologists have contributed to these lofty ideals through their research on race and gender issues within the services and

gender-stereotyped jobs

sexual harassment

177

by helping develop the training programs used to combat racism/sexism and improve human relations.

Unfortunately, a large proportion of the research done by military psychologists is never published in professional journals or books. This is not a function of the quality of the work, for much has very high technical merit, but rather because it tends to address issues of immediate practical importance, based on the needs of a particular sponsor (i.e., a senior manager in the service who needs to make a policy decision, a training institution that needs a methodological question answered, etc.), and occurs under the purview of a military laboratory, school, or other agency (e.g., the ARI, NPRDC, DEOMI, Naval Postgraduate School, and DoD Accession Policy Office). A great deal of this "fugitive" literature is accessible (using computer assisted searches) through the Defense Technical Information Center and the National Technical Information Service. (The author of this chapter is pleased to note a movement toward increased publication in standard professional journals and books—a trend that should be encouraged in the interest of enhancing scientific community and cross-utilization of military psychologists' work.) Military psychologists also frequently make presentations at meetings of professional societies (such as the American Psychological Association [APA], Society for Industrial and Organizational Psychology, and American Management Association). Certainly, creation of Division 19 (Military Psychology) and its associated journal (*Military Psychology*) of the APA has contributed substantially to the opportunity to share results from military research. In addition, specialized forums, such as the Applied Behavioral Sciences Symposium, sponsored by the USAF Academy, and the International Military Testing Association Annual Conference often feature presentations by military psychologists. Frequently, these professional activities include presentations or reports on diversity issues.

With a basic understanding of the context, let us now consider the first of the two major diversity issues in the military, the racial/ethnic concerns.

Racial/Ethnic Issues

A number of sociologists, psychologists, and other professionals have explored race and ethnic issues within the military. Many, spurred by concerns raised in the general society, have done so without formal sponsorship of the services. However, identification of major issues and initiation of most research are direct results of the efforts of behavioral scientists from military organizations. The research has covered many key areas, including racial/ethnic questions related to personnel (e.g., entrance testing, recruitment, assignment, performance, personnel evaluation, promotion, retention, and training), discipline, diversity training, and human relations (e.g., equal opportunity climate, fostering positive relations, group dynamics and cohesion). A comprehensive review of all these topics is beyond the scope of this chapter. However, the author shall attempt to discuss representative studies and give the reader some sense of the scope of the work that has been done.

Personnel. Much of the research on race and ethnicity involves personnel issues and has been conducted under the auspices of major military personnel research centers (e.g., ARI, NPRDC, or the Air Force's Armstrong Laboratory Human Resources Directorate—formerly the Human Resources Laboratory [HRL]). In many instances, broader studies of personnel issues (e.g., entrance testing, occupational assignment) addressed racial/ethnic questions, though the main thrust was not diversity concerns. Many such studies were sponsored through the Office of the Secretary of Defense, frequently by the Accessions Policy Office. For example, Eitelberg, Laurence, and Waters (1984) and Eitelberg (1988) offer extensive summaries on screening, selecting, and classifying personnel for military service. These reports include analyses based on race, ethnicity, and gender, in addition to a number of other relevant dimensions. A great deal of the information relates to data from the Armed Services Vocational Aptitude Battery (ASVAB), a test battery introduced in 1968 that has been used by all services since 1974 as a tool for screening enlistees and assigning them to occupations (Eitelberg, 1988).

Since the ASVAB (or its predecessors and composites, such as the Armed Forces Qualification Test, or AFQT) is used to make critical personnel decisions (i.e., entry qualification, occupational field, etc.), it has been the subject of much research on gender and race issues. Two notable incidents in the use of the ASVAB should be mentioned. The first is Project 100,000 (a result of the "Great Society" initiatives), in which 100,000 men who fell below current admission standards were allowed to enter the services annually. Many of the accessions were minority group members. The project was designed to provide employment for many of America's "subterranean poor" (Eitelberg, 1988). The second is the inadvertent misnorming of the AFQT between 1976 and 1980 (Sellman & Valentine, 1981). The **misnorming** resulted in the admission of nearly 360,000 men who actually fell below minimum standards (if the test had been normed properly). Clearly, both these incidents resulted in increased admission of minority group members, who traditionally had not done as well on standardized testing and may have lacked educational advantages.

For the interested reader, Eitelberg (1988) discusses both Project 100,000 and the misnorming in some detail. He reports that military commanders often complained of what they perceived as a diminished quality of the force during this period; however, statistical analysis indicates those scoring below the accepted standard did about as well as those in the lowest qualified group. Eitelberg's (1988) conclusion is noteworthy:

> It is probably safe to say that, from the standpoint of "military effectiveness" or "readiness," the jury is still out on both Project 100,000 and its unplanned replacement. Lower aptitude standards tend to be viewed more favorably when expedience demands it . . . [S]everal gross misconceptions about the two events have become a permanent part of the military's folklore. But Project 100,000 and the ASVAB misnorming did make at least one lasting contribution to defense manpower policy: they helped to raise some serious questions about the appropriateness of established selection standards and forced the military's policymakers to examine more closely the performance measures used to set those standards. (pp. 182–183)

Eitelberg's (1988) analysis of the status of women and minorities indicates significant progress by these groups in securing assignment to higher skilled jobs between the late 1970s and late 1980s. However, he points out there are still significant differences in the job distribution of minorities and women compared to white men (though the distribution in the military may be more equitable than in U.S. society as a whole). These issues (representation and distribution by occupation) continue to be of interest to social scientists exploring diversity issues in the military. For example, Moskos and Butler (1996) outline the significant rise of African Americans into leadership positions within the Army. Between 1970 and 1990, the proportion of senior noncommissioned officers who were black grew from 14% to 31%;

during the same period, black officers increased from 3% to 11% of commissioned officers; and by the mid-1980s, 7% of the Army's generals were black.

Another area of inquiry is the propensity of American youth to join the military. Major attitude studies by race, ethnicity, and gender are common. For example, Segal and Bachman (1994) report results from high school seniors surveyed as part of the *Monitoring the Future* project at the University of Michigan's Institute for Social Research. The DoD also sponsors the annual *Youth Attitude Tracking Study* (YATS; Ramsberger, 1993; Nieva, 1991), generating a number of analyses of attitudes toward the military. Such efforts provide many opportunities for psychologists to explore attitudinal data and behavioral propensities of minority group members and women who are potential recruits.

Many other personnel issues related to race have been researched in the military. A few
institutional examples: Thomas (1988) describes a number of reports dealing with institutional
discrimination discrimination issues (assignment, promotion, selection for schooling, access to reenlistment, retention, etc.); Dansby (1989) and St. Pierre (1991) look at retention by race (and gender); Rosenfeld, Thomas, Edwards, Thomas, and Thomas (1991) provide a historical review of the Navy's research on race (and gender) issues, including personnel research such as studies of Hispanic representation and turnover in the Navy's civilian work force; Greene and Culbertson (1995) present an analysis of potential racial bias in promotion fitness reports within the Navy; Smither & Houston (1991) explore issues related to fairness and redress of grievances within the military. In short, personnel issues within the military have been, and continue to be, a rich source for racial/ethnic (and gender) research.

Discipline. Another major topic for racial/ethnic research within the services is military discipline and punishment (i.e., military justice). There are three broad categories of
administrative discipline within the military: administrative discipline (e.g., separation from the service),
discipline nonjudicial punishment (NJP; minor punishments and fines administered by a local commander), and formal legal proceedings conducted under the Uniform Code of Military
nonjudicial Justice (UCMJ). Military psychologists have explored all three.
punishment

To a large degree, the research interest in race and military justice is the result of questions of equity raised by black civilian leaders. For example, in 1971 the National Association for the Advancement of Colored People (NAACP) and the Urban League challenged the racial fairness of military justice in the Army's European theater (Nalty, 1986). Members of these organizations, along with others, served on a task force commissioned by Secretary of Defense Melvin Laird to investigate the allegations. The majority of the task
discrimination force concluded that racial discrimination existed in the administration of military justice, and that it resulted from pervasive discrimination in American society (Nalty, 1986). Nalty summarizes the report as follows:

> Statistics indicated that blacks were more likely than whites to run afoul of regulations. Furthermore, black servicemen had the greater probability of undergoing trial by court-martial, being confined before the trial, being convicted, and receiving long sentences. Blacks were also more likely to receive . . . administrative discharge . . . and . . . nonjudicial punishment, administered on the authority of the commanding officer for comparatively minor offenses. The statistical disparity . . . existed despite what the task force agreed was a genuine effort on the part of the armed forces to ensure equal treatment. (p. 329)

The report of the task force was administrative and lacked the rigor of a scientific study. Yet the basic disparity in rates of military discipline for white and black soldiers was not in dispute. A statistical disparity persists, though the absolute numbers of disciplinary actions (especially for UCMJ offenses) has declined dramatically since the 1970s. (See, for example, Walker, 1992, who reports that while the total number of court-martial convictions across the DoD fell from 15,739 in fiscal year [FY] 1987 to 7,485 in FY1991, the proportionate overrepresentation of black servicemembers actually increased during the same period because of the faster decline in white convictions as compared to black convictions.) Landis and Dansby (1994) report that between 1988 and 1993, blacks were overrepresented in NJP (by 20–40%, depending on service) and UCMJ actions (by as much as 100%). While the overrepresentation of blacks in military prisons is less than half what it is in the penal system at large (Moskos & Butler, 1996), it is still an issue of some concern. Why does the disparity exist? This important question, raised by Secretary Laird's task force in the 1970s, remains to a large degree unanswered and continues to be the basis of many studies within the military.

disparity

Dansby (1992) outlines a structure for systematically addressing the issue. Using the analogy of a tree, three "roots" in the soil of society are discussed as potential exogenous (i.e., outside the military) sources leading to black overrepresentation in the military justice system: psychological, physiological, and sociological variables. Within the military several "branches" are discussed as endogenous factors: differential treatment by race, differential involvement in crimes by race, and statistical artifacts. Landis and Dansby (1994) continue to develop this model.

After a review of numerous empirical studies, Dansby (1992) concludes that a substantial portion of the disparity can be traced to exogenous sociological factors. In a study of incarcerated military members, Knouse (1993) presents empirical evidence of racial differences in exogenous variables such as socioeconomic, familial, and personality factors. Based on available results, it appears racial discrimination within the military justice system is not the major cause of the disparity. Consequently, Dansby (1992) and Landis and Dansby (1994) recommend an action research strategy addressing the sociological factors. Two possible techniques are special training programs: one for recruits and another for military supervisors. These programs are described as the *inoculation approach* and the *cultural assimilation model* (see also Knouse, 1994a, and Knouse & Dansby, 1994). Neither strategy has been fully developed or validated as of this writing.

inoculation approach

cultural assimilation model

Numerous studies relating to discipline have been conducted or sponsored by the NPRDC (see Rosenfeld et al., 1991, and Culbertson & Magnusson, 1992, for reviews), ARI and Department of the Army (e.g., Bauer, Stout, & Holz, 1976; Bell & Holz, 1975; Nordlie, Sevilla, Edmonds, & White, 1979; and Verdugo, 1994), Office of the Secretary of Defense (e.g., Flyer, 1990; Flyer, 1993; Flyer & Curran, 1993), and DEOMI (see Knouse, 1996a, for a review). Despite the large number of studies, much remains to be explored (Dansby, 1992).

Diversity training. Hope (1979), Day (1983), Thomas (1988), and Dansby and Landis (1996) offer reviews of the structure, policies, and research for diversity training in the military. Many of the studies in this area are evaluation research relating to the impact of the training on servicemembers in general or the effectiveness of DEOMI's training programs in preparing equal opportunity advisors and instructors (also see Knouse, 1996a; Johnson, 1995; and Johnson, 1996).

Day (1983) offers an extensive review of the effectiveness of diversity training in the military, especially in the Army. In general, there are many problems associated with evaluation research, making it difficult to determine the effectiveness of the DEOMI training or the training offered by the services to their members in general. Evidence reviewed by Day (1983) indicates the training varies in effectiveness (based on numerous criteria such as whether it was implemented seriously or just as a "paper-and-pencil" exercise). The overall program appears to have improved the race relations climate in the services (Day, 1983). Hope (1979) summarizes a number of studies concerning the effectiveness of the DEOMI training as measured by several approaches, including attitude change; cognitive change; and the opinions of the faculty, graduates, and commanders the graduates served. All sources indicated positive impact of the training, both personally and organizationally. Johnson (1995, 1996) found similar positive results in her surveys of commanders and supervisors of DEOMI graduates.

In addition to analysis of the impact of diversity training, many studies have considered the adequacy of the content of the DEOMI training to prepare equal opportunity advisors for their jobs. Most of the studies were occupational analyses conducted by the services (e.g., the Air Force Occupational Measurement Squadron) or contractors (e.g., Kinton & Associates). The results have been used to refine the curriculum to make sure graduates are adequately prepared to meet the needs of the field commands. Such studies have led to evolutionary changes in DEOMI's curriculum (such as an increased emphasis on gender issues and the addition of a writing and speaking program).

Human relations. Fostering positive human relations is a key concern in military organizations. Without mutual respect and recognition of each person's dignity and contributions to the organization, it is difficult to develop a cohesive, smoothly functioning team. The DoD Human Goals emphasize the "infinite dignity and worth" of each individual in the service, and military commanders generally acknowledge the need for a positive human relations climate in order for their units to function effectively. Much of the diversity research conducted by military psychologists has focused on this issue.

racial separatism

The range of research in this area is quite extensive (see Thomas, 1988, for examples of research on cross-cultural communications, racial separatism, unit race relations, cultural assimilation, and training for racial harmony). For the sake of brevity, we will focus on general EO climate issues and confine our consideration to two major sources of EO climate research: DEOMI and NPRDC. The work of each institution has been documented in numerous articles and technical reports (for the DEOMI program, see Dansby & Landis, 1991; Landis, Dansby, & Faley, 1993; Landis, Dansby, & Tallarigo, 1996; Dansby, 1995; and the summary in Knouse, 1996a; for information on the NPRDC effort, see Rosenfeld et al., 1991; Culbertson & Rosenfeld, 1993; Rosenfeld & Culbertson, 1993; Rosenfeld & Edwards, 1994; and Newell, Rosenfeld, & Culbertson, 1995).

organizational effectiveness

At DEOMI, most EO climate research involves the Military Equal Opportunity Climate Survey (MEOCS) or its various derivatives (i.e., the small unit, equal employment opportunity, senior leader, and short versions). The MEOCS is an organizational development survey and analysis service provided free to military commanders at all levels to help them assess EO and organizational effectiveness (OE) within their units and plan strategies to improve the organizational climate. There are four EO climate sections in the MEOCS: perceptions of unit EO behaviors, OE perceptions, general EO perceptions (i.e., in the total context and not just unit conditions), and a global measure called *Overall EO Climate*. Issues addressed include race/gender discrimination, racist/sexist behaviors, sexual harassment, positive EO behaviors, the so-called "reverse" discrimination, desire for racial separatism, job satisfaction, commitment to the organization, and perceptions of work group effectiveness (Dansby, 1994a). Commanders voluntarily request MEOCS for their units, and DEOMI keeps unit results confidential. After a commander receives his or her report, identifying information is stripped from the unit's data and the data are added to an overall database used for comparison and research purposes (i.e., to allow each command to compare their results to their service's results and those from all services). In addition, occasional probability samples are gathered across entire services to compare results to the overall unit database results. Since the program began in June of 1990, over 4,000 units have participated and a database of well over half a million respondents has been accumulated.

Results from the MEOCS have demonstrated several consistent patterns. Although most demographic subgroups rate the EO climate as "average" or better for their organizations, the perceptions vary by group. In general, the most favorable climate ratings are from majority men, and the least favorable are from minority women (Dansby, 1994a; Moskos & Butler, 1996). More specifically, the most favorable ratings are from majority officer men and the least favorable are from minority officer women (Dansby, 1994a). (In a recent article, Dansby & Landis, in process, present evidence that one reason minority officer women rate the climate less favorably than minority enlisted women is because of their smaller representation in units.) Results from the Senior Leader Equal Opportunity Climate Survey (Dansby, 1996), a version of MEOCS developed for the general/admiral/Senior Executive Service level, and from the normal MEOCS indicate that high ranking white men have the most favorable ratings of all.

The MEOCS database has been the source of many empirical studies on diversity issues, including the relationship between EO climate and Total Quality Management (Knouse, 1994b, 1996b), group cohesiveness and performance (Niebuhr, Knouse, Dansby, & Niebuhr, 1996), career commitment (Landis, Dansby, & Faley, 1994), demographic representation within the organization (Dansby & Landis, in process), organizational characteristics (Tallarigo & Landis, 1995), and acceptance of diversity (Niebuhr, 1994a). In addition, the MEOCS program has inspired numerous studies on EO climate survey development and improvement (e.g., Dansby & Landis, 1991; Landis et al., 1993; Dansby, 1994b; Niebuhr, 1994a; Albright & McIntyre, 1995; McIntyre, 1995; McIntyre, Albright, & Dansby, 1996; Dansby, 1996).

The NPRDC's extensive EO climate program includes both the Navy and the Marine Corps. The Navy assessed EO climate as part of its Human Resources Management organizational survey program as far back as 1975 (Rosenfeld & Edwards, 1994). There were also efforts to measure race relations in the Navy during the 1970s (Rosenfeld et al., 1991) that were similar to the Army's efforts at ARI (Thomas, 1988). The current NPRDC program involves measurement at two levels: the Command Assessment Team Survey System (CATSYS), at the individual unit level, and service-wide probability sample surveys (the Navy Equal Opportunity/Sexual Harassment Survey [NEOSH] and the Marine Corps Equal Opportunity Survey [MCEOS]) (Rosenfeld & Edwards, 1994; Newell et al., 1995; Thomas & Le, 1996).

The NEOSH (and MCEOS) is composed of nine modules, addressing a number of areas where discrimination might occur. The modules include: assignments, training, leadership, communications, interpersonal relations, grievances, discipline, performance evaluation, and Navy satisfaction (Rosenfeld, Culbertson, Booth-Kewley, & Magnusson, 1992). Results on each module are compared by demographic group, much as MEOCS results are compared by group for MEOCS scale scores.

For the most part, results of the NPRDC EO climate program are very similar to those for the MEOCS. Newell et al. (1995) summarize the EO results of the 1991 NEOSH as follows:

> 1) Navy personnel generally had positive EO climate perceptions; 2) men had more positive EO climate perceptions than women; 3) White male officers had the most positive EO climate perceptions, African-Americans, particularly African-American women had the least positive perceptions; and 4) Hispanics' EO perceptions typically fell between those of Whites and African-Americans and most often were closer to the perceptions of Whites. (p. 160)

The close correspondence between NEOSH and MEOCS, which use different instruments and methodologies, supports the construct validity of the two surveys and increases confidence in the findings.

The NEOSH and MCEOS are divided into two major sections: EO climate issues and sexual harassment issues (Rosenfeld & Edwards, 1994; Thomas & Le, 1996). The fact that a single EO issue (sexual harassment) carries such weight in the NPRDC's assessment of EO climate indicates the importance of this issue in military research efforts. We now turn our attention to the this sensitive and often discussed topic, along with other gender issues in the military.

Gender Issues

In recent years, gender issues in the military have occupied a prominent place in the popular press as well as the scientific literature. Questions regarding the impact of women's increased accessions and entry into traditionally male military specialties, along with gender discrimination and sexual harassment concerns, have occupied newspaper and television reporters, movie makers, political pundits, feminists, special interest organizations, congressional oversight committees, senior leaders within the services and executive branch, and behavioral scientists alike. Perhaps the most often discussed issue is sexual harassment, so we address it first, followed by a brief synopsis of other gender issues of interest to military psychologists.

sexual
harassment

Sexual harassment. The DoD definition of sexual harassment parallels that of the Equal Employment Opportunity Commission (EEOC). DoD Directive 1350.2 states:

> Sexual harassment is a form of sex discrimination that involves unwelcome sexual advances, requests for sexual favors, and other verbal or physical conduct of a sexual nature when:
> (1) submission to such conduct is made either explicitly or implicitly a term or condition of a person's job, pay, or career, or
> (2) submission to or rejection of such conduct by a person is used as a basis for career employment decisions affecting that person, or
> (3) such conduct has the purpose or effect of unreasonably interfering with an individual's work performance or creates an intimidating, hostile, or offensive working environment.

This definition encompasses two basic forms of sexual harassment: *quid pro quo* (i.e., "something for something"), a more severe form in which sexual favors are demanded in exchange for job benefits (e.g., continued employment, advancement), and *hostile environment*, a less severe form in which sexual behavior in the work place interferes with a person's performance or creates an intimidating environment. (See Eisaguirre, 1993, for a lay person's discussion of the general EEOC definition, the two basic categories, legal issues, and a summary of studies concerning sexual harassment. Pryor, 1985, also provides an excellent summary of the issues.) Another key feature of the definition is that the sexual behavior is unwelcome (i.e., uninvited and unwanted; Farley, 1978; MacKinnon, 1979). Secretary of Defense William Perry (in a 1994 memorandum to the secretaries of the military services, joint chiefs of staff, and other defense agencies) further clarified the policy to indicate that hostile environment harassment "need not result in concrete psychological harm to the victim, but rather need only be so severe or pervasive that a reasonable person would perceive, and the victim does perceive, the work environment as hostile or abusive." Secretary Perry's memorandum also indicated that the definition applied both on or off duty for military members, and that anyone in DoD "who makes deliberate or repeated unwelcome verbal comments, gestures, or physical contact of a sexual nature in the workplace is also engaging in sexual harassment." Goldman (1995) offers an excellent summary of the issue as it is understood in the military, and Bastian, Lancaster, and Reyst (1996) give a succinct history of sexual harassment policy concerns and deliberations in the DoD.

Several major studies of sexual harassment in the military have been conducted in the last 20 years. Two, while not specifically focused at the military, included information concerning the sexual harassment of civilians working in every service. These studies were conducted by the U.S. Merit Systems Protection Board (MSPB) (1981, 1988) in 1980 and 1987. The MSPB defined several categories of sexual harassment, varying in severity. The *most severe* form was actual or attempted rape or assault; the *severe* forms were deliberate touching, pressure for sexual favors, and letters and calls; the *less severe* forms were sexual remarks, suggestive looks, and pressure for dates. The survey asked respondents to indicate whether they had experienced any of these forms of harassment during the 24 months prior to completion of the survey. The table below shows the reported incidence rate, by gender and service, for the two surveys.

Results of the MSPB Surveys of DoD Civilians:
Those Experiencing at Least One Incident in the Previous Two Years[3]

SERVICE	1980 Survey		1987 Survey	
	Men	Women	Men	Women
Air Force	12%	46%	16%	45%
Army	16%	41%	11%	44%
Navy/Marine Corps	14%	44%	14%	47%
Federal Government Average	15%	42%	14%	42%

As the table indicates, there was relatively little change in the incidence of sexual harassment of civilian employees of the military between 1980 and 1987.

Since the MSPB surveys did not address the rate of sexual harassment for military members, the DoD conducted its own studies, modeling its survey after the MSPB survey. The military surveys, conducted by DMDC in 1988 and 1995, used probability samples representing military personnel from all services (Martindale, 1990; Bastian, et al., 1996). The rates for military women were higher than those reported by civilians in the MSPB surveys, though the rates for military men were similar to those of the male civilian employees. A major finding of the 1995 survey was a significant decline in reported sexual harassment from the 1988 survey. While 64% of the women in the 1988 survey reported at least one incident during the preceding 12 months, 55% gave a similar report in 1995. Similarly, the rates for men dropped from 17% to 14%. This drop in reported incidents is consistent with perceptions reported in the MEOCS (see figure below) and other climate surveys, which indicate sexual harassment is on the decline in the services.

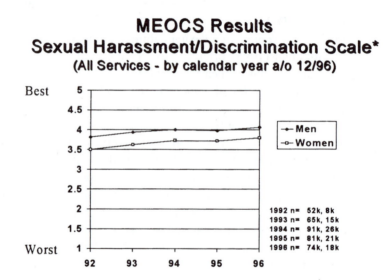

MEOCS Results
Sexual Harassment/Discrimination Scale*
(All Services - by calendar year a/o 12/96)

*Source: Defense Equal Opportunity Management Institute, Patrick AFB, FL

Though several notable incidents have brought sexual harassment in the military to the public's attention recently, the overall trend is toward an improved climate. However, DoD still has a long way to go before it reaches its goal of "zero tolerance" (Bastian et al., 1996).

In trying to understand the levels of sexual harassment, care must be taken in interpreting incidence rates. There are many different operational definitions used in measuring sexual harassment: perceptions of what has happened to others, reports of personal experience, and reports of formal complaints, to name a few (Eisaguirre, 1993; Culbertson & Rosenfeld, 1994). Furthermore, studies may use different time references (i.e., have you *ever* experienced, experienced in the *last year*, etc.), selection of participants may vary (i.e., self-selected or randomly selected), and the behaviors reported may not be considered harassing by the respondent. (For example, the 1995 DoD survey included a Form B listing additional behaviors not included in the original survey. Respondents were asked to report whether they considered the behaviors they experienced to be harassing. Many did not think of some of the incidents as sexual harassment.) A final consideration is differences in working conditions and exposure to sexual harassment. Although the military rates are higher than those reported by civilian employees in the MSPB surveys, the surveys do not consider the difference in working conditions for civilian and military employees (i.e., while civilians are typically "on duty" for normal 8-hour work days, military members are on duty 24 hours a day and often live in military quarters; this expands the opportunities for work-related harassment, as clarified by DoD's expansive definition of sexual harassment; also, the higher ratio of men to women— over six to one—in the military increases opportunities for military women to experience sexual harassment). However, even with consideration for all these caveats, sexual harassment remains an issue of major concern to the services (Bastian et al., 1996), and developing strategies for the prevention of sexual harassment will likely continue to be a high interest item.

Other findings of the 1995 DoD survey help paint a picture of sexual harassment similar to that in the civilian world: women are much more likely to experience harassment than men; junior personnel are more likely to be victims than senior personnel; black men experience harassment more frequently than white men; and the most likely perpetrators are coworkers, followed by superiors, then others in the service. Most incidents occurred on a military installation, at work, during duty hours. About a quarter of the victims reported the incident(s), and most of those who complained said action was taken (though 23% of women and 16% of men said their complaint was discounted or not taken seriously). Of those who did not report the incidents, over half said they took care of it themselves or didn't consider it important. About a quarter of the nonreporting women indicated they failed to report the incident(s) because they thought it would make work unpleasant, and 20% thought nothing would be done. A small but significant number of victims (20% of women and 9% of men) reported some degree (to a small, moderate, or large extent) of reprisal, but 80% of women and 85% of men said they felt free (to a small, moderate, or large extent) to report sexual harassment without the fear of something bad happening to them. About a third of those making complaints said they were dissatisfied with the resolution of the complaint (another third were neither satisfied nor dissatisfied, and the remaining third were satisfied). Eight in ten had received training about sexual harassment in the last year, and almost nine in ten knew how to report sexual harassment. (Bastian et al., 1996; cf. Eisaguirre, 1993)

Niebuhr (1997) summarizes recent research findings on sexual harassment in the military, though his work was published before the second DoD survey had been released. The table on the next page lists some of the findings reported by Niebuhr (1997).

As this brief overview has indicated, there is a wide range of research on sexual harassment in the military. While this topic remains the most prominent gender issue in the military, others have been considered as well. The next section provides a sampling of some of the other research.

Other gender issues. Besides sexual harassment, researchers have addressed a wide variety of other gender issues in the military. An early interest was the impact on the military of increased accessions of women after Congress removed the limits on numbers of women in the services. Binkin and Bach (1977) consider several such issues, including the economic impact (both on the military and for the women themselves), institutional attitudes (both within and outside the military), and military effectiveness; Holm (1992) discusses the

Finding	Researcher(s)	Date
1. Women in lower-status positions were more frequently harassed.	Fain & Anderton	1987
2. Women in male-dominated groups or who are gender pioneers in their jobs were more likely to be harassed.	Neibuhr & Oswald	1922a
3. Organizational climate concerning sexual harassment contributed to the degree of perceived harassment.	Niebuhur	1994b
4. More sexual harassment of women occurred in large units, where contact with larger numbers of men occurred.	Niebuhr & Oswald	1992b
5. Different survey designs may lead to substantially different estimates of sexual harassment incidence rates.	Culbertson & Rosenfeld	1994
6. Men and women may have different interpretations of what is sexual harassment.	Thomas	1995
7. There was a negative relationship between gender discrimination climate and both group cohesion and perceived group performance.	Niebuhr, Knouse, & Dansby	1994
8. The better the balance between the percentages of males and females in a work group, the greater their agreement in perceptions of how much sexual harassment occurs.	Niebuhr	1994b

expanding access to military occupations (i.e., job specialties); Shields (1988) explores sex roles in the military. Another interest has been the impact of pregnancy and single parenthood (e.g., Thomas & Thomas, 1992). Related to the latter is the questions of gender differences in absences from work (e.g., Thomas, Thomas, & Robertson, 1993; Thomas & Thomas, 1993a,b).

Gender equity in promotions and retention has been an issue of some concern as well (e.g., Firestone & Stewart, 1995). The data indicate women generally fare well in military promotions, doing about as well as their male cohorts in most promotion decisions. One interesting theme in the research on promotions has been the effort to determine how narrative descriptions in evaluation reports differ by gender (Thomas, Holmes, & Carroll, 1983; Greene & Culbertson, 1995). These studies have found that women's narratives may be shorter and have fewer dynamic, action words. Such deficiencies can have a negative impact on individuals, despite the equitable overall promotion statistics.

The following table summarizes some other gender issues that are recent research interests for military psychologists:

Finding	Researcher(s)	Date
1. Female aviators in the Air Force had more masculine or androgynous scores on a sex role scale (the Bem Sex Role Inventory) compared to other military and civilian women.	Dunivin, K. O.	1990
2. Younger, more highly educated respondents with better paying jobs were more likely to support women's participation in a variety of military roles	Matthews & Weaver	1990
3. Female Air Force aviators who planned to separate from the service were more likely to cite family concerns than job, career, or other concerns.	Schissel	1990
4. The relationship between stress, general well being, and job absences is more complex for women than for men.	Hendrix & Gibson	1992
5. Male service academy cadets were less approving of women serving in combat roles than their counterparts in a private civilian college.	Matthews	1992
6. Women had higher naval aviation qualifying test scores, yet had lower scores and completion rates for pre-flight training.	Baisden	1992
7. Compared to their male counterparts, women USAF pilots had higher extraversion, agreeableness, and conscientiousness personality scores.	King, McGlohn, & Retzlaff	1996
8. Scores on the Bem Sex Role Inventory were not predictive of ratings of military development for U.S. Military Academy cadets.	Stokan & Hah	1996
9. There was a negative correlation between sexism and both cohesion and group performance for an active duty military unit.	Niebuhr, Knouse, Dansby, & Niebuhr	1996

Gender issues will, no doubt, continue to be central to diversity concerns in the military for some time to come. Because the proportion of women recruits is on the rise, women will likely become a larger proportion of military personnel, even though the military has experienced a reduction in overall force levels. We can expect military psychologists will have more than a passing interest in this phenomenon and its related issues.

In our discussion of race and gender issues, we have covered the "big two" diversity concerns in the military. Yet, many see diversity as encompassing issues "beyond race and gender" (Thomas, 1991). In the next section we briefly address some of these issues.

Future Diversity Issues

What does the future hold for diversity issues within the military? We can safely project that race and gender will continue to be important; but what else will researchers address? We see several potential "hot topics" for the next few years, including the status of homosexuals in the military, racial identity issues (specifically, dealing with multiracial individuals), concerns about the "white male backlash" (i.e., reactions to affirmative action programs and perceptions of "reverse" discrimination), the conflict between individual rights and the need for cohesion in the military, and dealing with racial/ethnic/gender issues in foreign countries as the military takes on more humanitarian and peacekeeping missions.

Of all these issues, certainly the most prominent is the status of homosexuals in the military. Traditionally, military leaders have maintained that homosexuality is incompatible with military service because it is considered contrary to morale, good order and discipline, and cohesion within units. Congress and the courts have supported this position. Senate Bill 1337 (September 16, 1993) amends Chapter 37 of Title 10 (US Code) to clarify the policy (which has been characterized as "don't ask, don't tell, don't pursue") on issues of gender orientation. The law is clear that homosexual behavior is still forbidden within the military. However, homosexual orientation, per se, is not a reason for dismissal from the service, nor does it prevent entry into the service. So long as a person shows no *propensity to engage in homosexual acts*, does not openly state that he or she is homosexual or bisexual, and does not marry or attempt to marry a person of the same sex, the individual may serve in the military. The law reaffirms that military service is not a constitutional right. It also provides for continuation of a January 1993 policy that suspends questioning about homosexuality as a condition for entry into the military, but allows the Secretary of Defense to resume such questioning as necessary to implement the revised policy on homosexuality.

Very little research has been conducted on this issue. Since law and policy preclude homosexual behavior in the military, service laboratories are reluctant to sponsor research on the subject. At the request of the Congress, the United States General Accounting Office (1992) conducted an analysis summarizing previous research and incorporating findings of their own. The DoD contested the conclusions of the GAO's report, which indicated increasing support for permitting homosexuals in the services. The highly political and controversial nature of this issue will, no doubt, influence the course of research on the topic. Perhaps the most important question—the impact on military effectiveness—remains to be addressed from a scientific perspective.

As more and more interracial marriages occur, racial identity issues are becoming increasingly important in society at large, as well as in the military. The basic questions are how to identify racial categories in an increasingly multiracial society and what the impact of using multiracial (or nonracial) categories will be on various programs used to ensure racial equity in organizational decisions. Although more individuals are asking for a statistical category that reflects blended racial/ethnic background, the laws establish protected groups based on the traditional census categories. This is a very practical issue that should be addressed from a scientific perspective.

Another emerging trend is the "white male backlash" to affirmative action and other programs white males may perceive as discriminatory toward them. As the MEOCS data below indicate, servicemembers perceived more "reverse" discrimination as occurring between

1992 and 1995, with a possible reversal of the trend beginning in 1996. (Note that the scale has been abbreviated to better depict the difference in scores. Also note the overall high scores on the 5-point scale, indicating an overall favorable perception.)

Average MEOCS "Reverse" Discrimination Scale Scores, 1992-1996*
(Higher score indicates perception of better climate; all services, a/o 12/96)

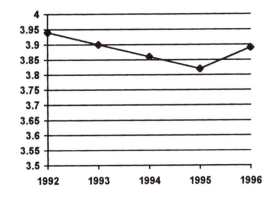

*Source: Defense Equal Opportunity Management Institute, Patrick AFB, FL

While the issues of homosexuality, racial classification, and white male backlash may become more important in the future, other diversity issues are also likely to emerge. We toss two more onto the table: the conflict between individual rights and the need for group cohesion and preparation for dealing with the potential racial/gender/ethnic conflicts in other parts of the world. As the services become more diverse, we can expect more conflicts between individual rights and organizational requirements. For example, as the military recruits more individuals from diverse religious backgrounds, we may see increasing demands for accommodation of particular religious views and practices (i.e., Sabbath observances, clothing requirements, etc.). Issues of religious discrimination may also become of more concern. Finally, as American troops take on more peacekeeping and humanitarian duties, they are likely to be stationed in countries where their primary duties require close contact with people of different racial/ethnic backgrounds and customs. Dealing with such diversity may be a problem for some military members. Therefore, we should research methods to prepare and train servicemembers to face such issues.

As this chapter has demonstrated, the military psychologist has a wide range of diversity issues to research in the military. The field has a rich heritage and challenging future. As long as our services consist of people from across America's diverse population, one has difficulty imagining a time when some issue related to diversity will not be "on the front burner."

References

Albright, R. R., & McIntyre, R. M. (1995). *Measuring equal opportunity climate in small units: Development of scales to evaluate the acceptance of diversity* (DEOMI RSP 95–10). Patrick Air Force Base, FL: Defense Equal Opportunity Management Institute.

Baisden, A. G. (1992). Gender and performance in naval aviation training. *Proceedings, Psychology in the Department of Defense 13th Symposium* (pp. 217–220). Colorado Springs, CO: USAF Academy.

Bastian, L. D., Lancaster, A. R., & Reyst, H. E. (1996). *Department of Defense 1995 sexual harassment survey.* Arlington, VA: Defense Manpower Data Center.

Bauer, R. G., Stout, R. L., & Holz, R. F. (1976). *Predicting military delinquency* (Research Problem Review 76–4). Alexandria, VA: U.S. Army Research Institute for the Behavioral and Social Sciences.

Bell, D. B., & Holz, R. F. (1975). *Summary of ARI research on military delinquency* (Research Report 1185). Alexandria, VA: U.S. Army Research Institute for the Behavioral and Social Sciences.

Binkin, M., & Bach, S. J. (1977). *Women and the military.* Washington, DC: The Brookings Institution.

Binkin, M., & Eitelberg, M. J. (1982). *Blacks and the military.* Washington, DC: The Brookings Institution.

Culbertson, A. L., & Magnusson, P. (1992). *An investigation into equity in Navy discipline* (TR-92-17). San Diego, CA: Navy Personnel Research and Development Center.

Culbertson, A. L., & Rosenfeld, P. (1993). Understanding sexual harassment through organizational surveys. In P. Rosenfeld, J. E. Edwards, and M. D. Thomas (Eds.), *Improving Organizational Surveys* (pp. 164–187). Newbury Park, CA: Sage.

Culbertson, A. L., & Rosenfeld, P. (1994). Assessment of sexual harassment in the active-duty Navy. *Military Psychology, 6,* 69–93.

Dansby, M. R. (1989). Military retention by gender, racial/ethnic group, and personnel category. *Proceedings, 31st Annual Conference of the Military Testing Association* (pp. 744–749). San Antonio, TX: Flamingo.

Dansby, M. R. (1992). *Racial disparities in military incarceration rates: An overview and research strategy* (DEOMI RSP 92–3). Patrick Air Force Base, FL: Defense Equal Opportunity Management Institute.

Dansby, M. R. (1994a, December). The Military Equal Opportunity Climate Survey (MEOCS). Presentation to the Equal Opportunity Research Symposium, Cocoa Beach, FL.

Dansby, M. R. (1994b). Revising the MEOCS: A methodology for updating the Military Equal Opportunity Climate Survey. *Proceedings, Applied Behavioral Sciences 14th Symposium* (pp. 59–64). Colorado Springs, CO: USAF Academy.

Dansby, M. R. (1995). *Using Military Equal Opportunity Climate Surveys for organizational development* (DEOMI RSP 95–11). Patrick Air Force Base, FL: Defense Equal Opportunity Management Institute.

Dansby, M. R. (1996). The Senior Leader Equal Opportunity Climate Survey: What do the bosses believe? *Proceedings, Applied Behavioral Sciences 15th Symposium* (pp. 310–316). Colorado Springs, CO: USAF Academy.

Dansby, M. R., & Landis, D. (1991). Measuring equal opportunity climate in the military environment. *International Journal of Intercultural Relations, 15*, 389–405.

Dansby, M. R., & Landis, D. (1996). Intercultural training in the military. In D. Landis & R. S. Bhagat (Eds.), *Handbook of Intercultural Training, Second Edition* (pp. 203–215). Thousand Oaks, CA: Sage.

Dansby, M. R., & Landis, D. (in process). Race, gender, and representation index as predictors of equal opportunity climate in military organizations. *Journal of Military Psychology.*

Day, H. R. (1983). Race relations training in the U.S. military. In Landis, D., & Brislin, R. W. (eds.), *Handbook of intercultural training (volume II): Issues in training methodology* (pp. 241–289). New York: Pergamon Press.

DEOMI. (1996). *Semi-annual race/ethnic/gender profile of the Department of Defense forces (active and reserve), the United States Coast Guard, and Department of Defense civilians* (DEOMI SSP 96–4). Patrick Air Force Base, FL: Defense Equal Opportunity Management Institute.

Dunivin, K. O. (1990). Gender identity among Air Force female aviators. *Proceedings, Psychology in the Department of Defense 12th Symposium* (pp. 71–76). Colorado Springs, CO: USAF Academy.

Eisaguirre, L. (1993). *Sexual harassment: A reference handbook.* Denver, CO: ABC-CLIO.

Eitelberg, M. J. (1988). *Manpower for military occupations.* Alexandria, VA: Human Resources Research Organization.

Eitelberg, M. J., Laurence, J. H., & Waters, L. S. (1984). *Screening for service: Aptitude and education criteria for military entry.* Alexandria, VA: Human Resources Research Organization.

Fain, T. C., & Anderton, D. L. (1987). Sexual harassment: Organizational context and diffuse status. *Sex Roles, 5* (291–311).

Farley, L. (1978). *Sexual shakedown.* New York: McGraw-Hill.

Firestone, J. M., & Stewart, J. B. (1995). Trends in gender and racial equity in retention and promotion of officers. *Proceedings: Equal Opportunity Research Symposium* (pp. 43–49). Patrick Air Force Base, FL: Defense Equal Opportunity Management Institute.

Flyer, E. (1990). *Characteristics and behavior of recruits entering military service with an offense history.* Arlington, VA: Defense Manpower Data Center.

Flyer, E. (1993, August). Inservice criminal behavior: Another measure of adjustment to military life. Paper presented at the 101st Annual Convention of the American Psychological Association, Toronto, Canada.

Flyer, E., & Curran, C. (1993, August). Relationships between preservice offense histories and unsuitability discharge. Paper presented at the 101st Annual Convention of the American Psychological Association, Toronto, Canada.

Goldman, J. L. (1995). *The issue is . . . sexual harassment* (DEOMI STSP 95–1). Patrick Air Force Base, FL: Defense Equal Opportunity Management Institute.

Greene, E. L., & Culbertson, A. L. (1995). Officer fitness report ratings: Using quantitative and qualitative methods to examine potential bias. *Proceedings: Equal Opportunity Research Symposium* (pp. 31–41). Patrick Air Force Base, FL: Defense Equal Opportunity Management Institute.

Hendricks, W. H., & Gibson, G. S. (1992). Gender differences: Stress effects on general well-being and absenteeism. *Proceedings, Psychology in the Department of Defense 13th Symposium* (pp. 202–206). Colorado Springs, CO: USAF Academy.

Holm, J. (1992). *Women in the military: An unfinished revolution*. Novato, CA: Presidio.

Hope, R. O. (1979). *Racial strife in the U.S. military: Toward the elimination of discrimination*. New York: Praeger.

Johnson, J. L. (1995). *A preliminary investigation into DEOMI training effectiveness* (DEOMI RSP 95–8). Patrick Air Force Base, FL: Defense Equal Opportunity Management Institute.

Johnson, J. L. (1996). *Local effects and global impact of DEOMI training* (DEOMI RSP 96–7). Patrick Air Force Base, FL: Defense Equal Opportunity Management Institute.

King, R. E., McGlohn, S. E., & Retzlaff, P. D. (1996). Assessment of psychological factors in female and male United States Air Force pilots. *Proceedings, Applied Behavioral Sciences 15th Symposium* (pp. 1–5). Colorado Springs, CO: USAF Academy.

Knouse, S. B. (1993). *Differences between black and white military offenders: A study of socioeconomic, familial, personality, and military characteristics of inmates at the United States Disciplinary Barracks at Fort Leavenworth* (DEOMI RSP 93–2). Patrick Air Force Base, FL: Defense Equal Opportunity Management Institute.

Knouse, S. B. (1994a). *Preliminary development of the military socialization inoculator: A means of reducing discipline problems by early socialization into appropriate military behaviors* (DEOMI RSP 94–9). Patrick Air Force Base, FL: Defense Equal Opportunity Management Institute.

Knouse, S. B. (1994b). *Equal opportunity climate and Total Quality Management: A preliminary study* (DEOMI RSP 94–3). Patrick Air Force Base, FL: Defense Equal Opportunity Management Institute.

Knouse, S. B. (1996a). *Recent diversity research at the Defense Equal Opportunity Management Institute (DEOMI): 1992–1996* (DEOMI RSP 96–14). Patrick Air Force Base, FL: Defense Equal Opportunity Management Institute.

Knouse, S. B. (1996b). *Diversity, organizational factors, group effectiveness, and total quality: An analysis of relationships in the MEOCS-EEO test version 3.1* (DEOMI RSP 96–6). Patrick Air Force Base, FL: Defense Equal Opportunity Management Institute.

Knouse, S. B., & Dansby, M. R. (1994). The organizational socialization inoculator. *Proceedings of the Southwest Decision Sciences Institute, 25*, 35–39.

Landis, D., & Dansby, M. R. (1994). *Race and the military justice system: Design for a program of action research* (DEOMI RSP 94–3). Patrick Air Force Base, FL: Defense Equal Opportunity Management Institute.

Landis, D., Dansby, M. R., & Faley, R. H. (1993). The Military Equal Opportunity Climate Survey: An example of surveying in organizations. In P. Rosenfeld, J. E. Edwards, and M. D. Thomas (Eds.), *Improving Organizational Surveys.* (pp. 210–239). Newbury Park, CA: Sage.

Landis, D., Dansby, M. R., & Faley, R. H. (1994). The relationship of equal opportunity climate to military career commitment: An analysis of individual differences using latent variables. *Proceedings, Applied Behavioral Sciences 14th Symposium* (pp. 65–70). Colorado Springs, CO: USAF Academy.

Landis, D., Dansby, M. R., & Tallarigo, R. S. (1996). The use of equal opportunity climate in intercultural training. In D. Landis & R. S. Bhagat (Eds.), *Handbook of Intercultural Training, Second Edition* (pp. 244–263). Thousand Oaks, CA: Sage.

MacGregor, M. J. (1981). *Integration of the Armed Forces, 1940–1965.* Washington, DC: Center of Military History, United States Army.

MacKinnon, C. (1979). *Sexual harassment of working women.* New Haven, CT: Yale University.

Martindale, M. (1990). *Sexual harassment in the military: 1988.* Arlington, VA: Defense Manpower Data Center.

Matthews, M. D. (1992). Women in the military: Comparison of attitudes and knowledge of service academy cadets versus private college students. *Proceedings, Psychology in the Department of Defense 13th Symposium* (pp. 212–216). Colorado Springs, CO: USAF Academy.

Matthews, M. D., & Weaver, C. N. (1990). Demographic and attitudinal correlates of women's role in the military. *Proceedings, Psychology in the Department of Defense 12th Symposium* (pp. 77–81). Colorado Springs, CO: USAF Academy.

McIntyre, R. M. (1995). *Examination of the psychometric properties of the Senior Leader Equal Opportunity Climate Survey: Equal opportunity perceptions* (DEOMI RSP 95–6). Patrick Air Force Base, FL: Defense Equal Opportunity Management Institute.

McIntyre, R. M., Albright, R. R., & Dansby, M. R. (1996). *The development and construct validation of the Small Unit Equal Opportunity Climate Survey* (DEOMI RSP 96–8). Patrick Air Force Base, FL: Defense Equal Opportunity Management Institute.

Moskos, C. C., & Butler, J. S. (1996*). All that we can be: Black leadership and racial integration in the Army.* New York: Basic Books.

Nalty, B. C. (1986). *Strength for the fight: A history of black Americans in the military.* New York: The Free Press.

Newell, C. E., Rosenfeld, P., & Culbertson, A. L. (1995). Sexual harassment experiences and equal opportunity perceptions of Navy women. *Sex Roles, 32,* 159–168.

Niebuhr, R. E. (1994a). *Measuring equal opportunity climate in organizations: Development of scales to evaluate the acceptance of diversity* (DEOMI RSP 94–5). Patrick Air Force Base, FL: Defense Equal Opportunity Management Institute.

Niebuhr, R. E. (1994b). *The relationship between organizational characteristics and sexual harassment.* Patrick Air Force Base, FL: Defense Equal Opportunity Management Institute.

Niebuhr, R. E. (1997). Sexual harassment in the military. In W. O'Donohue (Ed.), *Sexual harassment: Theory, research, and treatment.* Boston: Allyn & Bacon.

Niebuhr, R. E., & Oswald, S. L. (1992a). The impact of workgroup composition and other work unit/victim characteristics on perceptions of sexual harassment. *Applied H. R. M. Research, 3,* 30–47.

Niebuhr, R. E., & Oswald, S. L. (1992b, April). The influence of workgroup composition on sexual harassment among military personnel. Paper presented at the Psychology in the Department of Defense 13th Symposium, Colorado Springs, CO.

Niebuhr, R. E., Knouse, S. B., & Dansby, M. R. (1994). *Workgroup climates for acceptance of diversity: Relationship to group cohesiveness and performance* (DEOMI RSP 94–4). Patrick Air Force Base, FL: Defense Equal Opportunity Management Institute.

Niebuhr, R. E., Knouse, S. B., Dansby, M. R., & Niebuhr, K. E. (1996). The relationship between racism/sexism and group cohesiveness and performance. *Proceedings, Applied Behavioral Sciences 15th Symposium* (pp. 322–327). Colorado Springs, CO: USAF Academy.

Nieva, V. F. (1991). *Youth Attitude Tracking Study: Propensity report (final).* Rockville, MD: Westat, Inc.

Nordlie, P. G., Sevilla, E. R., Edmonds, W. S., & White, S. J. (1979). *A study of racial factors in the Army's justice and discharge system.* McLean, VA: Human Sciences Research.

Pryor, J. B. (1985). The lay person's understanding of sexual harassment. *Sex Roles, 13,* 273–286.

Ramsberger, P. F. (1993). *Influences on the military enlistment decision-making process: Findings from the 1991 Youth Attitude Tracking Study* (DMDC 93–004). Arlington, VA: Defense Manpower Data Center.

Rosenfeld, P., & Culbertson, A. L. (1993, August). Assessing equal opportunity climate: Integrating Navy-wide and command-level approaches. Paper presented at the 101st Annual Convention of the American Psychological Association, Toronto, Canada.

Rosenfeld, P., & Edwards, J. E. (1994, September). Automated system assesses equal opportunity. *Personnel Journal,* 99–101.

Rosenfeld, P., Culbertson, A. L., Booth-Kewley, S., & Magnusson, P. (1992). *Assessment of equal opportunity climate: Results of the 1989 Navy-wide survey* (NPRDC TR-92-14). San Diego, CA: Navy Personnel Research and Development Center.

Rosenfeld, P., Thomas, M. D., Edwards, J. E., Thomas, P. J., & Thomas, E. D. (1991). Navy research into race, ethnicity, and gender issues: A historical review. *International Journal of Intercultural Relations, 15*, 407–426.

Schissel, B. L. (1990). Factors affecting career decisions of Air Force female aviators. *Proceedings, Psychology in the Department of Defense 12th Symposium* (pp. 162–166). Colorado Springs, CO: USAF Academy.

Segal, D. R., & Bachman, J. G. (1994). Change in the all-volunteer force: Reflections in youth attitudes. In Eitelberg, M. J., & Mehay, S. J. (Eds.), *Marching toward the 21st century: Military manpower and recruiting* (pp. 149–166). Westport, CT: Greenwood.

Sellman, W. S., & Valentine, L. D. (1981, August). Aptitude testing, enlistment standards, and recruit quality. Paper presented at the 89th Annual Convention of the American Psychological Association, Los Angeles, CA.

Shields, P. M. (1988). Sex roles in the military. In Moskos, C. C., & Wood, F. R., *The military: More than just a job?* (pp. 99–113). New York: Pergamon-Brassey's.

Smither, R. D., & Houston, M. R. (1991). Racial discrimination and forms of redress in the military. *International Journal of Intercultural Relations, 15*, 459–468.

St. Pierre, M. (1991). Accession and retention of minorities: Implications for the future. *International Journal of Intercultural Relations, 15*, 469–489.

Stokan, L. A., & Hah, S. (1996). Psychological androgyny and its relationship to leadership grades of cadets at the United States Military Academy. *Proceedings, Applied Behavioral Sciences 15th Symposium* (pp. 97–102). Colorado Springs, CO: USAF Academy.

Tallarigo, R. S., & Landis, D. (1995). *Organizational distance scaling: Exploring climates across organizations* (DEOMI RSP 95–13). Patrick Air Force Base, FL: Defense Equal Opportunity Management Institute.

Thomas, J. A. (1988*). Race relations research in the U.S. Army in the 1970s: A collection of selected readings*. Alexandria, VA: U.S. Army Research Institute for the Behavioral and Social Sciences.

Thomas, M. D. (1995). *Gender differences in conceptualizing sexual harassment* (NPRDC TR–95–5). San Diego, CA: Navy Personnel Research and Development Center.

Thomas, P. J., & Le, S. K. (1996). *Sexual harassment in the Marine Corps: Results of a 1994 survey* (NPRDC TN-96-44). San Diego, CA: Navy Personnel Research and Development Center.

Thomas, P. J., & Thomas, M. D. (1992). *Impact of pregnant women and single parents upon Navy personnel systems* (NPRDC TN-92-8). San Diego, CA: Navy Personnel Research and Development Center.

Thomas, P. J., & Thomas, M. D. (1993a). Mothers in uniform. In F. Kaslow (Ed.), *The military family in peace and war.* (pp. 25–47). New York: Springer.

Thomas, P. J., & Thomas, M. D. (1993b). Surveying pregnancy and single parenthood: The Navy experience. In P. Rosenfeld, J. E. Edwards, and M. D. Thomas (Eds.), *Improving Organizational Surveys.* (pp. 145–163). Newbury Park, CA: Sage.

Thomas, P. J., Holmes, B. L., & Carroll, L. (1983). *Gender differences in the evaluations of narratives in officer performance ratings* (NPRDC TR 83-14). San Diego, CA: Navy Personnel Research and Development Center.

Thomas, P. J., Thomas, M. D., & Robertson, P. (1993). *Absences of Navy enlisted personnel: A search for gender differences* (NPRDC TN-93-3). San Diego, CA: Navy Personnel Research and Development Center.

Thomas, R. R. (1991). *Beyond race and gender*. New York: AMACOM.

United States General Accounting Office. (1993). *DoD's policy on homosexuality: Report to Congressional requesters on defense force management* (GAO/NSIAD -92-98). Washington, DC: United States General Accounting Office.

U.S. Merit Systems Protection Board. (1981). *Sexual harassment in the federal government: An update*. Washington, DC: U.S. Government Printing Office.

U.S. Merit Systems Protection Board. (1988). *Sexual harassment in the federal workplace: Is it a problem?* Washington, DC: U.S. Government Printing Office.

Verdugo, N. (1994, December). Research on the Army's criminal justice system. Presentation to the Equal Opportunity Research Symposium, Cocoa Beach, FL.

Walker, M. R. (1992). *An analysis of discipline rates among racial/ethnic groups in the U.S. military: Fiscal years 1987–1991* (DEOMI RSP 92-4). Patrick Air Force Base, FL: Defense Equal Opportunity Management Institute.

Endnotes

[1] The views expressed in this chapter are those of the author and do not necessarily reflect the views of the Department of Defense or any of its agencies.

[2] Indeed, in the foreword to Dr. Thomas' book, the Technical Director of ARI may have presaged this view when he indicated, "Dr. Thomas provides a proven model for future research, *should it be needed*." (emphasis mine)

[3] Adapted from U.S. Merit Systems Protection Board, 1981, 1988.

The National Naval Medical Center in Bethesda, Maryland. Credit: Globe Photos, Inc.

10

Military Psychologists: Training and Practice

W. Brad Johnson, Ph.D.

* Introduction * Training * Application and Selection * Indoctrination
* Structure of Military Psychology Internships * Unique Training
Opportunities * Unique Stresses of the Military Internship * Postdoctoral
Training * Practice * Common Practice Components for Military
Psychologists * Range of Duty Assignments for Military Psychologists
* Drawbacks to Practicing in the Military * Chapter Summary * References

Introduction

American psychology has traditionally enjoyed an excellent relationship with the military (Allen, Chatelier, Clark, & Sorenson, 1982; Mullins, 1992). Although psychologists practicing in military settings perform functions which closely parallel those of civilian psychologists, such as assessment, biofeedback, consultation and psychotherapy, several components of military training and practice are clearly unique to the military environment.

First, the military is highly mission oriented. That mission is to defend the country and to be able to fight and win a war if necessary. Everything that does not contribute to this mission is considered secondary in importance (Driskell & Olmstead, 1989). For a psychologist in the military, the overriding goal must be the application of behavioral science knowledge and principles as well as research techniques to aid the armed services in achieving and maintaining advantage over any enemy. In keeping with this mission, the military psychologist enjoys considerable influence over the working environment and living situation of military personnel the psychologist serves. Such control of the client's "total environment" is unheard of in civilian outpatient practice.

Second, the military environment is often more serious than civilian counterparts as military personnel must be prepared to fight for their country and die in support of its defense (Driskell & Olmstead, 1989; O'Hearn, 1991). Military personnel may be sent into hostile situations with minimal warning, thus it is essential that they meet the highest level of physical and psychological health and fitness. The consequences of inadequate selection and classification procedures, poor training or bad equipment design are obviously very serious.

Training of psychologists in the military is unique for several reasons. Rath and Norton (1991) noted three unique features of military preparation of psychologists: (a) the military only trains those psychologists it plans to retain for further duties in the military, (b) military psychologists must be prepared to function both independently as a solo practitioner and as a member of a larger team, and (c) psychologists trained for military duty must be taught broad, basic skills in assessment and treatment rather than rapidly becoming specialists. It is also true that the psychological services most likely to be effective in combat situations are quite different from the kinds of treatment traditionally offered by psychologists. Military psychologists must be highly proficient with both traditional clinical skills and very unique combat related psychological approaches.

post-doctoral

This chapter will highlight the process of training psychologists for military service with particular focus on internship training in the military, as well as post-doctoral training opportunities. The unique features of practicing psychology as a clinical or counseling psychologist in uniform will also be explored. Military psychologists enjoy some of the most exciting and challenging practice opportunities to be found in the profession. They also face some difficult problems in balancing their roles as military officer and practicing psychologist.

Training

To become a psychologist in the military, one must complete a four year college degree and
Doctor of Psychology
then gain admission to a civilian Doctor of Psychology (Psy.D.) or Doctor of Philosophy (Ph.D.) program in clinical or counseling psychology in a professional school of psychology or university based doctoral program. These programs require a total of four (Psy.D.) or five
internship
(Ph.D.) years to complete. At this time, the military does not subsidize graduate training for psychologists until the final year of the doctoral program which is the formal internship year. In
clinical training
the three or four years prior to the internship, the psychologist in training attempts to complete all required course work and clinical training as well as a doctoral dissertation (major research
doctoral
project). During this year, the student applies for a formal clinical internship in psychology.
dissertation

The internship is an important milestone in the development of the clinical psychologist. During the internship year, the psychology graduate student is "socialized" into the profession and prepared to function as an independent psychologist (Sturgis, Verstegen, Randolph, & Garvin, 1980). The military currently maintains nine internship training sites fully approved by the American Psychological Association (APA). When applying for a military internship, the psychology graduate student has his or her first interactions with medical officer recruiters and the directors of training at the sites themselves. Although little has been written about military internships (Johnson & Wilson, 1993), this chapter will outline the process of
indoctrination
application and selection to military internships, the indoctrination process, the structure of training in military settings, the unique opportunities in military internships and the stressors common of training in the military.

Application and Selection

Before deciding to apply to a military internship, it is important for the student to clarify long and short term professional goals, core values, geographic preferences and the potential for a
pay-back
good fit within the military environment (Brill, Wolkin, & McKeel, 1985). Currently, pay-back

commitment following the year of internship is three years. Obviously, the student must be willing to commit to practice as a military psychologist for a minimum of three years. Research has suggested that most graduate students look for factors such as APA approval, desirable geographic location, diversity in training and a range of theoretical orientations among faculty when selecting an internship (Tedesco, 1979). Each of these factors is present in military settings with the added enticements of a substantial salary and full military benefits for the intern and family members (Eggert, Laughlin, Hutzell, Stedman, Solway & Carrington, 1987).

APA approval

The selection process for military internships is more taxing than that for civilian sites. A committee of high ranking psychologists within the specific military service meets to review application packages which typically include: detailed application forms, physical examination results, undergraduate and graduate transcripts, letters of recommendation from professors, written evaluations by those who have interviewed the applicant, and a personal statement from the applicant regarding serving in the military. Military selection committees will, like other committees (Tipton, Watkins, & Ritz, 1991) value excellent clinical training, solid research experience and high performance academically in graduate school. In addition, they will assess potential for service as a military officer by evaluating factors such as, "proper military bearing" confidence, and political views not obviously incongruent with military service. Those intern applicants with the greatest potential for selection are those who have a wide range of clinical experiences and appear likely to become good "generalist" psychologists. This means they will likely be able to function well in a wide range of different circumstances with nearly any kind of client and client problem.

generalist psychologists

Indoctrination

Officer indoctrination occurs in the summer preceding the internship (Air Force and Navy) or in the summer following the internship year (Army). Although the intern is a fully commissioned 0-3 officer before officer indoctrination, indoctrination training marks his or her first real contact with the military system. Officer orientation or indoctrination courses range from two and one half (Air Force) to fourteen (Army) weeks. Naval officer Indoctrination School in Newport Rhode Islands lasts for six weeks. The purpose of these schools is to orient new psychologist officers and to focus their attention on their new roles in the military (Rath & Norton, 1991). The indoctrination experience consists of learning to wear the uniform and understand military structure and protocol. There is a good deal of physical training and many hours spent in lectures on topics ranging from military law to combat casualty management. Frequent examinations, inoculations, watch standing and training unique to the specific military branch should also be anticipated. When indoctrination precedes the internship, it serves to assist the psychology graduate student in managing the very significant adjustment caused by leaving graduate school and entering the very foreign world of a large military medical center as an officer.

inoculations

Structure of Military Psychology Internships

Currently, the Army, Air-Force and Navy each run three large and fully accredited internships in psychology. The Army internships are located at Walter Reed Army Medical Center in Washington DC, Dwight D. Eisenhower Army Medical Center at Fort Gordon (Georgia) and

Tripler Army Medical Center (Honolulu, Hawaii). Air Force internships are located at Malcom Grow Medical Center at Andrews Air Force Base (Maryland), Wilford Hall Medical Center at Lackland Air Force Base (San Antonio, Texas), and Wright-Patterson Medical Center at Wright-Patterson Air Force Base (Dayton, Ohio). The Navy internships are located at the National Naval Medical Center (Bethesda, Maryland), Portsmouth Naval Hospital (Portsmouth, Virginia) and San Diego Naval Hospital (San Diego, California).

Each military internship is designed to teach basic clinical and professional skills and to help the intern learn how to apply those skills effectively in a military environment (Rath & Norton, 1991). Experience on the part of military psychologists has demonstrated that military intern graduates typically report to billets requiring a higher level of practice than their civilian counterparts. As a result of the great reliability and accountability demanded in the military environment, the professional skills of military interns are more closely tracked. This means that training is often experienced by military interns as quite rigorous and intensive. Although each of the internship training sites is unique in structure, faculty expertise and possible training experiences for interns, all of them share certain common features. Each internship is attached to an independent department of psychology within a major regional medical center. Each places great emphasis on thorough generalist training as graduates are likely to be stationed in "solo" duty stations immediately following graduation. This is particularly true for Air Force psychologists who often serve as the only mental health professional on small and comparatively isolated installations. In addition to broad clinical skills, each site also places particular emphasis on practice skills critical to the military
critical environment. These include consultation with military commands, combat psychology,
incident critical incident stress debriefing following traumatic events, suicide prevention and treatment of post-traumatic stress disorder. In addition to their role as psychologist in training, military
stress psychology interns are also required to carry out various administrative duties common of military medical officers. These include rotation on the administrative officer watch bill and
debriefing serving as a member of medical boards which evaluate service members for fitness for duty.

rotation Each of the internships is quite similar in operating with a rotation structure in which interns rotate between several distinct training moduals. Each site offers several standard rotations (typically three to four months each) and most offer several special elective rotations of shorter duration. The standard rotations include outpatient, inpatient, behavioral medicine/ health psychology and assessment. The outpatient rotation focuses on rapid evaluation and brief treatment of patients with a wide variety of presenting problems. On the medical psychology rotation, interns respond to consults from other services within the hospital, such as neurology, oncology, dentistry, anesthesia, endocrinology and internal medicine. These consults request behavioral medicine diagnosis, treatment and/or recommendations regarding fitness for duty for patients with a wide variety of psychophysiological problems such as headaches, back pain or other physical discomfort which seems to stem from primarily psychological causes. On inpatient rotations, interns participate in and are responsible for admission, diagnosis and treatment of patients with severe psychological disorders. As part of the multidisciplinary team, interns are direct providers and carry clinical loads and responsibilities similar to those of psychiatry residents. During the psychological assessment
psychological rotation, interns respond to requests from around the medical center for psychological
evaluations evaluations of various types of clients. In addition to detailed interviews, interns administer full

batteries of psychological tests and integrate this data in concise written psychological reports which assist the referral source in understanding the patient's diagnosis, psychological structure and treatment needs. Specific sites may also offer rotations in child psychology, neuropsychology, community/military psychology, clinical research, alcohol rehabilitation and family practice.

In addition to these major rotations, military interns also have several year long or "cross-rotational" requirements. These include carrying long-term individual outpatient psychotherapy cases, long term family and/or marital counseling cases, serving as co-leader of one or more groups during the year and attending several hours each week of supervision with various psychologists on faculty. Interns also spend approximately 150 hours in required weekly seminars and additional time attending specialty workshops and seminars offered on topics relevant to their practice as psychologists and practice in the military in particular. Topics might include assessment techniques, post-traumatic stress disorder and group or child psychotherapy. Finally, interns undergo a limited amount of "operational training" during the year in which they may visit military installations similar to those on which they might work as psychologists. These experiences are meant to further familiarize the intern with the various military communities, the needs of commanding officers for consultation and politics and structure of life in the military.

cross-rotational requirements

operational training

Unique Training Opportunities

Military internships appear to be among the most comprehensive and high caliber internships offered anywhere in the United States (Johnson & Wilson, 1993). There are several reasons for this. Stipends or pay for military internships are among the highest in the country (Kurz, Fuchs, Dabek, Kurtz, & Helfrich, 1982) and the pay and numerous military benefits often allow the intern in the military to live quite comfortably following a very difficult period of graduate training.

"Adventure" is another important feature of training in the military. Unlike interns in other settings, military psychology interns are routinely exposed to a wide range of operational training and make frequent visits to combatant communities ranging from submarines to fighter-bomber squadrons. Experiences during indoctrination school and opportunities for qualification on various weapons and military specialties may also serve to keep the military intern pleasantly interested in his or her training.

Military interns also receive considerably more experience with direct consultation than do many other interns (Kurz et al., 1982). Interns offer direct consultation to referral sources (physicians, psychologists, social workers, chaplains, etc.) and member's commanding officers regarding fitness for duty and potential for long-term service to the military as well as deployability and need for any ongoing psychological services. Interns are frequently called to testify in military court proceedings and may be asked to assist with evaluations of a command's environment and/or morale from a psychological perspective. Also, interns often participate actively in selection of service members for highly sensitive roles in the military. For example, in the Navy, interns will screen members who apply for service to Antarctica. Interns in all branches will conduct evaluations for security clearances when necessary.

Unique Stresses of the Military Internship

Internship training in psychology is quite stressful for interns in most settings. In the military, the intern is likely to face additional sources of stress which may serve to make the military internship considerably more taxing. Most interns report that the most stressful client behaviors include suicidal threats, reports of criminal behavior and expressions of anger or rage toward the intern. Military interns are likely to confront a large number of suicidal patients and patients reporting various illegal behaviors. These may be authentic or used by the military members to escape military service. Active duty personnel are also likely to become angry with military interns who must often return them to full duty or separate them from the service against their will.

Military interns are also likely to carry higher than average case loads and work longer hours than interns in other settings. Part of the reason for this is the need to prepare the military psychologist for the experience of "intensity" in practice likely to be faced often in practice as a military psychologist. For example, when an air-craft carrier battle group pulls into a port, it is not uncommon for a Navy mental health department (sometimes consisting of a single psychologist) to have five to ten "emergency" cases to evaluate for immediate disposition and care. Military interns may also be aware of a great deal of scrutiny and evaluation by supervisors who are tasked with carefully evaluating the intern's performance as a psychologist under stressful circumstances as well as determining the extent to which the intern functions effectively as an officer in the military. Many military interns also experience some degree of loss of identity and social support during the internship year. Although successful in graduate school, interns have often relocated, left comfortable surroundings behind and been confronted with the demands of a new identity as a military officer while trying to perform well under very stressful training conditions. Toward the end of the internship year, interns within each branch of the military will be assigned to military medical facilities and installations for their three-year period of pay-back. Given the wide range of differences in geography and types of practice offered by different billets, interns may feel strongly about where they should be stationed and, like other military personnel, will have to ultimately defer to the needs of the military.

Military interns must also consider the implications of an armed conflict for their training experience. While psychologists at the intern stage are rarely deployed to combat areas, this remains a possibility. Immediately following graduation of the internship, and particularly after receiving state licensure (typically one year following graduation), the recent graduate is fully eligible for deployment anywhere in the world. Even in the case of internship interruption, however, the military would be obligated to allow the intern to complete training at the earliest time feasible.

Finally, it is important to briefly consider the implications of failing a military internship in psychology. Given that military internships are uniquely demanding personally and professionally, and that interns are simultaneously evaluated on the basis of psychological and military skills, it is not surprising that military interns may more frequently be placed on probation and perhaps even excused from the internship (Johnson & Wilson, 1993). Dismissal from a military internship carries unique and often critically negative career implications for psychologists in training. Most significantly, the ex-intern (like any other military member) remains open to scrutiny from prospective employers who may question the nature of his or her military discharge. Also, as commissioned officers, ex-interns may be required to remain on active duty for some time following dismissal during which they may be tasked with administrative duties unrelated to their training in psychology. Finally, failure of an internship may cause the ex-intern to be placed in a probationary status, or perhaps even excused from his or her graduate program in psychology altogether.

Postdoctoral Training

Following graduation from the internship, and if the dissertation project has been completed, the military intern typically is awarded the doctoral degree (Psy.D. or Ph.D.) by his or her graduate school. They will then be assigned to an initial clinical billet for a period of three years. Whether the psychologist plans to exit the military following this initial commitment or continue on for another tour or for a career, there will be requirements for various kinds of continuing training as a psychologist. Military psychologists often need special training at specific times in their careers (Rath and Norton, 1991). These might include assignment to a new billet with a different client population or which requires new clinical skills. There may also be times when a psychologist must become familiar with a new area of clinical concern (ex. cognitive changes associated with progression of HIV or the treatment needs of children of service members stationed overseas). Additionally, a psychologist is required by state laws to periodically upgrade his or her clinical skills through continuing education.

To address these training needs, military psychologists are often invited to take part in operational training within the military. Developing various military proficiencies such as sea or air going qualifications make the psychologist better able to serve unique military units. Specific educational programs are also made available within the military for psychologists. These might include training in managing casualties, critical-incident stress debriefing and leadership within the military (a requirement for promotion to certain ranks). Like other military personnel, psychologists must compete for attendance at those "schools" necessary for promotion and improvement in clinical skill. It is also routine for military psychologists to regularly attend professional civilian conventions and conferences designed to update skills and offer leading edge information in the field. It is important for military psychologists to maintain close professional ties with the larger community of psychologists. This is important for two reasons. First, it is imperative that civilian psychologists and governing bodies understand the unique practice demands of psychologists in military settings. Second, such liaisons facilitate the military's long-standing reputation as a center for cutting-edge research and training in psychology.

Finally, clinical and counseling psychologists in the military are eligible for sub specialty or "fellowship" training if they remain on active duty beyond the initial three year commitment. Sub specialty training is referred to as "postdoc" or fellowship training. Currently, military fellowships, like internships, are one year in length and are located in major military medical centers. In addition, military psychologists, if granted a postdoctoral fellowship, may elect to apply for a fellowship in a major civilian university or medical center. During the fellowship year, psychologists participate in supervised clinical experience and graduate level course work in a specialty area within clinical psychology. Currently, the Air Force, Army and Navy offer fellowships in subspecialties including pediatric/child psychology, Aviation psychology, community psychology, Health psychology-behavior medicine, Forensic psychology and

fellowship

postdoctoral

psychopharmacology

Neuropsychology. During the past five years, several military psychologists have also been selected each year for participation in the psychopharmacology fellowship sponsored by the Department of Defense as a demonstration project. Graduates of this unique fellowship have prescription privileges similar to psychiatrists and may provide medications as well as psychological interventions for patients when needed.

Practice

Professional psychologists in the military enjoy some of the most unusual and exciting practice opportunities to be found anywhere. An Air Force psychologist may spend the morning providing counseling to several regular service member clients and the afternoon training jet pilots in strategies for achieving greater relaxation and concentration in the cockpit. A Navy psychologist might spend two days "underway" in a nuclear submarine for the purpose of consulting with the commanding officer regarding morale or stress effects onboard. An Army psychologist might accompany her battalion on a maneuver in the field and later that day serve as the primary psychologist in the emergency room at a large Army medical center. In this section, we will explore those components of practice common to most billets for clinical psychologists in the military. We will then highlight the wide variety in primary billets open to military psychologists with an emphasis on the unique nature of the job demands in each. Finally, this chapter will explore some of the difficulties associated with military practice. In several respects, the professional roles of psychologist and military officer may be incompatible, creating unique sources of stress for the military psychologist.

Common Practice Components for Military Psychologists

A survey of military psychologists in 1984 (Mangelsdorff, 1989) found that most tended to remain on active duty as a psychologist after the initial payback period and on to retirement if they reported three experiences. Psychologists who found themselves committed to the military (and its mission), those who experienced substantial opportunity for personal accomplishment and those who believed the military offered good potential for professional development tended to remain on active duty. A review of the major components in the practice of most military clinical psychologists should show why so many of these psychologists experience significant professional development and personal accomplishment.

Whether the psychologist functions in a major medical center or an isolated operational community, it is likely that he or she will carry out certain duties common to nearly all clinical psychology billets (Rath & Norton, 1991). Most military psychologists will conduct both routine and emergency psychological evaluations for a mental health unit or directly at the request of a commanding officer in the field. There can be a wide range of reasons for routine psychological evaluations. Most commonly, an active duty or other eligible service member or dependent family member will simply request mental health care, leading them to the facility's mental health unit and the clinical psychologist. Common reasons for self-referral among military personnel include problems adjusting to life circumstances (change of job, end of relationship, etc.), problems with depression or anxiety, problems with relationships, as well as interest in personal growth or development. For example, a Senior Chief in the Navy who has completed 20 years on active duty and is suddenly confronted with the probability of forced retirement within three years. He experiences some anxiety (i.e. sleep disturbance, problems with concentration, and stomach upset) as he thinks more and more about this. He wonders if he can "make it" outside the military, whether he will have any marketable job skills and where he will finally settle after moving constantly for over 20 years. He also has some fear about how his marriage will withstand such a radical shift. The Senior Chief will be seen by a psychologist who may diagnose him as having a very common "Adjustment Disorder" in response to a predictable life transition. Most likely, the psychologist will meet with the service member over several weeks or several months time to explore his experiences and concerns. The Chief will be assisted with "normalizing" his experience and reducing his distress by modifying the way he thinks about or evaluates these circumstances. He may be encouraged to explore his fears about retirement. Marital counseling and vocational/job placement testing might also be helpful. If the Senior Chief's symptoms worsen and he becomes significantly depressed, the psychologist might refer him for a medication consultation and a trial of antidepressant medicine may be started.

    In contrast to those who "self-refer" for psychological services, those active duty members who are referred by other sources for an evaluation typically present with more serious symptoms or life difficulties which may require more intensive or administrative intervention on the part of the psychologist. When a psychologist receives a referral from a unit commanding officer, the service member has typically demonstrated some problem behavior publicly. The member may present with rather serious symptoms of a psychiatric disorder, may be "acting-out" (refusing to comply with orders, assaulting others, etc.) or may be performing so poorly on the job, the commanding officer has reason to suspect a problem with intelligence, substance abuse or emotional difficulty.

    In such cases, the military psychologist is tasked with making an accurate diagnosis of the member's condition and recommending appropriate treatment and/or administrative disposition. Most psychologists in military settings have become quite familiar with the two primary recommendations on the part of a psychologist which may cause a service member to be discharged from military service. The first is the recommendation for administrative separation. Active duty personnel may be separated from service for a variety of conditions which may come to the attention of a clinical psychologist. For example, sleep-walking, bed-wetting, obesity, motion sickness or evidence of fraudulent enlistment (providing false information at the time of entry into the military) are all just cause for separation from military service. Each of these may first come to light during a thorough psychological evaluation. The psychologist must then consider whether such a condition warrants separation from the military and the potential costs or hazards to the member and the military if he or she remains on active duty.

    The most frequent cause for a military psychologist recommending administrative separation is the existence of a personality disorder. A personality disorder is a chronic pattern of behavior beginning early in adulthood which interferes with social and/or occupational functioning. For example, a service member with an Antisocial Personality Disorder will likely have a long history of delinquency as an adolescent. As an adult, he or she will be manipulative, exploit others, rebel against authority and be generally irresponsible. Obviously, such a pattern of behavior would make the service member a poor match for military service when success in the military depends on traits such as ability to follow orders,

Adjustment Disorder

normalizing

antidepressant

administrative separation

personality disorder

honesty and a cooperative approach to tasks. When a psychologist determines that a service member has a personality disorder and that this disorder is interfering with performance of duty, a decision to administratively separate the member from the service may be made. In those cases where the disorder is severe, the service member may be expeditously separated. Separation of a service member from duty may be quite unpleasant for the psychologist, particularly when the service member is opposed to separation. Nonetheless, commanding officers will depend on psychologists to separate those members likely to pose a long-term risk to the functioning or safety of their units.

Medical
Board

The second type of recommendation which a military psychologist might make which will likely lead to termination from service is a recommendation for a Medical Board. When a service member presents with a serious psychiatric disorder such as Major Depression, Panic Disorder or a psychosis, attempts are first made to provide effective treatment for the disorder. This might include medications, hospitalization and intensive psychotherapy. During such treatment, the psychologist may recommend a period of limited-duty. If treatment is not effective, however, or if there is substantial risk of a reoccurrence of the disorder, a recommendation may be made for a medical discharge from active duty. In such cases the service member may be entitled to some benefits after discharge, particularly if the disorder was related to military service.

When treating a service member for a serious psychological problem, it is critical that the military psychologist remain aware of how his or her diagnosis and recommendation may permanently impact the member's career. For example, a pilot or submarine sailor may be immediately disqualified from performing their routine duties when any documentation is made of a mental health concern. Psychologists must also keep in mind that separation or medical disqualification from the military may produce new psychological problems in the service member. For example, there have been numerous cases of active duty members committing suicide during the time they are being discharged from duty (Fragala & McCaughey, 1991). This is not surprising when one considers that the career military person may suffer a profound loss of identity and career when suddenly discharged after years of service.

Military psychologists are as likely to practice in major regional medical centers as they are to practice in very isolated rural bases (Rath & Norton, 1991). In both settings, psychologists routinely engage in a wide range of professional activities. These include conducting evaluations, psychological testing, screening for various duty stations or clearances, conducting both individual and group psychotherapy and counseling, providing emergency services (assessment and diagnosis) as an on-call mental health specialist, providing educational lectures to active duty members and other health-care professionals, and consulting with various leaders regarding management and personnel concerns.

In their role as consultant, psychologists perform several critical functions for the combatant community. First, a psychologist is often asked to assist a command with individual or group personnel problems. As a consultant, the psychologist may provide treatment if indicated, recommend a separation or medical board for severe psychological problems, provide necessary education and training, or recommend structural changes or personnel reassignments within the command in order to improve functioning and reduce the presenting problem. Psychologists are also frequently tasked with performing evaluations and expert testimony for military court martials. Here the psychologist will be asked to render an opinion regarding the mental stability of a service member, his or her ability to appreciate the wrongfulness of any criminal behavior and his or her ability to participate in a legal defense.

It is also true that many military clinical psychologists participate in applied research projects. Often, psychologists will have more extensive training in scientific research than any other active-duty professional. For this reason, various commands may request assistance from local psychologists in developing and carrying out research for the purpose of evaluating the value of various programs or increasing knowledge in a unique or emerging area within the military. For example, selecting active duty members for specific roles in the military is a critical task. When members are assigned to tasks for which they are poorly matched, there is an increased chance they will be unsuccessful and perhaps endanger the success or safety of their entire unit. Recently, Navy psychologists have been involved in conducting research on the personality factors which appear to lead to successful performance on Submarines (Moes, Lall, & Johnson, 1996) and in Navy dive units (Beckman, Lall & Johnson, 1996). Moes et al. (1996) found that highly successful submarine personnel tended to be detached (reported few strong emotions and generally kept to themselves), proper (concerned with appropriate standards of conduct), and workaholic (generally place work above all else). These findings make sense when one considers that submarine personnel must work extremely long hours in confined quarters. Successful Navy divers were generally controlling, highly individual, analytical, willing to modify their environments and generally quite optimistic (Beckman et al., 1996). Again, these traits make sense when one considers frequent dangers and demands for rapid decision-making required by the diving profession. By determining those personality factors most predictive of success in a military specialty, psychologists may then contribute to selection and assignment procedures and increase the chances that new recruits will be assigned to those jobs for which they are well suited.

Military clinical psychologists are also called to frequently participate in the development of new and innovative programs in applied psychology. Leshner and DeLeon (1996) recently highlighted three initiatives or pilot programs being conducted by military psychologists. First, the DOD "LEAN" program is a three-week inpatient treatment program **LEAN** for obesity and other behavioral health disorders. Here psychologists emphasize exposure, training and commitment to healthy lifestyles, emotions, exercise, attitudes and nutrition. Second, several psychologists have developed a stress-inoculation program for military pilots. Pilots are trained in biofeedback assisted self-regulation to help them reduce stress and therefore perform more effectively in the air. Finally, a military psychologist who recently graduated from the psychopharmacology residency is serving as a consultant to a military pain clinic. Here he administers and interprets psychological tests, makes diagnoses, provides treatment, promotes health programs and prescribes medications as needed.

Range of Duty Assignments for Military Psychologists

While each of the activities of military clinical psychologists described above will undoubtedly be familiar to the psychologist serving in most clinical/medical settings, there are many billets for psychologists which afford quite different opportunities and demands. It is important to understand the wide range of duty assignments available for psychologists in the military today. In this section we will briefly highlight some of these assignments.

1. MAJOR MEDICAL CENTER

Many Psychologists are stationed in very large teaching medical centers. These centers typically have internship and residency training in psychology and numerous medical specialties. Very often there is an independent department of psychology with specialists in neuropsychology, behavioral health psychology, pediatric psychology and assessment. In these settings, psychologists will be involved in evaluation and treatment of both outpatients and inpatients within the medical center. They will frequently be asked to consult with other departments in the medical center and will typically serve emergency room watch, offering emergency evaluations as needed around the clock.

2. MENTAL HEALTH CLINICS

Psychologists on military bases without a medical center will typically function in a smaller mental health clinic with psychiatrists and social workers. In these settings, psychologists are more likely to provide outpatient assessments and treatment to members of the line community. There will typically be significant opportunity to consult directly with commanding officers and to enjoy a high level of participation with operational units to increase familiarity with the demands of such roles for service members.

3. ISOLATED BASES

Quite often, a military psychologist may find him or herself functioning as the only mental health professional for an entire military installation in a remote area. Here the psychologist must provide care for all active duty personnel and their family members. The psychologist in such situations must be broadly skilled in all areas of evaluation, treatment and consultation. Clients may range from children to elderly adults and clinical work will range from suicide emergencies to long-term counseling. The psychologist in isolated communities must work more closely with commanding officers and must be particularly careful to guard the privacy of those clients served.

4. SERVICE ACADEMIES

Military psychologists hold doctoral degrees. As a result, they are uniquely prepared to serve as instructors in the military. Currently, there are several psychologists assigned to each of the three military academies. In these settings psychologists function much like other college professors, teaching classes and advising students. There are also many additional senior service schools and post-graduate programs which utilize military psychologists as instructors. For example, the Uniform Services University of the Health Sciences (USUHS) typically has several military psychologists on its faculty. Their combined experience as scholars and service providers to the line community (military combat personnel) make psychologists particularly well suited for training and research. In addition, psychologists may serve as directors of unique training sites within the military. For example, a Navy

SEAR   psychologist heads the SEAR school in rural Maine which is tasked with training Navy flight crew members to avoid capture and resist torture if their aircraft were to go down in enemy territory.

5. PROGRAM DIRECTORS

Psychologists are often assigned to leadership of special mental health related programs in the military. At this time, psychologists are tasked with leadership of alcohol rehabilitation units, special programs for handicapped or learning disordered children and unique branch specific health programs such as suicide awareness, weight control, health and fitness and smoking cessation. In these roles, military psychologists combine their skills as program developers, administrators, personnel managers and professionals familiar with a wide range of research literature in the area of their specialty program.

6. SPECIAL FORCES

In each branch of the military, clinical psychologists are assigned to larger special forces or special operations units. In this role, the psychologist participates in selection of personnel to special forces teams based on mental health and psychological fitness for arduous and dangerous duty. Psychologists also serve as part of the support teams for these units and remain on full time stand by to provide support and debriefing after difficult missions or operations in which traumatic events occur. Psychologists with such units are frequently required to conduct critical-incident stress debriefing in which unit members work through difficult losses shortly following such events in order to reduce the probability of long-term effects.

7. MILITARY INTELLIGENCE

Several psychologists are assigned to intelligence gathering units in the various military branches. The Air Force, for example, has several psychologists assigned to the National Security Agency. In these billets, psychologists may collect data on various world leaders. This data may be of assistance in development of psychological profiles of world leaders to be used in crisis or conflict situations. Collection of psychological data with bearing on terrorist behavior may also be important as is offering reasoned assessments of the threat posed by those viewed as hostile to our country.

8. INVESTIGATIONS

Military psychologists may be assigned to investigative units of various kinds. Air Force psychologists may serve as investigators of flight accidents and may offer reports concerning the human and psychological factors related to such mishaps. A Navy psychologist is assigned to Naval Investigative Services. In this billet the psychologist serves as the primary investigator in cases where an active duty member dies under unusual or questionable circumstances. Reconstruction of the events leading to the member's death and post-mortem analysis of his or her previous psychological functioning are the primary objectives. Such billets require the military psychologist to become a forensic specialist.

9. CORRECTIONAL FACILITIES

Military psychologists may be assigned to one of several military correctional facilities. Army psychologists may be assigned to the federal penitentiary at Leavenworth Kansas. Here psychologists provide ongoing evaluation and treatment for incarcerated active duty members. Common presenting problems among inmates include difficulty with adjustment to prison life, depression, anxiety and existential or philosophical concerns related to hope and continuing with life in the face of profound distress and disappointment.

Drawbacks to Practicing Psychology in the Military

dual role

stigma

Military psychologists occupy two distinct roles as both psychologist and commissioned military officer. At times, this dual role may create conflicts for both psychologists and the active duty members they serve. In this final section, the most pressing difficulties faced by military psychologists will be examined.

In the military community, those who seek mental health care may face stigma and rejection (or at least substantial questioning and scrutiny) by their peers and superiors. This stigma creates an obvious obstacle for those who need and/or desire assistance for a behavioral or emotional problem (O'Hearn, 1991). Military officers in particular tend to avoid psychological services offered within the military. It is also well known that those in specialties which require mental health clearance generally avoid mental health care or seek it only outside the military network. While often in violation of military policy, such utilization of civilian mental health care is seen by some as the only sure way to safeguard their careers and simultaneously secure the assistance they need. For example, personnel with top secret security clearances, or those tasked with operation of nuclear power plants may believe that any contact with a military psychologist will certainly lead to questions about their stability and therefore jeopardize their careers.

Military psychologists must therefore work diligently at safeguarding the privacy and confidentiality of active duty members whenever possible while also serving the needs of the military. One of the most effective roles for the psychologist in this regard is that of educator. Reducing stigma associated with mental health care will require considerable education about the actual role of the psychologist and the potential benefits of mental health care. Offering highly publicized psychoeducational lectures, groups or focused services for common difficulties (eg. smoking-cessation, weight-control and stress-reduction) serve as excellent methods for increasing the comfort with and acceptance of psychology within the military.

One recent study of mental health stigma in the military (Porter & Johnson, 1994) found that among 138 commanding and executive officers in the Navy and Marine Corps, most reported very neutral attitudes regarding the reliability and competence of service members who had previously received psychological services. Although these findings are encouraging and suggest that psychological stigma may be declining in the service, these same senior officers felt strongly that a service member's commanding officer should have full awareness of all services received by the member. Thus military leadership is unlikely to recognize an active duty member's "right" to fully confidential mental health care.

Psychologists on active duty must maintain a very delicate balance between their simultaneous allegiance to the Department of Defense (DOD) and the profession of psychology, particularly the American Psychological Association (APA) and the APA ethical standards (APA, 1992). Military psychologists, like other active duty officers, are required to place the needs and interests of the military foremost in supporting their commanding officers and the larger military mission. At the same time, they are required to maintain licensure as professionals, which requires them to adhere to the standards and principles of the APA. Such dual allegiance sometimes creates conflicts in practice which seem to lack easy solutions (Johnson, 1995). Although the APA has encouraged military psychologists to exercise good judgment in balancing these two missions (APA, 1994), such balance is not always possible. For example, Jeffrey, Rankin and Jeffrey (1992) described two actual cases in which military psychologists were sanctioned for attempting to adhere to either military or APA guidelines. In the first case, a military psychologist was sanctioned by APA for failure to maintain the confidentiality of a service member's record long after the psychologist had been stationed elsewhere. In the second, a psychologist, adhering to the principle of confidentiality, was sanctioned by his commanding officer for failure to reveal an alleged violation of the Uniform Code of Military Justice by a third party.

confidentiality

Camp (1993) has referred to this predicament as the "double agent" status of military psychologists and psychiatrists. Camp has vividly described the ethical problems faced by battlefield psychologists during the Vietnam War. During this conflict, treatment for those with battle fatigue consisted of very brief supportive techniques which deemphasized patienthood and encouraged a very rapid return to combat. The obvious conflict for many practitioners was whether it was ethical for a psychologist or psychiatrist to encourage a soldier with symptoms of stress and trauma to return to action when doing so might very well result in death. Can a military psychologist ethically engage in such practice when the APA code of ethics (APA, 1992) requires psychologists to always serve their client's "best interest?"

double agent

ethics

Part of the difficulty with practice in the military is that the psychologist may not be clear about precisely "who" his or her client actually is in any given situation. Is the "client" the individual service member to whom the the psychologist might offer services (this would be the view supported by APA)? Or is the client the larger military organization which has hired and commissioned the psychologist? Crosby and Hall (1992) found evidence of this dual-role problem in their analysis of referral patterns at an Air Force mental health clinic. The authors discovered that among active duty personnel referred for evaluation and treatment by their commanding officers, 88% had serious occupational or legal problems. In these cases, the primary client seems to be the military and recommendations are often made for separation from the service. On the other hand, those members who self-referred to the clinic had very few job-related problems and psychologists rarely recommended administrative action. In the case of self-referral then, the primary client appears to be the individual service member.

As a result of their military allegiance, there are two areas in which military psychologists are particularly vulnerable to violating APA (APA, 1992) ethical standards of practice. These include (a) maintaining confidentiality and (b) avoiding multiple kinds of relationships with clients. Below, each of these will be discussed with actual case examples to highlight how each might be problematic for the client and potentially for the psychologist.

Current DOD directives clearly allow access to confidential material by any federal employee with a "need for the record in the performance of their duties" (Jeffrey et al., 1992, p. 91). Further, military case law generally does not recognize a service member's right to confidentiality. As a result, military psychologists often have little authority when it comes to deciding if a client's records should be released to his or her command, and if records are to be released, which data in the record are appropriate for review by others. Active duty service members may therefore be surprised, angered and embarrassed to learn that persons other than their psychologist have had access to their mental health record. Johnson (1995) offered the following case study to highlight the problem of minimal assurance of confidentiality in the case of a Marine:

> An enlisted Marine who had been treated over a period of one month for depressive symptoms became acutely suicidal and was admitted to a military medical center. During admission, the treating psychologist informed the inpatient physician that the member's increasing fear of being discovered as a homosexual as well as a problematic relationship had been the primary treatment issues and appeared to have directly caused his suicidal thoughts. Later, the member's commanding officer demanded and received copies of the inpatient treatment records which contained reference to the outpatient treatment. The commanding officer then obtained copies of the outpatient records without the psychologist's permission. These were used to begin an investigation of the Marine's possible homosexual behavior and he was harassed considerably by peers. As a result, he stopped receiving treatment and agreed to be separated from the military.

This case shows how psychologists on active duty may have difficulty protecting the confidentiality of the clients they serve. For this reason, it is critical to give military clients clear "informed consent" or full disclosure at the start of therapy regarding who may eventually gain access to their records. This case also highlights the difficulty faced by military psychologists in caring for gay or lesbian clients. The psychologist may find him or herself in a bind when considering that while homosexual behavior is officially prohibited by military personnel, scientific data shows no evidence that homosexual service members are less capable, reliable or effective in their job performance (Herek, 1993). In addition, the psychologists' ethical code requires humane and professional care of all clients regardless of sexual orientation.

informed consent

In addition to difficulty protecting client confidentiality, military psychologists may find themselves in uncomfortable dual relationships with clients which are required by their military role. For example, a psychologist stationed at a small rural Army base may by necessity live next door to, attend church with and perhaps serve directly under those for whom he or she will also provide psychological services. Psychologists may also have difficulty defining the exact nature of their role with clients. The following case study (Johnson, 1995) highlights the difficulty which may follow when a psychologist is required to switch from a primarily "therapeutic" role to an investigative or forensic role:

> A female Army officer was treated for more than a year by a military psychologist to deal with childhood sexual abuse and an eating disorder. Although the officer showed significant improvement and had a good relationship with her psychologist, she continued to have some mood swings, occasional suicidal thoughts and minor episodes of bulimia. As a result, her commanding officer requested a full psychological evaluation to determine whether the officer should retain her top secret security clearance. Although the psychologist protested and asked that another professional be assigned to complete the evaluation, the psychologist was ordered to complete the evaluation. The psychologist felt there was no alternative but to recommend against the security clearance while the officer was experiencing such turmoil. As a result, the clearance was denied and the officer rather angrily ended her

therapy. Her military career was negatively impacted and her symptoms were noted to worsen.

In this case, the service member appears to have been negatively effected by participation in military mental health care. Because the psychologist was unable to prevent a shift in her primary role with the client (from therapist to evaluator), the client was poorly prepared for this new "relationship" with the psychologist and felt somewhat betrayed as a result.

Chapter Summary

In this chapter, we have summarized the nature of training and practice for psychologists in the military. Military internship and post-doctoral experiences offer some of the most competitive and cutting-edge training experiences available to psychologists anywhere. There are also unique stresses and benefits to military internship training. Few environments offer clinical and counseling psychologists the range of jobs or professional activities currently available to military psychologists. Psychologists in all three branches remain on the cutting-edge in their field when it comes to training, clinical practice and applied research. It is also true that practicing psychology in the military presents the practitioner with several unique ethical quandaries. The essential conflict between the psychologist's military and professional/ethical allegiances may pose problems for both the psychologist and the clients he or she serves.

References

Allen, J. P., Chatelier, P., Clark, H. J., & Sorenson, R. (1982). Behavioral science in the military: Research trends for the eighties. *Professional Psychology, 13,* 918–929.

American Psychological Association (1992). Ethical principles of psychologists and code of conduct. *American Psychologist, 47,* 1597–1611.

American Psychological Association (1994). Report of the ethics committee, 1993. *American Psychologist, 49,* 659–666.

Beckman, T. J., Lall, R., & Johnson, W. B. (1996). Salient personality characteristics among Navy divers. *Military Medicine, 161,* 717–719.

Brill, R., Wolkin, J., & McKeel, N. (1985). Strategies for selecting and securing the predoctoral clinical internship of choice. *Professional Psychology: Research and Practice, 16,* 3–6.

Camp, N. M. (1993). The Vietnam war and the ethics of combat psychiatry. *American Journal of Psychiatry, 150,* 1000–1010.

Crosby, R. M., & Hall, M. J. (1992). Psychiatric evaluation of self-referred and non-self-referred active duty military members. *Military Medicine, 157,* 224–229.

Driskell, J. E., & Olmstead, B. (1989). Psychology and the military: Research applications and trends. *American Psychologist, 44,* 43–54.

Eggert, M. A., Laughlin, P. R., Hutzell, R. R., Stedman, J. M., Solway, K. S., & Carrington, C. H. (1987). The psychology internship marketplace today. *Professional Psychology: Research and Practice, 18,* 165–171.

Fragala, M. R., & McCaughey, B. G. (1991). Suicide following medical/physical evaluation boards: A complication unique to military psychiatry. *Military Medicine, 156,* 206–209.

Herek, G. M. (1993). Sexual orientation and military service: A social science perspective. *American Psychologist, 48,* 538–549.

Jeffrey, T. B., Rankin, R. J., & Jeffrey, L. K. (1992). In service of two masters: The ethical-legal dilemma faced by military psychologists. *Professional Psychology: Research and Practice, 23,* 91–95.

Johnson, W. B. (1995). Perennial ethical quandaries in military psychology: Toward American Psychological Association-Department of Defense collaboration. *Professional Psychology: Research and Practice, 26,* 281–287.

Johnson, W. B., & Wilson, K. (1993). The military internship: A retrospective analysis. *Professional Psychology: Research and Practice, 24,* 312–318.

Kurz, R. B., Fuchs, M., Dabek, R. F., Kurtz, S. M. S., & Helfrich, W. T. (1982). Characteristics of predoctoral internships in professional psychology. *American Psychologist, 37,* 1213–1220.

Leshner, A. I., & DeLeon, P. H. (1996). The power of science in furthering professional practice: Opportunities for those with vision. *Professional Psychology: Research and Practice, 27,* 219–220.

Mangelsdorff, A. D. (1989). A cross-validation study of factors affecting military psychologists' decisions to remain in service: The 1984 active duty psychologists survey. *Military Psychology, 1,* 241–251.

Moes, G. S., Lall, R., & Johnson, W. B. (1996). Personality traits of successful Navy submarine personnel. *Military Medicine, 161,* 239–242.

Mullins, F. A. (1992). Recent advances in Navy clinical psychology. *Navy Medicine, 83,* 10–12.

O'hearn, T. P. (1991). Psychotherapy and behavior change. In R. Gal and A. D. Mangelsdorff (Eds.), *Handbook of military psychology* (pp. 607–623). New York: Wiley.

Porter, T. L., & Johnson, W. B. (1994). Psychiatric stigma in the military. *Military Medicine, 159,* 602–605.

Rath, F. H., & Norton, F. E. (1991). Education and training: Professional and paraprofessional. In R. Gal and A. D. Mangelsdorff (Eds.), *Handbook of military psychology* (pp. 593–606). New York: Wiley.

Sturgis, D. K., Verstegen, P., Randolph, D. L., & Garvin, R. B. (1980). Professional psychology internships. *Professional Psychology, 11,* 567–573.

Tedesco, J. F. (1979). Factors involved in the selection of doctoral internships in clinical psychology. *Professional Psychology, 10,* 852–858.

Tipton, R. M., Watkins, C. E., & Ritz, S. (1991). Selection, training and career preparation of predoctoral interns in psychology. *Professional Psychology: Research and Practice, 22,* 60–67.

11

Counseling/Clinical Psychologists' Role in the Military

Ronald Ballenger, Ph.D.

* Introduction * Overview of Alcohol and Drug Abuse Prevention and Control Program * Structure of the ADAPCP * Overview of the Exceptional Family Member Program * Structure of the Exceptional Family Member Program * Psychological Assessment Procedures in EFMP * Overview of Neuropsychiatry and Mental Health * Structure of the Neuropsychiatry and Mental Health Department * Professional Development * Ethical Issues Confronting Psychologists within the Military Environment *Issues of Confidentiality * How to Seek Employment within the Federal Government * References

Introduction

The United States government's Office of Personnel Management employs psychologists to provide psychological services and/or administrative services in a variety of programs established by the United States Department of Defense in the various military services. These programs include, but are not limited to, the Alcohol and Drug Abuse Prevention and Control Program; the Exceptional Family Member Program; and Neuropsychiatry and Mental Health Department. These programs are designed to ensure that a high state of operational combat readiness is maintained for all service personnel. This chapter will highlight the roles that psychologists may play in these various programs to support the military member and family members.

Overview of Alcohol and Drug Abuse Prevention and Control Program

treatment

The ADAPCP was developed to provide evaluation and treatment for soldiers identified to have used drugs illegally and to have misused alcohol. The program is governed by Army Regulation 600–85. The purpose of this chapter, with regard to the ADAPCP, is to describe the current programs provided and to explain how the current programs have developed. Role that psychologists play in this program are described. Also, what may be the involvement of psychologists in the future in the Department of Army is discussed. Evaluation of appropriate force size and configuration in light of the demise of the Soviet Union and the rise of terrorist threats may provide different roles and duties for psychologists to ensure the readiness of military personnel.

Structure of the ADAPCP

The Deputy Chief of Staff for Personnel (DCSPER) is the proponent for the ADAPCP and has the responsibility for plans, policy, programs, budget formulations, and behavioral research pertaining to alcohol and other drug abuse in the Army (AR 600–85).

The Office of the Surgeon General provides the guidance for development of required professional services and resources to support the ADAPCP. Psychologists who are employed in the ADAPCP as a clinical director of a Community Counseling Center (CCC), the name given to the ADAPCP facility by 600–85, 1–9b, that provides treatment to individuals having drug/alcohol problems, must be credentialed by the Army hospital serving the area in which the post is located. The credentialing process indicates the types of therapeutic interventions that can be used in the clinic by the therapists providing the direct treatment under the clinic director's credentialed privileges and supervision. These interventions can include cognitive, behavioral, insight, and family therapy as well as others. The Clinical Director, whether a psychologist or social worker, normally is required to be licensed to practice psychology or social work in one of the fifty states of the United States.

credentialing

cognitive

Without the state licensure the credentialing committee of the hospital providing medical support to the post where the CCC is located will, usually, not credential the Clinical Director. Therefore, no treatment can be provided either by the Clinical Director or by the staff that works under the Clinical Director's credentialing privileges. The Clinical Director works under the supervision of and in consultation with a medical doctor located in the hospital designated to provide services to the area where the CCC is located.

Another component of the ADAPCP is the urinalysis testing program designed to monitor all military members, no matter their Military Occupational Speciality (MOS) or their place of work and all civilian workers involved in the ADAPCP as to whether or not any individual is using illegal drugs. The Biochemical Testing Coordinator (BTC), who works under the supervision of the Alcohol and Drug Coordinator (ADCO), is responsible for monitoring all urine testing in the military community. The ADCO oversees all ADAPCP activities and programs in a particular military area. The military Unit Alcohol and Drug Coordinator (UADC), a military person designated by the unit commander, conducts the random selection of the military members to be urinalysis tested and oversees the urine collection and processing, ensuring that the chain of custody is secure and inviolate until the samples are delivered to the BTC for shipping to the testing laboratories. The ADCO position is sometimes filled by a licensed psychologist.

Individuals can be identified and referred for evaluation at the Community Counseling Center through five different methods. The most desired way is for the individual, whether active duty or civilian, to self refer to the local CCC. Voluntary referral, if not done subsequent to a felonious or misdemeanor offense, is, of course, encouraged and command policies are in place that discourage or prohibit adverse actions being taken against one who self refers.

Another referral source is from the military unit commander. If the unit commander becomes aware of a military member's or civilian's substance misuse, a referral to the CCC for screening and assessment can be made. Attendance at the CCC is mandatory on the part of the military member referred for assessment; it is not if the individual is a civilian. Procedures provided by the Civilian Personnel Office (CPO) must be followed when a civilian is thought to be misusing psychoactive substances.

If a soldier provides a urinalysis sample that is found to be positive, a referral must be made to the ADAPCP for an assessment as to the extent of the abuse. Also, should an alcohol breath test show a blood alcohol level that exceeds the legally prescribed level, a referral is mandatory on the part of the unit commander of that individual to the CCC.

Upon routine or emergency examination by a medical doctor who determines that alcohol and/or drug abuse may be present, a written referral to the CCC may be done. Upon receipt of the referral, the Clinical Director or designee will notify the patient's unit commander of the medical officer's referral.

Should a military member be involved in a drug or alcohol related incident investigated either through military or civilian law enforcement officials, the unit commander must refer the individual to the CCC for an initial screening. Incidents that could initiate a referral include driving offenses, spousal conflict, fighting, and other disorderly behaviors that could involve law enforcement personnel who have reason to believe that drugs and/or alcohol were involved.

Once a military member is screened by the CCC and the unit commander develops a belief, predicated upon a review of the military member's overall service record and the recommendations of the CCC staff presented in the Rehabilitation Team Meeting (RTM), that the individual has no potential for future service in the military, the individual will be considered for processing for separation from the military service. The separation will be grounded in various regulatory procedures available to the unit commander. However, the unit commander may receive other recommendations from the CCC Clinical Director. Referral to another agency may be recommended if evidence is uncovered of, for example, psychiatric limitations that are afflicting the service member or physical ailments that have not been appropriately diagnosed and/or treated.

Even if the Clinical Director is a psychologist in such a situation where the patient is determined to be exhibiting a psychiatric condition, the patient is referred to the department of psychiatry of the medical facility providing medical support to the area where the CCC is located for evaluation and treatment, if needed, and referral to another facililty, if resources are not sufficient.

During the assessment phase of a patient referred to the CCC, the intake counselor conducts an emotional and behavioral evaluation and enters these obtained data into the patient's record. The assessment will include, but not necessarily be limited to obtaining a history of previous emotional, behavioral, and substance abuse problems and treatment as well as current emotional and behavioral functioning. Assessment for suicidality and personality characteristics of such a nature that could seriously affect the treatment or rehabilitation of the patient is made. An evaluation of the current social environment is attempted and, when indicated, a mental status examination and/or psychological testing to include intellectual, projective, and personality testing is performed.

Should the Clinical Director be a psychologist credentialed to perform these assessments by the local hospital credentialing committee, the evaluations can be performed locally. However, if the Clinical Director is not a psychologist, then the patient is referred to a psychologist in the division of psychiatry at the local military supporting hospital for such psychological evaluations that may be needed.

Most often, though, the recommendation will be made that the individual be enrolled in the Alcohol and Drug Abuse Prevention Training (ADAPT) course. This recommendation does not require that the individual be formally enrolled into the ADAPCP. This course has the goals of imparting awareness of the effects of drug abuse and alcohol misuse; the impact of substance use upon one's career, family and other relationships; creating an atmosphere wherein the military member can have an opportunity to evaluate one's personal drinking habits and patterns; and developing alternatives to substance use. Attempting to change undesirable attitudes, values, and behavior associated with alcohol or other drug use is of particular importance during this educational experience.

After completion of the ADAPT, the patient is seen again six months hence to ensure that the patient has continued to implement the information that was gained during ADAPT. Referral to outpatient or to outpatient rehabilitation or other appropriate services is a viable option at any point during the active phase of the ADAPT or during the six month period following its completion. The Education Coordinator (EDCO) is responsible for the overall design and management of the ADAPT services. This individual functions under the guidance of the EDCO. The EDCO is usually educated to the bachelor level and has obtained additional training in substance abuse through programs offered by the ADAPCP. The authority to enroll a military member into outpatient/inpatient treatment lies with the CCC Clinical Director. The unit commander has the prerogative of accepting or rejecting the recommendations provided during the RTM. An appeal procedure is in place that permits the commander to appeal the decision of the Clinical Director to the Department of Psychiatry located in the local servicing hospital.

Should the screening reveal, however, that the educational approach encompassed in the ADAPT is insufficient to meet the requirements presented by the patient's needs, then enrollment in the ADAPCP on an outpatient basis may be recommended. This enrollment may consist of individual, group, or family counseling for, at least thirty days and no more than 360 days. This enrollment is categorized as outpatient. The outpatient enrollment often consists of, at least, nine intensive group sessions and some individual sessions, if issues arise that are not seen to be involved with the reasons for referral. The military person enrolled in outpatient has, as a place of duty, participation in the three hour, thrice weekly session at the CCC.

These sessions are closed group sessions which means that the same individuals participate from the beginning to the end and new members are not allowed to enter the group in the middle of the group process. Closure of the group provides a chance for the group members to build trust in each other and to be honest with oneself. Group members are not segregated according to rank. Privates through generals could be treated in the same group. More typically, though, enlisted personnel and junior officers, captains and under, are found in outpatient treatment groups. Failure to attend is grounds for disciplinary action by the unit commander or grounds for the CCC personnel to declare that the member has failed the treatment designated. This could, then, result in the individual being discharged from the military under a Chapter nine regulation indicating failure in the rehabilitation program. During the outpatient experience, the patient is required to be totally abstinent from the use of psychoactive substances. Involvement with an Alcoholics Anonymous meeting may be recommended as well as the addition of Antabuse therapy if the CCC counselor, with the concurrence of the Clinical Director, believe that the patient is at risk of relapse. Antabuse is a drug that blocks the metabolizing of alcohol by the liver and creates a toxic reaction when one consumes or contacts alcohol in practically any form from any source in conjunction with its use. Close consultation between the Clinical Director and the medical facility supporting the CCC program is necessary to ensure appropriate treatment for the patient when Antabuse is felt to be needed and the patient agrees to such treatment.

Should a patient enrolled in the outpatient program not respond favorably to the outpatient treatment or if it becomes apparent that there is a long term dependency on substances that did not surface during the outpatient initial assessment, then an intensive residential rehabilitation regimen may be implemented. This regimen includes six weeks of inpatient treatment with mandatory non residential follow up period for a total treatment program of one year. The inpatient portion of treatment is under the supervision of a medical doctor. Provision of direct treatment may be by psychologists, social workers, or other providers with expertise in treating individuals with substance dependence issues.

Group therapy is the primary treatment modality with individual, family and other modalities employed as determined necessary by the treatment team in the rehabilitation treatment facility (RTF). Cognitive-Behavioral intervention strategies are the most used interventions in residential treatment. Group members are placed in role playing situations similar to actual situations that the members have been involved in prior to coming to treatment. The role of the psychologist in this situation is one of developing the creative powers of the group members. Group development and maintenance of such can founder on the inability of the therapist to draw out the members in these role playing/role creating dramas. It is well, especially in this type of therapy milieu, to have colleagues with whom to consult and help generate interventions that will bind the group members to behaviors that will help them to fend off the pleas of their peers to imbibe in alcoholic beverages.

The psychologist, as the clinical supervisor, must be able to provide novel, creative suggestions that will maintain the morale and investment of the therapist providing the counseling service. Often, these counselors/therapists have limited training and experience having received but six months of training in the Military Occupational Speciality (MOS) of Behavioral Specialist. The skill, patience, and endurance of the psychologist may be tested in facilities employing the 91X, the MOS designation of the individual rated as a Behavioral Specialist, as the 91X employs techniques, interventions, and personal beliefs within the therapy group.

Members are instructed to develop coping responses to these situations. For example, many of the group members are military members between the ages of eighteen and twenty-five. Much entertainment in and around the military environments consists of going to local establishments that serve alcoholic beverages and engaging in contests, say, of which individual can consume the most alcohol in the shortest amount of time and still be able to perform a behavior such as balancing two shots of liquor on the back of each hand and then drinking the shots without spilling them or dropping the glasses.

Or, playing games such as "Indian" which consists of each individual having a particular sign that will be shown to the rest of the group. Each individual memorizes the others' signs.

Then the group, in rhythm, hits upon the table with hands. One member has been designated as the "chief" and has the duty of showing his or her sign and then two beats later, to show the sign of another group member. The member whose sign has been given must recognize that the sign has been given and then show that sign and then show the sign of another group member and so on. If a member playing this game fails to recognize that a relevant sign has been given to him or her and does not, therefore, maintain the momentum of the game, then that member is obligated to drink the alcoholic drink that one has. Needless to say, drunkenness follows fairly rapidly in these games. At least one source, however, indicates that drinking games seem to have diminished in recent years and have been replaced with solitary or very small group drinking situations (McCollum, 1997).

The individuals in the treatment group would develop strategies for resisting the peer influence and pressure to engage in this or similar type of behavior once they have returned to their duty stations. These strategies might include developing oral responses that would reduce or stop the insistence of peers to engage in the drinking of alcohol while still being able to engage in the activities that the group developed or developing other ways to be engaged in group activities without having the consumption of alcohol as a central focus. Other strategies may include stress management techniques, education about the physiological effects of alcohol as well as psychological and behavioral, and intensive evaluation of core values upon which group members ground their lives. These interventions become salient issues for discussion in the group therapy sessions.

A determination will be made during the first two weeks of inpatient treatment regarding a patient's progress in treatment. If a patient is not progressing in treatment during the first two weeks of inpatient treatment, as judged, for example, by the patient insisting that the behavior engaged in that prompted the inpatient referral was not inappropriate or that peers drank as much or more, the patient will be discharged from the facility back to the referring military unit and CCC. The unit commander can make a determination, in conjunction with CCC personnel, if desired, whether further progress is likely and, if not, a discharge from the military can be implemented.

After successful completion of treatment in the RTF and return to the referring military unit, the patient is encouraged to become involved in Alcoholics Anonymous (AA) or Alanon and Alateen, as appropriate, for family members. Participation in AA is encouraged but cannot be mandated and non-participation cannot be used as criteria to measure success or failure in any of the ADAPCP tracks.

The ADAPCP is responsible for providing counseling services to civilian employees, family members, and retirees. There is no cost to the patient for outpatient ADAPCP services or for medical evaluations performed under ADAPCP designation. Civilians may self refer to the ADAPCP for evaluation and treatment or, if a governmental employee, be referred by a

confidentiality   supervisor. Civilians referred to the ADAPCP are afforded a very high level of confidentiality. Neither the referring supervisor, nor anyone else, can obtain any information about the diagnosis, treatment, or prognosis of an employee unless the employee signs a release of information. The supervisor cannot even obtain information from the treating agency as to whether the employee entered into treatment unless a release of information is signed. Of course, if the employee refuses to sign such a release, the employer is free to terminate the individual's employment.

A police official, such as a member of the Criminal Investigation Division (CID), cannot obtain access to an individual's chart, civilian or military, without executing a request form to obtain specific information from a specific individual's ADAPCP chart. This request form is then given to the Chief of Psychiatry in the servicing hospital for the ADAPCP CCC providing treatment to the individual for a decision as to whether that information can be released. Alternatively, the Patient Affairs Division (PAD) of that hospital may have the duty of determining if and to whom specific information should be released. Information in the charts can be obtained by subpoena if the judicial authority to whom the plea is made is satisfied there are sufficient legal grounds to grant it (USAREUR Staff Judge Adjutant General).

Adolescents, between the ages of twelve and nineteen years of age who are enrolled in a Department of Defense Dependent School (DoDDS), and who exhibit substance use issues, may be referred by school authorities, parents, Community Civilian Misconduct Authority (CCMA) or other individuals or agencies to the Adolescent Substance Abuse Counselor (ASAC). This program is exclusively for American Department of Defense schools located outside of the United States. In the continental United States, community resources are engaged to assess and treat adolescents who exhibit substance use issues. The ASAC is an employee contracted by the U.S. government to provide assessment and treatment to, primarily, adolescents referred, usually, by school officials or the Community Civilian Misconduct Authority (CCMA) for help.

The Adolescent Substance Abuse Counselor's Services (ASACS) are confidential and cannot be revealed to the referring officials, parents or other agencies without informed consent on the part of the patient. Oftentimes, information that has been requested by an outside agency may be supplied to the patient by the Clinical Director of the treatment facility and the patient, then, has the burden and responsibility of deciding to release it to that requesting source. Psychologists and social workers are the primary professionals employed to either supply the services or supervise the provision of services to this patient population.

Recommendations may be made to the governing officials of the local hospital to send a patient to residential treatment or, if the patient is located in an overseas setting, to return the patient to the United States for treatment. In locations where United States psychological and medical services are limited and host nation resources are unavailable or inadequate, the psychologist who is in that location must make recommendations without consultations with colleagues. This unique aspect of psychological evaluation of substance abusers in this type of remote site places unusual stresses on the psychologist. The person serving in this location must be able to function independently and with confidence in one's judgment. — host nation

The ASAC is often confronted not only with alcohol use but also with illegal drug use on the part of the patient referred to treatment. Also, the adolescent may have a history of anti social behavior such that a diagnosis of Conduct Disorder may be appropriate. Given a referral by the local misconduct authorities, of, perhaps, a fourteen year old female who has been arrested for shoplifting hard liquor in the local Class VI store, whose mother is an untreated alcoholic and a stepfather who has little investment in the well being of the family, the ASAC has the extremely difficult task of attempting to engage this girl in a treatment in which she has no interest whatever other than to satisfy the local misconduct authority. How can treatment proceed with these kinds of factors to overcome? What is the role of the psychologist/therapist in this type of referral? The most salient intervention in this case could — Conduct Disorder be to develop a monitoring structure to oversee the girl's daily behavior since it is improbable that she has had sufficient parental monitoring from either her mother or her stepfather. Further investigation into the girl's history reveals that her biological father was sentenced to prison, where he remains, for armed robbery. The biological mother and father divorced after the father was sentenced to prison and the girl's mother and stepfather married subsequently.

Thus, the psychologist's major intervention may be to implement a daily monitoring program, initially. This may be done in conjunction with the school authorities who could be requested to ensure that the child attends school by tasking the stepfather to bring her to school each day and to pick her up at the end of each day. The stepfather could be encouraged to ask for an intervention upon his wife by the local Community Counseling Center staff. This may have the effect of stopping or reducing the use of alcohol on the part of the mother such that she might be able to provide monitoring and attention to her daughter. Also, the marital unit may be strengthened to the extent that the stepfather feels that emotionally satisfying attachments can be developed within the family.

However, if, for example, to further complicate a case such as this, it is found that the stepfather has sexually molested his stepdaughter, then it becomes incumbent upon agencies such as the Army Community Services' (ACS) divisions of Family Advocacy Program (FAP) and Social Work Services (SWS), to remove the daughter from the home and place her in a foster home until more suitable and/or permanent living environment can be secured for the child. Marshaling the resources to meet this child's needs, especially in an overseas setting, consumes enormous energy and time. Often, there are no foster families available to provide temporary care for the child. The psychologist involved must, then, find appropriate facilities in the United States that can provide care and protection for the child. Coping with the myriad issues involved in this case, or one similar to it, can tax the psychologist or mental health provider's inner resources enormously. The initial case must be handled properly and the health care provider's mental and physical well being must be properly attended. The agencies involved should have procedures in place to assess what resources are needed by the provider involved to complete the tasks required and to preserve and protect the provider's well being.

The task of the ASAC is fraught with pitfalls, as is obvious in the description provided. Even with parental involvement of any kind in this or similar case, the interactions are likely to be resistive, inconsistent, and noncompliant on the part of the parents and the child. The therapist may provide enormous energy in attempting to restructure the family and to motivate parents and child to make behavioral changes only, in the end, to see all efforts thoroughly made inconsequential. Thus, the psychologist/therapist in the role of ASAC, must be made of good temperament and indefatigable energy.

Overview of the Exceptional Family Member Program (EFMP)

The EFMP serves to provide medically related services to family members of soldiers and civilians. The EFMP is governed by Army Regulation 608–75. The program helps identify the medical needs of the Exceptional Family Member (EFM) during the assignment process, whether the assignment is within the continental United States (CONUS) or outside the United States (OCONUS), and to ensure that the special needs of the EFM can be accommodated in that assignment location. These needs include, but are not limited to, neurological disorders, asthma, mental disorders, and other disabilities that may need specialized attention.

Structure of the Exceptional Family Member Program

The Assistant Chief of Staff for Installation Management is the proponent for the EFMP and has the responsibility of ensuring the implementation of federal law pertaining to the education of disabled individuals. The EFMP was created in 1981 as a result of a Department of Defense (DoD) directive to the Department of Defense Dependent Schools (DoDDS) to provide a free, appropriate public education to school age children with disabilities. The military medical departments were directed by the DoD to develop agencies to provide medically related services (MRS) to children found eligible for special education services. The United States Army, as well as other military services, developed agencies to implement DoD guidance. The Army developed the EFMP to implement DoD directives.

Enrollment in the EFMP is mandatory if the care needed exceeds that which can be provided by a family physician or routine pediatrics/internal medicine or obstetrics. The enrollment of a family member in the EFMP supports the military mission of the Army and the individual soldier, whose family member is in need of special services, because the family will be placed in a setting where the military sponsor will not be burdened with the thought that medical care will not be available. Enrollment in the EFMP does not have an impact upon favorable personnel actions for the soldier or military member but it may have a decided impact upon where the soldier or military member and one's family is assigned. Therefore, travel destinations may be affected if a family member is enrolled in the EFMP.

The EFMP is mandated to care for children with medical educationally related and developmental difficulties. These are children who need the care of subspecialists. These are providers having skills and knowledge beyond that of a family practitioner or primary care provider. The subspecialists located in the EFMP include clinical child psychologists, speech therapists, physical therapists, developmental pediatricians, occupational therapists and early childhood educators. Social workers are employed as program coordinators but also provide social work services to early intervention cases.

Clinical Child Psychologists in the EFMP have the duty to provide evaluation and, if necessary and appropriate, treatment of children referred to the EFMP by the Case Study Committee(CSC) of DoDDS. The referring CSC is obligated to specifically state the questions that are to be answered by the evaluating subspecialists. The CSC may request to know what kinds of interventions could be of help to the child, family and educators. The clinical psychologist, upon receipt of the referral form, may elect to assess the child by use of a variety of assessment tools. These may include projective devices, behavior checklists, personality assessments, achievement tests, intelligence tests, as well as clinical interviews and observations. Observation may also be done in the classroom setting. This is especially so if the referral is in reference to presumed hyperactivity and/or attentional issues noted in the classroom setting.

The psychologist has a specific time period in which to respond to the CSC referral. This period is forty-five days from the time the parents sign the permission form to allow the assessment to begin. The school psychologist, a DoDDS employee, will, in most cases, carry out the initial evaluation for learning disorders or behavioral concerns. Once these assessments are completed or, sometimes concurrently, the child is referred to the EFMP clinical child psychologist for further, in depth, assessments regarding the child's psychological functioning.

If the time period has lapsed by the time the referral reaches the EFMP subspecialist, the school is notified that a new permission to assess the child is required. To prevent this time consuming event from occurring, the CSC chair will often notify the subspecialist(s) to be involved in the assessment that a referral is being sent. The EFMP subspecialists can then make appointments to see the child prior to the actual paperwork arriving at the clinic and, thus, expedite the assessment process. The actual functioning of this referral process is dependent, in many cases, upon the working relationships among the various CSC chairs and the subspecialists who provide the assessment and written response to the CSC chairs.

The clinical child psychologist must work closely with the CSC chair to insure that the chair understands what referrals are appropriate to make to the EFMP for evaluation and those which should be made to the local medical treatment facility (MTF). The task of discriminating as to which agency a referral should be made is sometimes difficult. Usually, if the referral is regarding a child who is a behavioral problem in school and/or in the community, and may have an adjustment disorder, attention deficit/hyperactivity disorder (ADHD), or depression which may improve with medication, the referral should go to the local medical treatment facility for initial triage. If the EFMP psychologist should, for some reason, continue with a non-CSC referred case, then an issue of which agency should pay for the service provided arises. If the EFMP psychologist should continue the service, perhaps because that service is the only service available in that specific area, then EFMP may bill CAPS or the regional military hospital responsible for mental health service provided by EFMP personnel. If facilities and resources are available, then the individual and/or family may be referred to the community mental health clinic or the child and adolescent psychiatry (CAPS) department. Since this a medical service that is deemed necessary and the services may not be appropriately available, the family may then be relocated to areas where the services do exist. The CAPS department is located, usually, in the military hospital servicing the particular post where the school is located.

There are now three military hospitals providing service to active duty Army members and civilians in the European theatre. These hospitals are located in Landstuhl, Germany, Heidelberg, Germany, and Wuerzburg, Germany. Clearly, since most posts do not have community mental health facilities nor an EFMP clinic or psychologist, a severe burden is placed upon CAPS and upon the patient who, if service is to be obtained, must drive to these locations.

The lack of referral facilities in OCONUS creates a unique environment for the psychologist working in that environment. Host nation facilities cannot be used in most instances because of language barrier difficulties as well as differences in treatment modalities. For example, with regard to the treatment of attention deficit/hyperactivity disorder (ADHD), most physicians in Germany will not prescribe pharmacological agents to treat the condition. Rather, the disorder is viewed as being a non neurological disorder best treated with family and/or individual intervention. Thus, in areas where military medical treatment resources are limited, the treatment of this very treatable disorder, becomes problematic resulting in continued academic and behavioral difficulties.

On a more serious note, when an adolescent or child makes suicidal statements or attempts, there are no residential treatment facilities available for acute or long term care in the European setting. The psychologist treating that child must create an ad hoc observation team for the patient, which may consist of family members, friends, or military unit members,

Adjustment Disorder

Attention Deficit Hyperactivity Disorder

triage

to ensure the patient's safety until arrangements can be made to transport the patient and, if necessary, the patient's family to CONUS. This may take up to seventy two hours to orchestrate and accomplish.

The critical focus of attention for the psychologist who is attending this or similar crisis situation, is to ensure that families are kept together, if at all possible, and to identify the best locations for the patient's well being. If the military or civilian sponsor of the patient had provided appropriate information concerning the patient prior to developing plans to come to Europe, then this kind of emergency could have been avoided. However, when the disabling condition arises, de novo, then the burden on the attending provider, usually a psychologist or psychiatrist, becomes one of assessing the risk of suicidality and activating the process for emergency medical return to the CONUS.   *de novo*

The development of appropriate responses to a threat or attempt of suicide in locations where there are few, if any, host nation facilities with secure wards and no American military facilities with secure wards challenges even the most seasoned psychological practitioner. The clinical judgment and clinical and organizational skill of the psychologist to develop responses to this type of crisis will resound throughout the military theatre, especially if the judgment and skill exercised are found to be wanting in the final analysis.

The readiness of the whole military force in a particular theatre will be affected with the successful completion of a suicide, particularly of an adolescent family member. Psychologically, all military members who have family members will wonder if their family members are at the same or similar risk as the individual who did commit suicide. This kind of stressful event is compounded when soldiers are deployed to distant missions.

Military members always bear the stress inherent in being in a military environment that they may be put in harm's way, on very short notice. They do not wish to have the emotional stress to bear, too, of wondering if their loved ones are being properly attended to in the rear or in garrison. To reduce, if not eliminate, that stress is part of the job of the psychologist in the EFMP but more so of the mental health workers in the Department of Child and Adolescent Psychiatry. The EFMP psychologist can develop relationships with CAPS such that when an emergency situation does arise, CAPS can be contacted and their resources used to develop an intervention and transition plan.

Psychological Assessment Procedures in EFMP

The psychologist in the EFMP, according to Army Regulation 40–18, must hold a license appropriate to the discipline and have documented evidence of training and/or experience in providing services to special-needs children from birth to age twenty-one. The psychologist uses a variety of assessment tools to evaluate and diagnose children properly referred to the EFMP by the CSC.

Many of the referrals made by the CSC are based upon the behavior exhibited by the student in the academic setting. These behaviors may be characterized as being inattentive, unable to understand and follow directions, appearing sad, withdrawn, or acting belligerent vis a vis a peer and/or a teacher, or being unable to remain on task. The classroom teacher, suffice to say, believes that the child's academic development is being deterred to such an extent that an evaluation pursuant to special education placement is necessary.

The child and family are usually seen in the EFMP clinic for initial assessment. Much material gathered by the CSC chair and school psychologist will be available to the EFMP psychologist. History taking can be reduced and a more focused elaboration of issues noted in the history provided by the CSC can be pursued. This clinical elaboration of the parents' view of what is occurring to their child can help guide the psychologist's course of evaluation. For example, if the parents indicate that there has been serious strife within the marital unit for an extended period of time and this period of time coincides, to a reasonable degree, with the observed behaviors of the child that triggered the CSC referral, then projective tests may be given to the child. These tools may help reveal that the child is overly worried and concerned about one's sense of security, safety, and certainty within the family structure.

The reason that the child has not paid attention in class and/or has acted out against peers may be because of unresolved feelings of anger and fear secondary to the ongoing strife between the mother and father. Thus, the focus of treatment may become, not the child, but, rather, the marital unit. The EFMP psychologist may carry out this treatment if it is believed that it is the relational issues in the family that is having an impact on the academic functioning and development of the child and the psychologist has the space and time available to provide marital therapy. Again, the issue of allocation of costs for the services among the various agencies involved arises.

The diagnosis given in this example may be such that the child will not be found eligible for placement in special education. The Department of Defense Dependent Schools guidelines recognize only specific diagnoses that allow an Individual Education Plan (IEP) to be developed and the child to be given specific special education services. The EFMP psychologist, having made a diagnosis that is not within the special education guidelines of DoDDS, can provide services to the family on a space available basis or refer the family to the Department of Psychiatry for services.

As noted above, this type of referral is exceedingly problematical because of the probable distance required to access these facilities and the stigma attached to seeing a therapist within the Department of Psychiatry. The couple will not seek treatment and the child will continue to behave inappropriately. Thus, the EFMP psychologist is in an especially difficult circumstance with regard to providing what are clearly needed services but not having the regulatory mandate to provide them since the child did not qualify for special education services. Often efforts are made to find private providers in the local area that are approved TRICARE providers who are willing and able to provide services. TRICARE is a managed care or managed cost organization developed by the United States Department of Defense.

In some instances the EFMP psychologist is asked to assess for the presence of specific learning disabilities. Some of the most used instruments are the Wechsler Intelligence Scale for Intelligence-Third Edition (The Psychological Corporation, 1991) and the Wechsler Individual Achievement Test (The Psychological Corporation, 1992). The Wechsler tests are very useful because both tests use components of the same norm group to develop the scales on both instruments. Thus, one is able to determine severe discrepancies between abilities and predicted achievement and to, more definitively, determine if a child does have a learning disability in reading, writing or arithmetic.

The EFMP psychologist may use projective devices such as the Holtzman Inkblot Test, (Holtzman, 1961), the Roberts Apperception Test for Children (Roberts, 1982), the Draw A

Person, (Naglieri), or the Thematic Apperception Test (Murray, 1965) to assess for psychological functioning. The Minnesota Multiphasic Personality Inventory for Adolescents (MMPI-A) (University of Minnesota Press, 1992) may be used for personality assessment as well as the Personality Inventory for Children (Larcher, 1984). Both are well anchored in research and have excellent psychometric properties so that one can be fairly certain of the reliability and validity of the results obtained. This is essential when decisions are being made as to whether or not a child should be entered into special education services. Other tools that may be used for evaluations include the Achenbach Children's Behavior Checklist and Teacher Report Form. These forms provide T scale scores across several domains of behavior with clearly defined areas indicating if psychopathology may be extant. These checklists, in conjunction with clinical interviews and observation, and other psychological tools, help the psychologist develop a considered opinion concerning the child referred for evaluation for possible placement in special education services.

Thematic Apperception Test

T scale scores

Once an opinion is developed, a Medical Evaluation Report (MER) is written which states the methods used to develop the diagnoses, what the diagnoses are and on which Diagnostic and Statistical Manual IV (DSM-IV) axis each should be placed. Recommendations are always made for educational interventions and, if needed, for medically related services and provider.

If a diagnosis is made that requires the services of a provider beyond that of a family physician, then the individual must be enrolled into the EFMP. This is done by the health provider involved with the case. This provider may be a psychologist but also may be one of the other providers that may be noted on the Individual Education Plan (IEP). These providers could be an occupational therapist, a social worker, or physical therapist. Also, physicians at the medical treatment facilities can enroll individuals in the EFMP if that provider believes that the individual will need the services of a subspecialist beyond that of a family physician.

The Department of Army form 5862-R is used for the enrollment procedure. The provider is obligated to state the diagnosis, the level of severity, frequency of care, whether inpatient or outpatient, the type of care provider and the frequency of care by that provider. The patient does not have a choice as to whether or not one is enrolled into the EFMP. This decision is made by the health care provider involved. For example, an MTF physician enrolls a conduct disorder patient into the EFMP. The physician codes the patient and notes that the child should be seen two times a week by a clinical child psychologist and once per month by a clinical child psychiatrist. When the family is scheduled to move to its next duty assignment in Bamberg, Germany, for example, travel, most likely, would be denied by the European Regional Medical Command (ERMC) medical director headquartered in Landstuhl, Germany at the Landstuhl Army Medical Center. Had the coding indicated that the patient be seen every five weeks or longer, then travel may have been recommended. Should a patient requiring the level of care noted in this example arise de novo in the European theatre, then a compassionate reassignment to CONUS would, most probably, be initiated.

The compassionate reassignment is a request made by the service member through the chain of command of that service member. Supporting documentation from professional caregivers involved with the service member's care may accompany the request. The commander of the service member's unit has the ultimate responsibility of affirming or denying the request for compassionate reassignment.

Again, the integrating theme of psychological interventions and recommendations within the military community is that of mission readiness. If the commander believes that the mission would be ill served by the compassionate reassignment or even the medical reassignment of a service member, then that reassignment may be denied or applied only to the family member concerned.

This aspect of mission readiness is the singular difference between working as a psychological practitioner in a military setting as opposed to working in a civilian environment. The needs of the mission take precedence over individual needs, wants, or whims. The psychologist involved must understand the unique nature of military mission readiness in order to accept and appreciate decisions made by unit commanders, at various levels, that are contrary to what one might expect in a civilian setting. The psychologist coming from a civilian setting cannot retain, or, at least cannot act on, previously held values regarding individual needs and rights; the needs of the larger, military group are paramount in a military setting. Family needs are secondary.

Overview of Neuropsychiatry and Mental Health

combat
fatigue

The primary function of this division is to ensure that military personnel are combat ready and, in case of combat fatigue as a result of combat operations, are returned to duty as quickly as possible. Prevention of mental health issues is a primary goal of this division. This is attempted by training unit commanders and senior non commissioned officers in the recognition of mental health disorders and making appropriate referrals to mental health personnel. These personnel include psychologists, psychiatrists, counselors and social workers.

Structure of the Neuropsychiatry and Mental Health Department

The proponent agency for neuropsychiatry and mental health is the Office of the Surgeon General (OTSG). The Army Regulation governing the development, staffing, and purpose of this department is, primarily, 40–216. This regulation establishes that this department is to aid command in conserving and maintaining manpower at maximum efficiency through the application of sound mental health and neuropsychiatric principles.

The psychologist who is an active duty military individual and is a member of a military medical battalion that is in a combat environment, will have the duty of assessing the mental state of combat casualities. Combat fatigue is the term applied to negative combat stress reactions with uncomfortable feelings and performance degradation. These terms do not imply a mental disorder.

There are several degrees of battle fatigue according to AR 40–216. The psychologist or mental health provider in attendance must make triage decisions. The mental health professional involved with treating these individuals must make a decision as to the level of battle fatigue a military member is suffering. The three levels of battle fatigue are mild, moderate, and severe. Even a decision that the person is suffering from severe battle fatigue and is too disruptive to be managed in the military unit, does not mean that the combatant cannot recover within a fairly short period of time and be returned to combat duty.

Often, if an individual suffering from combat fatigue is provided rest, sleep, food, fluids, and a clean environment, the symptoms of fatigue disappear within a short period of time, usually two weeks or less. The psychologist must provide a demeanor that suggests to the military combatant the combat fatigue is transitory and a return to duty is expected and as soon as possible. If, in the restorative environment, the mental health professional believes that a patient is decompensating, then a decision must be made to retain the patient in that setting for further support and treatment or to return the patient to the rear.

The mental health officer must accept the responsibility of making this and similar decisions with regard to the severity of the patient's condition and the environment required in which to treat the person. The therapeutic objective is to return the patient to combat duty but the difficult decisions as to how to achieve that ultimate objective will lie in the judgment and care of the psychologist or mental health professional most involved with that patient. The psychologist must believe in one's own capabilities and skills to function in this type of stressful environment.

Other issues that confront the mental health professional in combat environments is that of anxiety during heavy shelling or intense fire fights and then depression that may ensue during lulls in combat or actual cease fires. The changes in the war environment will bring changes in the type of disorders being seen by mental health professionals and the provider must be able to recognize these changes and to provide treatment accordingly. Management of the trauma suffered by combatants becomes the primary function of the provider in a war environment with the therapeutic goal of return to combat.

Management of post traumatic stress disorder becomes primary subsequent to being removed from the war environment. Groups composed of combatants that allow them to speak about their common or not so common experiences and to share their stories and fears help the mental health healing process to, at least, begin and to initiate reintegration into everyday society.

In a non war fighting environment, psychologists perform functions more similar to that found in a civilian setting. Even though the environment may be defined as non combat, the military member, as well as family members, may suffer from a variety of psychological issues. The psychologist may be called upon to provide psychoeducational interventions to forestall more serious psychological conditions if a conflictual situation should occur as happened in the Arabian Gulf in the early part of the 1990s.

Presentations to military units regarding stress identification and management may be a function carried out by the psychologist, psychiatrist or social worker located in the Department of Psychiatry. In an era and environment that require varied deployments, these types of classes can be very helpful in educating individuals to identify stress factors in one's life and various methods to cope with these stresses.

Again, remembering the theme of mission readiness and combat effectiveness, the better the soldier and family members are able to cope with the stresses inherent in a military environment, the more effective the soldier and the unit will be when called upon to deploy and, perhaps, engage in combat. The family members will be better able to withstand the absence of the military member from the family and to draw upon their own inner resources to maintain psychological health.

psychological
autopsies

The Child and Adolescent Psychologist located in the Department of Psychiatry, is required to perform psychological autopsies when a service member or family member dies from self inflicted measures. A lengthy process of interviewing those close to the deceased to ascertain the state of mind of the decedent prior to the suicide is carried out. Information is gathered that will, it is hoped, help prevent future suicides by developing methods of heightening awareness among commanders and non commissioned officers as to signs of suicidality within the military members under their command.

Psychological autopsies are also carried out should a family member commit suicide. Again, this is for the purpose of creating better awareness of the behavioral signs that may precede a suicide attempt. Once these factors can be determined, interventions within the military community can be arranged and implemented with the expectation that future suicides and suicide attempts can be prevented or, at least, reduced.

Often the CAPS psychologist and less so the EFMP psychologist will be called upon by various agencies within the military community to provide courses or presentations regarding the possible factors that may lead to a suicide attempt. DoDDS may ask that presentations be given to school faculty and administrators at both the elementary and secondary level that will provide guidance to them as to how to recognize behavior that could be a precursor to a suicide attempt. Once having made that recognition, then the procedures to follow to prevent an attempt are addressed. These presentations create within the military community a sense that everyone in the community is important to each other. Even though military readiness and combat effectiveness is the sine qua non, the care and consideration of family members in a garrison environment is raised to a very high level of awareness.

The psychologist working in the CAPS department or in a mental health department in the military setting will be involved in providing information to patients regarding the development of behaviors that will, presumably, provide a healthier, more productive life. This information may concern the physical and psychological issues embedded in the use of products containing the drug nicotine or alcohol. Or issues regarding the inappropriate and/or unhealthy use of foods may be the focus of the information. The use of illegal substances is, of course, grounds for dismissal from the military.

Professional Development

Psychologists within the Department of Defense, whether military members or federal service civilians, have unique opportunities for professional development. Each agency has an annual conference where psychologists may obtain continuing education units. These conferences are usually held at conference sites that permit mental health providers from a variety of agencies to share information and to discuss common concerns. Attendance at these conferences is, often, mandatory and is funded by the department or agency for which one works. Also, conferences and congresses are available at various cities, for example, throughout Europe.

Often these conferences are under the aegis of organizations such as the Milton Erickson Foundation, the European Branch-American Counseling Association, or the World Council for Psychotherapy. Presenters of world reknown stature present papers and are available for small group discussion and interaction. Also, the attendees may submit papers for presentation thus allowing an opportunity to see and hear how psychological services are devised and delivered in a variety of cultures throughout the world. Often, attendance at these conferences will be funded by the agency or department for which one works if it can be demonstrated that the content of the material to be presented at the congress is germane to one's own professional work setting.

Ethical Issues Confronting Psychologists within the Military Environment

Issues of Confidentiality

The psychologist within the context of a military environment is confronted, even more vividly than in a civilian setting, with situations that may require the breaching of confidentiality. Ethical conflicts will arise within the military environment because, even though the psychologist is presumed to have a primary responsibility to one's patient, as stated earlier, the essential reason for being for a military organization is to be ready to defend the interests of the United States with armed strength. If a member of that organization would jeopardize that readiness due to mental instability, emotional distress, or other factors, and these, or other relevant factors are known to the psychologist, what affirmative duties must the psychologist carry out? Guidance is provided in various ethical codes developed by a variety of mental health associations such as the American Counseling Association, the American Psychological Association, the American Psychiatric Association, and the National Association for Social Workers.

The confidentiality for active duty military members within the context of a counseling relationship is circumscribed in several sources. These sources may be codes of ethics, statutory and case law as well as military regulations. Confidentiality, at bottom, is a right to privacy. The mental health professional will maintain the privacy of what is said by the patient or client within that relationship, revealing information only when there is a compelling reason to do so. The various codes of ethics as well as statutory and case law indicate the circumstances under which a mental health professional may be morally and/or legally obligated to reveal information obtained within a therapeutic relationship.

The basic philosophical issue that arises is the conflict between the need to maintain the secrecy of statements made to another person acting in a professional relationship such as a psychologist, social worker, or physician, and the release of that secret information if another's life or essential well being is believed to be at risk. When is the right to have secrets safeguarded canceled by a presumed higher good, the protection or preservation of the life of another? The continuum along which this dynamic conflict lies is relatively broad. Until quite recently, deciding when to release information was a decision that was dictated by the mental health professional's personal sense of when another individual was at risk from the patient involved in the therapist-patient relationship.

In general, mental health professionals now have an affirmative duty to warn identifiable victims of probable harm to themselves or to their property if the patient's behavior cannot be controlled. The mental health professional has a duty to prevent a patient who poses a clear danger to other individuals, even though unspecified, from having an opportunity to inflict harm.

affirmative
duty

Obviously, the mental health professional involved is in a dilemma not only with regard as to when confidentiality must be breached but also in having to determine when the threat is credible, whether the patient's behavior can be controlled, and does the patient pose a clear danger. Each member of a professional speciality must understand what reasonable level of judgment must be exercised under similar circumstances within that professional speciality when confronted with a patient who, potentially, may pose a harm to others.

For psychologists working within the military environment, whether they are active duty or civilians, it is imperative that guidance be sought from the local Staff Judge Adjutant General's office regarding policies and directives as to how to respond when confronted with a situation as noted. One must not rely, solely, on the ethical standards promulgated by one's professional organizations. The psychologist must be aware of military court rulings as well as military regulations that may differ from the ethical standards developed by professional organizations.

The issue of testifying in a court procedure may arise for the psychologist employed in a military setting. Whether or not a psychologist will be granted the privilege to not testify regarding the contents of counseling sessions may turn on issues of military readiness and ensuring the security of the United States. A court decision (Jaffee v. Redmond) indicating that licensed clinical social workers do not have to reveal the contents of counseling sessions may not be applicable in the military environment. Questions that may arise as to whether information provided to a mental health provider within the context of a military environment is privileged will have to be answered by further adjudication. The questions may also be delineated as to whether the issue is civil or criminal in nature. The psychologist working within the military context should seek guidance from professional associations and from the local Staff Judge Adjutant General's office.

How to Seek Employment within the Federal Government

One may write to the Office of Personnel Management (OPM) located in Washington DC to obtain current information regarding application procedures to be considered for employment within the federal government. The information provided by the applicant on the requisite forms will be evaluated and a decision rendered as to whether the information provided is sufficient to permit a rating classification to be made within one of the job series delineated by OPM.

Psychologists are rated within the job series numbered GS-180, Engineers and Scientists. The grade levels within that series range from GS-9 through GS-13. Each grade level has ten steps. Each grade level and step have associated salaries which may increase on a yearly level depending upon funding by the United States Congress. No matter where within the federal government one is applying for a government service position as a psychologist, the application procedure is the same. For examples, this applies to applications for positions in Veterans Affairs as a direct service provider, to the Department of Defense as a research psychologist, or to the Department of Navy as a clinical director of a facility providing mental health or alcohol/drug counseling.

References

Achenbach, T. (1991). *Teacher's Report Form for Ages 5–18.* University of Vermont. Burlington.

Department of the Army (1995). *Army Regulation 600–85 * Change 2, Alcohol and Drug Abuse Prevention and Control Program.* Washington, DC.

Department of the Army (1984). *Army Regulation 40–216, Neuropsychiatry and Mental Health.* Washington, DC.

Department of the Army (1996). *Army Regulation 608–75, Exceptional Family Member Program.* Washington, DC.

Department of the Army (1995). *Army Regulation 40–48, Nonphysician Health Care Providers.* Washington, DC.

Holtzman, W. (1961). *Holtzman Inkblot Technique. New York: The Psychological Corporation.*

Jaffee v. Redmond. 116 S. Ct. 1923 (1996).

McCollum, R. (1997). Personal communication.

McArthur, D. and Roberts, G. (1982). *Roberts Apperception Test for Children.* Los Angeles: Western Psychological Services.

Minnesota Multiphasic Personality Inventory-Adolescent (1992). Minneapolis: National Computer Systems.

Murray, H. A. (1965). Uses of the Thematic Apperception Test. In B.I. Murstein (Ed.), *Handbook of Projective Techniques.* New York: Basic Books.

Naglieri, J. A. (1988). *Draw A Person, A Quantitative Scoring System.* San Antonio: The Psychological Corporation, Harcourt Brace Jovanovich.

Wechsler Individual Achievement Test (1992). San Antonio: The Psychological Corporation, Harcourt Brace & Co.

Wechsler, D. (1991). *The Weschler Intelligence Scale for Children–Third Edition. San Antonio:* Harcourt Brace & Company.

Wirt, R., Larcher, D., Klinedinst, J., Seat, P. (1984). *The Personality Inventory for Children.* Los Angeles: Western Psychological Services.

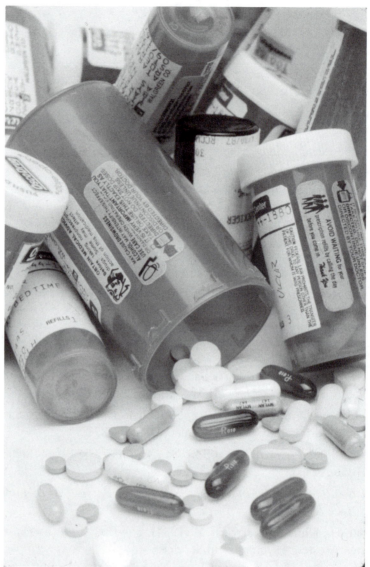

Copyright © G. M. Cassidy/The Picture Cube

12

Military Psychologists, Pioneers in Prescribing

John Sexton, Ph.D.*

Introduction

Military psychologists have been pioneers in a variety of territories like psychological testing, human factors, and posttraumatic stress. Psychologists prescribing is another new territory in which military psychologists have taken a pioneering role.

This chapter will include everything from 'the soup to the nuts' on the issue of psychologists prescribing. The chapter will begin by answering perhaps the most important question . . . "Why prescribe?" This question deserves attention; for if a sufficient answer did not exist, this chapter would never have been written. One of the key reasons why some psychologists should prescribe is to fill the considerable need that exists in various groups in our country. One of those groups happens to be the military. It is the unique role of the military in combat that makes it a group in need of appropriate mental health interventions. It was this fact that helped to set the stage for military psychologists to undertake this pioneering effort of being the first to prescribe.

The second section will review the history of the movement toward psychologists prescribing. It will take the reader from the early efforts in the 1970s through the Department of Defense's Psychopharmacology Demonstration Project (DoD PDP).

There will be a brief discussion of other curricula designs and the efforts by the American Psychological Association (APA) and other organizations in the quest for prescription privileges.

The final portion of the chapter will look to the future of psychologists prescribing. With the August 1995 endorsement by the APA's Council of Representatives of a resolution on

Psychopharmacology
Demonstration
Project

prescription
privileges

---

*The views expressed in this article are those of the author and do not reflect the official policy or position of the Department of the Navy, Department of Defense, or the United States Government.

prescription privileges, the issue has picked-up steam in many states, and each day new pages could be written for this chapter. While the timing is difficult to predict, one's crystal ball would likely reveal psychologists outside of the military prescribing according to their state's statutes in the very near future.

This chapter will hopefully be of interest to anyone studying military psychology; but it should be of significant interest to anyone contemplating a career in psychology, for this matter has a monumental impact on the future of mental health services.

Why Prescribe

Some psychologists should prescribe in order to provide cost-effective, quality care that could reach the underserved. Each one of these variables will be supported by evidence that will answer the question of why some psychologists should prescribe.

In a chapter entitled Expanding Roles in the Twenty-First Century, in the book *Health Psychology Through the Life Span*, Dr. Patrick De Leon, a psychologist and Administrative Assistant to Hawaii Senator Daniel Inouye, poignantly wrote about the responsibilities of mental health professionals in caring for the people of our country. He wrote, "Psychologists are convinced that to significantly improve the quality of life for individuals either who require psychotropic medications or, perhaps more important, for whom psychotropic medication has been inappropriately prescribed, behavioral scientists must become intimately involved in this clinical decision-making process. The power to prescribe is equally the power not to prescribe or the power to ensure that medications are appropriately used. There is considerable data indicating that, for example, children, the elderly, women, people of color, and those residing in nursing homes, rural America, and inner-city ghettos are often inappropriately medicated" (p 439, DeLeon, Howell, Newman, Brown, Keita, and Sexton, 1996).

Quality mental health care does not occur often enough. In general, psychiatrists and psychologists provide good quality mental health care; however, they tend to favor their strong suit and treat some patients with medication or psychotherapy when the other or both modalities of treatment are clinically indicated. Our concern with the quality of mental health care need not be with the licensed mental health professional, but more with the medical professional whose training and experience in mental health is insufficient to make accurate psychiatric diagnoses and effect modern psychotherapeutic or pharmacotherapeutic inter-ventions. The majority of all psychotropic drugs are prescribed by general practitioners. The percentage is greater than 75 percent (Senate Bill, 1995). General practitioners, who are reimbursed for greater than half of all out-patient mental health visits in the country, have been publicly criticized by their own American Medical Association (AMA) for their ability to prescribe psychotropics (Brief of the AMA, 1989). An article in the Journal of the American Medical Association noted "Primary care doctors were less likely to prescribe appropriate antidepressant medications or provide necessary counseling and they were more likely to prescribe tranquilizers, which are generally considered ineffective against major depression" (p 51; Sturn, and Wells, 1995).

There are many groups within our country who are underserved or inappropriately served. One of the underserved groups are those in rural America. Psychiatrists will not be found in 62% of the counties in the United States. According to the Bureau of Health, all

(margin terms: psychotropic, psychiatrists, psychotherapy, pharmacotherapeutic interventions, antidepressant medications)

states except Idaho and Missouri report some level of shortage of psychiatrists (American Psychological Association, 1992). Less than 10 percent of 2200 rural hospitals in the United States in 1994 had any form of mental health services (Enright, 1994).

The seriously mentally ill have become a group that is significantly underserved. To a large degree, this occurred with the de-institutionalization of the seriously mentally ill and the closing of many state and county mental health facilities all across our country. A 1990 report by the Public Citizen Health Research Group and the National Alliance for the Mentally Ill, recommended that psychologists, physician assistants, and nurse practitioners receive additional training in order to prescribe psychotropics and meet the needs of the seriously mentally ill (Torrey, Erdman, Wolfe, and Flynn, 1990).

nurse practitioners

A group whose mental health needs are not properly served is our children. One might call it an epidemic, viewing the dramatic increase in the diagnosis of Attention Deficit Hyperactivity Disorder (ADHD). With the diagnosis often comes stimulant therapy, before psychotherapeutic (usually behavioral) interventions are employed. There has been a significant increase in our country in the number of prescriptions written for stimulants. Child psychiatrists will make an appropriate diagnosis, and treat ADHD well; but there are too few child psychiatrists in some states. In 1990, Wyoming had only one child psychiatrist (American Psychological Association, 1992).

Attention Deficit Hyperactivity Disorder

stimulant therapy

Another group whose mental health needs are inappropriately served is women. Women receive 73 percent of all prescriptions for psychotropics, while they only account for 58 percent of physicians' visits. They receive 90% of the scripts for psychotropics when the physician is not a psychiatrist (Russo, 1985).

Elderly Americans are a group that is both underserved and overserved. The Inspector General of the U.S. Department of Health and Human Services has labeled the mismedication of the elderly our nation's "other drug problem" (Office of the Inspector General, 1989). The Health Care Financing Administration had seen the excessive inappropriate use of antipsychotics in elderly nursing home patients, and drafted federal guidelines to curtail the mismedication. With the elderly comprising approximately 12% of our population, it is hard to believe that they consume approximately 40% of all sedatives (DeLeon, Fox, and Graham, 1991).

Two other factors will worsen the current unmet mental health needs in our country. Factor one is the 'graying' of America. As baby-boomers move into their later years, there will be greater numbers of the aged who may be inappropriately prescribed psychotropics for behavioral control. Medication interactions, often a problem in the elderly, will be greater in number.

The second factor leading to more unmet mental health needs is the declining percentage of medical school graduates going into psychiatry as their chosen field.

America has had significant unmet mental health needs for many years. The forecast was for an even more gloomy future. Some influential people saw the growing need and did something about it. That is where the history of psychologists prescribing begins.

The Early History of Prescription Privileges

The first program known to train non-physician, doctoral level, mental health professionals to prescribe psychotropics was the Doctorate in Mental Health (DMH) program. This five year training program, which involved approximately two years of classroom instruction and three years of clinical training, was based in the University of California and San Francisco's Mount Zion Hospital from 1973 to 1986. Most of the courses were taught at U.C. Berkley. The program sprouted from the beliefs that the mental health needs were growing and the training of a full-service mental health professional was not efficiently done. Psychiatrists spent too much time in training in the basic sciences and too little time in psychotherapy training. Psychologists, on the other hand, were not trained to provide biological interventions, a significant portion of mental health treatment, and were overly trained in research methodology. The DMH program was designed to take the best of psychology and psychiatry training, discard the excess, and efficiently train the clinician in mental health diagnoses and treatments. Unfortunately, the 80 or so graduates were not granted licenses to prescribe medications, and many went on to medical schools or acquire licenses as psychologists (Sammons, Sexton, and Meredith, 1996).

One of the first high level public presentations supporting prescription privileges for psychologists occurred as Senator Daniel Inouye addressed members of the Hawaii Psychological Association at their annual convention. His message was clearly in concert with the theme of the convention, Psychology in the 80's: Transcending Traditional Boundaries. He suggested that "clients will be well-served" when psychologists obtain prescription privileges (DeLeon, Sammons, and Sexton, 1995).

The Hawaii Psychological Association pursued legislative action that would order a study of the feasibility of psychologists prescribing. State Resolution 159, 1985 was never reported forth from the legislative committee. Interested psychologists continued their efforts, and in 1989, introduced a far-reaching bill that suggested that the Board of Psychology be authorized to certify appropriately trained psychologists to prescribe from a limited formulary. A training program of no less than 60 semester hours of coursework and 400 hours of supervised experience was outlined. After extensive hearings, Hawaii State House Resolution 334–90, ordered a think tank, The Center for Alternative Dispute Resolution, to report on the mental health needs in Hawaii and whether training psychologists to prescribe was indicated (DeLeon, Fox, and Graham, 1991). A consensus opinion was unable to be reached, and further action did not occur.

The DoD PDP

In 1988, the U.S. Congress was prompted to examine the idea of psychologists prescribing, largely due to the efforts of Senator Inouye and his Administrative Assistant, Dr. Patrick DeLeon, who many consider to be the 'father of prescription privileges'. House-Senate conferees looked at a federal group that had a specific need. That group was the military, with its need for large scale mental health interventions during time of war. In particular, the conferees noted that a significant percentage (20 to 30 %) of combat casualties during war are purely psychiatric casualties. It has been estimated that during high intensity, long duration combat, that the percentage could double. The conferees knew that mental health professionals could return 75% of the psychiatric casualties to the front lines within 72 hours, if appropriate interventions occurred. One of those interventions would be the appropriate use of sedatives to aid sleep. This focus on addressing a real need in our society ultimately resulted in congressional language which gave the DoD its marching orders to establish a program to train military psychologists to prescribe (Sammons, Sexton, and Meredith, 1996).

House Resolution 4781 directed the DoD to establish a "demonstration project" to train military psychologists to prescribe psychotropics. The Assistant Secretary of Defense for Health Affairs then directed the Army to begin the demonstration project by September 15, 1988. Opposition to the project, led primarily by the American Psychiatric Association out of concern psychologists would be poorly trained and would cause harm to patients, delayed curriculum development and implementation. In 1989, the House-Senate conferees expressed considerable concern about the DoD's delay in beginning the project. The conference report stated "the Department cannot ignore directions from Congress and therefore should develop such a training program in fiscal year 1990" (H.R. 3072, H. Conf. Rep. No. 101-345, p. 30 in DeLeon, Fox, and Graham, 1991).

It took this second directive from Congress for the DoD to knuckle-down and begin planning a training program. The Army Surgeon General's Blue Ribbon Panel was assembled with representatives from the National Institute of Mental Health, the Food and Drug Administration, the Uniformed Services University of Health Sciences (USUHS), Psychiatry and Psychology Consultants to the Surgeons General, APA, and other medical groups. The panel examined a number of possible training options.

One of the training models examined by the Blue Ribbon Panel was one developed at Wright State University, School of Professional Psychology. This design was to produce "limited practice prescribers" who were trained in a specialty track within a predoctoral graduate program, a continuing education program or a postdoctoral fellowship. Fox et al. compared their 390 classroom hour design with that proposed by Balster, President of APA's Division 28 (Psychopharmacology). The 'Wright State University Model' is described in Fox, Schwelitz, and Barclay, 1992. The 'Balster Model' is described in Balster, 1990. See Table 1 for a comparison of these two curricula with an APA model and the DoD iteration #3 program. All tables show contact hours.

The 'Wright State University Model' did not stand up to the support received for what would be referred to as the 'USUHS Model', which would become the training program used in Iteration #1 of the Psychopharmacology Demonstration Project (PDP).

The problem with the roughly outlined USUHS Model was that many details remained to be worked-through, and the course could not possibly begin until the start of the next medical student class in the Fall of 1991. In an effort to follow the orders of Congress and establish a program in Fiscal Year 1990, the DoD sent two Army psychologists to the physician assistant course at the Army's Academy of Health Sciences at Fort Sam Houston, Texas in September 1990. The Army psychologists were to receive coursework in anatomy, physiology, chemistry, labs, pharmacology and physical examinations, before progressing to an 8-month supervised clinical practicum in an inpatient psychiatry service at either the   practicum
National Naval Medical Center in Bethesda, Maryland or Walter Reed Army Medical Center (WRAMC) in Washington, DC. After completing less than half of the didactic coursework, one psychologist who had continued in the program, was ordered to stop attending classes and

Table 1

CURRICULA COMPARISON*
(contact hours)

	Balster	Fox	DoD #3	APA
Anatomy			48	
Biochemistry	45	20	57	45
Neurosciences		60	54	
Physiology	75	60	39	75
Pathophysiology			60	
Clinical Medicine			277	
Pharmacology	90	20	83	90
Clinical Pharmacology	45	90	21	75
Psychopharmacology	45	80	21	45
Pharm. Professional	15	20		15
Bio. Basis of Beh.	45			45
Psychopath/Addiction		40		
Total	360	390	660	390

(*Some course titles were compressed into commonly named courses)

return to WRAMC. The early termination of this training program was the result of the well-orchestrated efforts of organized psychiatry. Political pressure was felt by Congress as the membership of the American Psychiatric Association responded to an action-alert letter that used phrases describing the prescription privilege training program like "dangerous precedent…endangering military patient care…making military personnel nothing more than medical research subjects". The House-Senate conferees directed the DoD to cease the current training program and follow the recommendations of the Blue Ribbon Panel, which was the USUHS training model (DeLeon, Fox, and Graham, 1991).

The lead agent, the Army, hammered-out the details for what would come to be known as the first iteration of the DoD Psychopharmacology Demonstration Project (PDP). At the same time, the search was on for candidates for the fellowship. Only the Army and the Navy decided to send candidates to this first iteration. They chose to each send two licensed psychologists who had completed at least one tour of duty as a military psychologist after their psychology internships. Other than the few months of medical training experienced by one of the Army psychologists who attended a small portion of the physician assistant training, none of the candidates had ever completed any formal medical training or pre-med coursework.

The four candidates came from the four corners of the earth to the Psychology Department of Walter Reed Army Medical Center (WRAMC) in July 1991. We all were very surprised to see a curriculum that looked like medical school. Some of us had been told that we'd take a few classes at USUHS and do a practicum at WRAMC over a two year period, and did not expect to complete approximately 75% of the two-year classroom and laboratory coursework of the medical students and do an extensive clinical practicum at WRAMC at the same time. From the first day at WRAMC, our protests were heard. It appeared that we were doomed for failure in having to pass the medical school coursework with no basic science preparation and complete a demanding practicum concurrently.

During the first day at WRAMC, we looked ahead to one of our first courses, a 160 hour course in biochemistry. None of us had ever completed a course in inorganic, let alone organic chemistry, typical prerequisites for biochemistry. My last chemistry course was in propulsion chemistry at the US Naval Academy nearly 20 years earlier. To ease our anxiety (a bit), a tutor helped us acquire some chemistry basics in the few weeks before the classes began at USUHS in August 1991.

The two year didactic curriculum, totaling 1339 hours in the classroom and lab, was a very demanding one. (See table #2). The anatomy course began with histology; not the study of hists, but microscopic examination of the human anatomy. Anatomy continued with gross anatomy; and it was truly that…gross. Nearly six months were spent in full body dissection of a cadaver. The clinical medicine course taught us to do head to toe physical examinations. We started this course by getting to know our classmates in ways we wish we hadn't. Pathology was both important and interesting, for we have found that the psychologist who prescribes needs to know about all sorts of physical symptoms and pathology. The clinical concepts course was an interesting diagnostic guessing game, where students were given pieces of a medical history and laboratory, electrographic, and radiographic findings, and asked to guess the usually obscure diagnosis. The medical students took several courses of moderate significance like parasitology (40 hours), microbiology (160 hours) and preventive medicine (47 hours), and a handful of less significant and consequently much shorter courses like the history of medicine, biostatistics, ethics, radiology, and two psychology courses.

Table 2

PDP Iteration #1

Years	Course	Contact Hours
'91–'92	Anatomy	341
	Biochemistry	160
	Physiology	172
	Clinical Medicine	86
'92–'93	Pathology	215
	Clinical Medicine	132
	Clinical Concepts	100
	Pharmacology	86
	Clinical Pharmacology	47
Total		1339

Taking-on material and learning during medical school can be compared to trying to get a drink from a fire hydrant. The material is so deep and comes at such a high rate of speed, one must become an expert test-taker in order to survive. We took 46 examinations, a dozen of which were hour-long attempts to identify tissues using a microscope. Teaching clinically experienced psychologists how to prescribe a limited number of psychotropics by sending them through traditional medical student coursework was absurd. We understand that the depth and breadth of training for a physician-to-be must be extensive, but we did not need that much training. Because we psychopharm fellows had to memorize a considerable amount of truly extraneous material, in order to perform adequately on examinations, some of the important information was hard to retain.

It is important to note that on top of the medical school coursework, we had to do clinical work at WRAMC. With the few hours a day that we were not in the classroom or sleeping (maybe it would be more accurate to say "and/or" sleeping), we stood night call with psychiatry residents, did psychiatric chart reviews, and did some brief work in consultation-liaison psychiatry.

The absurdity of this demanding schedule was finally acknowledged; and during our second year at the medical school, the clinical work at WRAMC was discontinued. A third, purely clinical year, was added to the PDP in order to give us this very valuable and intense training.

Three of the initial four fellows finished the second year of the program. Major Dale Levandowski enjoyed medical school so much, he decided to leave the PDP and trade in his oak leaf rank insignia for a 'butter-bar' (Second Lieutenant) rank insignia of a new medical student at USUHS. Navy Lieutenant Morgan Sammons and I went on to the third year of the PDP.

Our clinical year mirrored the typical second year of a psychiatry residency, a year of inpatient psychiatry work. Dr. Sammons and I stood in line with the four second-year psychiatry residents at WRAMC, and took the next new patient in turn as the patient was admitted to the ward. We completed intake physical exams and ordered labs and meds under the supervision of a staff psychiatrist. We were responsible for all aspects of the patient's care during their hospitalization, to include the treatment of medical problems or an appropriate consultation with a specialist. We prepared the patient for discharge and occasionally followed them as an outpatient. We had the opportunity to treat more than 100 patients each.

antimanic   Great learning occurred as we quickly applied high doses of antipsychotic and antimanic medications to those patients with serious mental illnesses.

The learning environment at WRAMC was phenomenal. The staff psychiatrists were knowledgeable and skilled teachers. The psychiatry residents were able to learn group psychotherapy as they co-facilitated groups with us, and we in turn were able to learn medicine as they helped us with our patient's medical problems.

Following our June 1994 graduation, Lieutenant Commander Morgan Sammons traveled a few miles to his assignment with the Psychology Department at the National Naval Medical Center in Bethesda, Maryland. I traveled south to the Psychology Department at the Naval Medical Center, Portsmouth, Virginia. Since no psychologist had ever been given the independent privilege to prescribe, the process of acquiring the clinical privilege took many months instead of a few weeks. Additionally, there was a difference in the degree of physician oversight initially required at the two Naval hospitals. Because each medical facility's commander decides the extent of the clinical privileges of the members of his or her medical staff, the privileges granted a prescribing psychologist will likely vary from facility to facility.

Meanwhile, back at WRAMC, the two iteration # 2 fellows finished their one year of didactics and were entrenched in their clinical practicum. This second iteration was significantly different from the first, due to the efforts of many. The training director of the PDP, Army Colonel Marvin Oleshansky, a superb psychiatrist and mentor, had caved-in to the complaining he heard from the first iteration fellows, and designed a didactic program that was half the length of the first iteration. Colonel Oleshansky convinced the program director of the PDP, Army Colonel and psychologist, Gregory Laskow, to run the new design up the flag pole and see how it was viewed. Indeed, we had a group to view it. Members of the American College of Neuropsychopharmacology (ACNP), an august group of nationally recognized experts in psychiatry, psychology, neurology, and pharmacology, were awarded a government contract to evaluate the performance of the fellows and the design of the PDP. They saluted the new curriculum, which cut the number of didactic hours by 698, down to a more reasonable total of 640 hours. (See table #3.)

The ACNP evaluation panel, made up of a balanced group of three psychiatrists and three psychologists, is very knowledgeable about training programs in psychopharmacology. Quarterly visits by the panel resulted in the examination of the many facets of the program. They assessed the degree of learning of the fellows by interviews and comprehensive two-day oral and written examinations given to the first iteration fellows on three occasions during their fellowship. The panel recommended the significant revision of the PDP, which has remained essentially the same from iteration #2 through iteration #4.

psychopharmacology

There have been minor adjustments in both the didactic and clinical portions of the fellowship since the major change from iteration #1 to #2. The number of didactic hours has increased by 20 hours between iterations 2 and 4; but more significantly, the type of other students in the classroom has changed. During iteration #1, the fellows completed all of their didactics with medical students. In the second iteration, the fellows had a major portion of their coursework presented only to the two of them in the classroom. This training was much more focused, cutting away the extensive amount of superfluous material for a psychologist who is training to prescribe psychotropics. Iterations 3 and 4 have linked-up with nurses at USUHS who are being trained as family practice nurse practitioners. Together they take core courses in the basic sciences.

Table 3

PDP Iteration #2

Course	Contact Hours	Differ Fr #1
Anatomy	54	↓287
Biochemistry	54	↓106
Neurosciences	42	included
Physiology	39	↓133
Pathology	102	↓113
Clinical Medicine	124	↓ 7
Clinical Concepts	100	0
Pharmacology	83	↓ 3
Clinical Pharmacology	21	↓ 26
Psychopharmacology	21	↑ 21
Total	640	↓698

The clinical practica have undergone a favorable transition. Iteration #1 was primarily a psychiatry inpatient experience. This had its advantage in the fellows being able to rapidly apply high doses of psychotropics to a large number of seriously mentally ill patients, who might also have concomitant medical illnesses. The downside of the first iteration's training was the lack of training in what we primarily do after the fellowship, out-patient work. What was initially believed to be valuable training in medicine with consultation and liaison psychiatrists, did not prove to be an efficient learning format. More training in medicine is being attempted by having the fellows work alongside physicians in family care or medicine services. The more recent practicum design essentially splits the one year experience between psychiatry inpatient and outpatient experiences.

Air Force Major James Meredith was the sole graduate of iteration #2 in June 1995. He hit the deck running at Malcolm Grow Medical Center at Andrews Air Force Base, Maryland, by being granted the privilege to prescribe under supervision very soon after his arrival.

Iteration number three, which began in July 1994 and ended in June 1996, made history by having the largest class, five fellows, and the first women. CDR Jim Parker was assigned to Naval Hospital Bremerton, Washington. Army Major Anita Brown, who in her pre-fellowship life was the assistant head of both the Practice Directorate and Education Directorate of APA, was transferred to Fort Hood, Texas. Air Force officers Major Elaine Rush went to Keesler Air Force Base, Mississippi, and CAPT Brian Pfeiffer went to Eglin Air Force Base, Florida.

Iteration number 4, possibly the last of the DoD PDPs, should graduate three fellows in June 1997. Those fellows are Army Major Debra Dunivan, Army CPT Tim Duke, and Navy LT Gilbert Seda. This will bring the total number of prescribing psychologists to 10.

I'll discuss the possible demise of the PDP later in the chapter, now turning your attention to some of the efforts by the APA and other groups to establish curricula and move toward prescriptive authority.

Other Prescription Privilege Efforts

The DoD's efforts to create some psychologists who prescribe ran parallel to efforts by the APA and other organizations. Senior APA officials, and most notably APA Presidents like Jack Wiggins, Ron Fox and Bob Resnick, gave strong support to the effort in the DoD and in the APA.

The issue of psychologists prescribing made its formal debut with the APA in August 1990 as the Council of Representatives voted 118 to 2 to establish a task force to study the issue. Many were surprised to discover that APA's highest policy body went on record supporting this highly controversial issue that might lead to the expansion of state practice acts (DeLeon, Fox, and Graham, 1991).

The seven member Ad Hoc Task Force on Psychopharmacology, along with a number of consultants and liaison personnel, thoroughly examined for greater than 18 months, the desirability and feasibility of psychopharmacology training for psychologists. They considered such diverse things as the changing paradigms of mental illness, educational and practice patterns of psychologists, research on the efficacy of psychological plus pharmacological therapies, and the implications for psychologists and consumers of psychologists prescribing psychotropics. The report included data on the distribution of psychologists and psychiatrists in the US, and groups with unmet mental health needs. In their conclusions, the task force recommended that all psychologists be trained to one of three levels regarding psychopharmacology. Level 1, known as Basic Psychopharmacology Education, would typically be taught to all students in psychology graduate schools, and cover the biological basis of behavior and an introduction to psychotropics. Level 2, labeled Collaborative Practice, would require a doctoral degree, and would add to level 1 training in-depth coursework on psychotropics as well as very basic knowledge of physical assessment and pathophysiology. This training would possibly permit a psychologist to help a prescribing clinician manage the psychotropic medications for their joint patient. In order to prescribe psychotropics, a licensed psychologist would need to complete the extensive level 3 training, known as Prescription Privileges. The Ad Hoc Task Force recommended that APA: oversee the development of the three different training programs; support legislation for licensing prescribers; encourage research on combination medication plus psychotherapy; work with various agencies to develop employment strategies to meet mental health needs (American Psychological Association, 1992).

With guidance and financial support from the APA, some states built prescription privilege task forces who would work on the many varied aspects of acquiring prescriptive authority. The California Psychological Association, which had been working on the issue of psychologists prescribing since the late 1980s, joined forces with the California School of Professional Psychology, our nation's largest producer of doctoral level psychologists, to create a model curriculum for prescription privileges. It was believed by many in the area of prescription privileges, that for legislative approval to occur, a well-accepted curriculum would have to be off the drawing boards, and possibly have been completed by some students. More than a dozen experts from varied disciplines related to medicine, pharmacology and psychology assembled in the Spring of 1994 to design the curriculum. While there was a good consensus about which courses needed to be taught, there was considerable disagreement about the number of hours for each course. The final product of this Blue Ribbon Panel of the Professional Education Task Force gave a range of the number of hours needed for each course. The five core areas of instruction and subcourses, listed later in this chapter, would eventually be adopted by the APA as its model curriculum (California Psychological Association and the California School of Professional Psychology—Los Angeles, 1995).

In addition to the APA's Ad Hoc Task Force on Psychopharmacology, its Committee on the Advancement for Professional Practice (CAPP), established a Task Force in April 1994 to examine the rationale for prescription privileges and curriculum designs in order to present up-to-date recommendations that might be used by the APA's Board of Directors. The CAPP Task Force endorsed the model curriculum developed by the California Blue Ribbon Panel, suggesting that the coursework required a minimum of 300 classroom hours. The Task Force proposed a model of 350 contact hours followed by a supervised practicum of at least 100 pharmacotherapy patients (Sammons, Sexton, and Meredith, 1996). The Task Force recommended that CAPP and APA provide support to State Psychological Associations (especially rural states), educate and work with organizations inside and outside of psychology, and develop legislative initiatives to advance prescription privileges.

A monumental occurrence for the prescription privilege effort occurred when the APA's Council of Representatives adopted a resolution on prescription privileges. An August 10, 1995 news bulletin from APA announced the resolution, which stated in part, that the practice of psychology encompasses the alteration of behavior by both physical as well as psychological interventions. The resolution stated that the association supports activities in seeking prescription privileges. The Council asked for the formation of a Prescription Privilege Conference Committee from a variety of committees and organizations like CAPP, and the APA's Board of Educational Affairs (BEA) and Board of Scientific Affairs (BSA). This committee considered curricula and guidelines developed by the DoD's PDP, the California Blue Ribbon Panel, and CAPP.

On August 12, 1996, the APA Council of Representatives approved two significant documents produced by the Prescription Privilege Conference Committee. The first was the Recommended Postdoctoral Training in Psychopharmacology for Prescription Privileges. This seminal document serves as the basis by which state psychological associations and other organizations can begin an 'approved' training curriculum. The prerequisites for this postdoctoral program are a doctoral degree in psychology and a current state license and listing as a health services provider psychologist. The didactic training should be provided by a regionally-accredited training institution and include at least 300 contact hours in five core content areas. The recommended subcourses and hours are listed in table 4. This didactic training must be followed by a clinical practicum which involves the psychopharmacological assessment and treatment of at least 100 patients under physician supervision (American Psychological Association, 1996).

The second important document approved by the APA Council of Representatives was Model Legislation for Prescriptive Authority. States may cut and paste the carefully-constructed legal language provided in this document to seek legislative authority for psychologists to prescribe. A key footnote suggests that the only group of mental health professionals who might be 'grandfathered' into prescriptive authority are those psychologists who have completed the DoD PDP.

Table 4

APA Curriculum

Course	Contact Hours
Neuroanatomy	25
Neurophysiology	25
Neurochemistry	25
Pharmacology	30
Clinical Pharmacology	30
Psychopharmacology	45
Developmental Pharmacology	10
Clinical Dependence; Pain	15
Pathophysiology	60
Physical Assessment and Labs	45
Pharmacotherapeutics	40
Total	350

The Future

Support for prescription privileges continues to grow. Surveys done in the mid to late 1980s found that a little over one-third of psychologists supported the notion that appropriately trained psychologists could safely and effectively prescribe. As psychologists become better informed of need that exists in our country and how psychologists certified to prescribe might fill it, the percentage of those psychologists who support this concept has grown. Examining surveys across a number of states indicates that the vast majority of psychologists, possibly 70 to 75%, support the idea of some psychologists prescribing. This has added fuel to the fire and prompted the banding-together of psychologists within their states to establish a state prescription privilege task force. At last count, 25 states had a task force. Some state psychological associations have formally endorsed their task force, while others might give it a 'special interest' status. These task forces might involve themselves in various political-educational efforts, curricula-implementing efforts and legislative action. Bills seeking statutory approval for psychologists to prescribe have been introduced in Hawaii, California, Indiana, Montana and Oregon (DeLeon, Sammons, and Sexton, 1995). According to recent communications with the APA's Practice Directorate, six states intend to introduce legislation for prescription privileges in 1997, and five states or regions are in the process of establishing prescription privilege (Level 3) training programs.

On January 11, 1997, students from the Illinois School of Professional Psychology, a member of the American Schools of Professional Psychology, sat in a classroom to hear a lecture on biochemistry. This was the first of 324 contact hours in the first-ever offering of a training program modeled after the APA-approved curriculum. Upon completion of the 324 classroom hours, the 'difficult-to-orchestrate' field practicum will begin. It is anticipated that students will assess and prescribe psychotropics 15 hours per week for one year under the supervision of board certified psychiatrists. This level 3 program appears to be the most promising one for producing the first prescribing psychologist outside of the pioneer military prescribing psychologists.

Level 1 and 2 training programs will soon follow. The Working Group on Psycho-pharmacology of the APA's Board of Educational Affairs has published a (level 1) Introductory Curriculum for Training in Psychopharmacology which includes nine modules that could be taught in approximately three college credit hours. Several state psychological associations plan offerings of this training.

The Working Group continues to refine the (level 2) curriculum for Psychophar-macology Training for Collaborative Practice with Specific Populations. It is anticipated that psychologists with this level of training, who work with special populations like the aged, the seriously mentally ill, children, and the developmentally disabled, will possibly be able to manage the psychotropic medications prescribed by another clinician for their joint patient. The Illinois School of Professional Psychology training program would permit the attainment of level 2 training enroute to level 3 (prescription privilege training).

Basic training in psychopharmacology, or more advanced training that sets the stage for prescribing, will likely become available in many venues. Free-standing training groups will advance the effort. Prescribing Psychologists Register, which has already given many hours of training to hundreds of psychologists, is improving their curriculum to more closely

Prescribing
Psychologists
Register

approximate the APA-approved curriculum. The movement for civilian psychologists to prescribe will probably snowball as a result of the efforts of these training institutions.

Unfortunately, the progress of the DoD training program has been slowed, but it is not dead-in-the-water. As is true about much in life, money and politics have considerable influence. The American Psychiatric Association's massive effort to stop the DoD PDP resulted in the U.S. House of Representatives passing a bill that would have stopped the PDP and stopped the graduates, who had already been prescribing, from continuing to prescribe. A Senate committee disagreed, which prompted a conference between the two houses of Congress. A compromise was reached between the Senate Armed Services Committee and the House National Security Committee. According to Section 742 of the House's Congressional Record of December 13, 1995, no new students will be enrolled, current PDP graduates may continue to prescribe, and the Comptroller General of the United States shall submit to Congress by April 1, 1997 a report evaluating the success of the PDP. The report was to include a cost-benefit analysis of the program, a discussion of utilization requirements, and a recommendation about whether or not the PDP should be extended.

On April 1, 1997, the U.S. General Accounting Office (GAO) sent to Congress their evaluation of the PDP. The title of their report (Defense Health Care, Need For More Prescribing Psychologists Is Not Adequately Justified) could serve as a one-line summary of their evaluation. The GAO report stated that the military health system spent $6.1 million to train the ten prescribing psychologists, even though the military has more psychiatrists than it needs. This stands in opposition to other data provided by military officials, the APA, and an extensive evaluation done by the Vector Research Group.

NBC news found this topic interesting enough to use it in its 'Fleecing of America' segment on Tom Brokaw's NBC Nightly News on April 30, 1997. As one might expect, the news segment had a prominent list to one side (to borrow a Naval term). It concluded that our nation's taxpayers were 'fleeced' in training the ten military psychologists to prescribe.

The future of the DoD PDP will be determined by Congress. The data could show that training military psychologists to prescribe is a cost effective way of providing quality mental healthcare. Although this may be true, money and politics will be powerful variables in the equation of whether or not the training program continues.

The ending of this chapter that you have just read, must occur. The ending of the chapter on military psychologists prescribing will hopefully never occur. The need for cost effective, quality mental healthcare that could reach the many who are underserved, was apparent to senior officials. They turned to those who are often pioneers. Those who frequently respond with "can do" (or for the Navy/Marine types "aye aye"). Military psychology rose to the occasion. This effort by military psychology may go down in history as a fine contribution to the welfare of mankind.

References

American Psychological Association. (1992). *Report of the Ad Hoc Task Force on Psychopharmacology of the American Psychological Association*. Washington, DC: Author.

American Psychological Association. (1996). *Recommended postdoctoral training in psychopharmacology for prescription privileges*. Unpublished document.

Balster, R. L. (1990). Predoctoral psychopharmacology training for clinical/counseling psychologists. *Psychopharmacology Newsletter, Division 28, American Psychological Association,* 23, 1.

Brief of the American Medical Association in Capp vs. Rank, Footnote No. 35 (1989).

California Psychological Association and the California School of Professional Psychology— Los Angeles. (1995). *Report of the Blue Ribbon Panel of the Professional Education Task Force.* Los Angeles: Author.

DeLeon, P. H., Fox, R. E., and Graham, S. R. (1991). Prescription privileges: Psychology's next frontier? *American Psychologist*, 46, 384–393.

DeLeon, P. H., Howell, W. C., Newman, R., Brown, A. B., Keita, G. P., and Sexton, J. L. (1996). Expanding roles in the twenty-first century. In R. J. Resnick and R. H. Rozensky (Eds.), *Health Psychology Through the Life Span* (pp 427–453). Washington, DC: American Psychological Association.

DeLeon, P. H., Sammons, M. T., and Sexton, J. L. (1995). Focusing on society's real needs: Prescription privileges for psychology? *American Psychologist,* 50, 1022–1032.

Enright, M. (1994). American Psychological Association responds. *Rural Health.* 1, 1–4.

Fox, R. E., Schwelitz, F. D., and Barclay, A. G. (1992). A proposed curriculum for psychopharmacology training for professional psychologists. *Professional Psychology: Research and Practice*, 23 (3), 216–219.

Office of the Inspector General. (1989). *Medicare drug utilization review.* (OAI–01–88–00980). Washington, DC: US Department of Health and Human Services.

Russo, N. F. (Ed.). (1985). *A woman's mental health agenda.* Washington, DC: American Psychological Association.

Sammons, M. T., Sexton, J. L., and Meredith, J. M. (1996). Basic science training in psycho-pharmacology: How much is enough? *American Psychologist*, 51, 230–234.

Senate Bill. (1995). *State of California SB777.* February 23, 1995.

Sturn, R. and Wells, K. B. (1995). How can care for depression become more cost effective? *Journal of the American Medical Association,* 273, 51–58.

Torrey, E. F., Erdman, K., Wolfe, S. M., and Flynn, L. M. (1990). *Care of the seriously mentally ill: A rating of state programs.* Arlington, VA: National Alliance for the Mentally Ill.

Reuters/Robert York/Archive Photos

13

Careers in Military Psychology

Martin Wiskoff, Ph.D.

* Introduction * Academic and Other Preparation * Work Settings for Military Psychologists * Professional Linkages * Range of Opportunities and Typical Careers * Employment Outlook and Financial Compensation * Advantages and Disadvantages of a Career in Military Psychology * Why I Chose Military Psychology * References

Introduction

Why consider a career in military psychology? As I will describe in this chapter, military psychology is a microcosm of psychology and consequently offers opportunities to psychologists of all persuasions, including those who wish to spend their career or a portion of it in a military uniform. It is also a discipline that crosses international boundaries, with military psychologists found in many countries. While my discussion is focused on military psychology in the United States, the problems addressed by the research are of concern to the militaries of all nations and there is the potential for cross-national research efforts, technical exchanges, teaching in overseas extension programs and even assignments to serve jointly with military forces of other nations.

Perhaps of greatest importance, military psychology offers the opportunity to make a significant difference in the lives of individuals and in the stability of our nation. A small sample of the types of contributions that can be made by military psychologists include (a) working in mental health or family counseling clinics to improve the lives of service personnel and their families; (b) performing research on the effects of battlefield environmental factors on soldiers in order to prevent or reduce battlefield casualties; and (c) analyses of humanitarian and peacekeeping missions to determine procedures that could save military and civilian lives.

I will first review the academic preparation for the field and the work settings in which civilian and uniformed military psychologists are employed. Subsequent sections cover professional linkages, the range of opportunities and typical careers, and the employment outlook including financial compensation. I conclude the chapter by presenting my view of the advantages and disadvantages of a career in military psychology as well as a personal perspective from my long association with the profession.

Academic and Other Preparation

Military psychology is unique in that being a microcosm of psychology, it includes many entry paths. Only a few individuals enter graduate school with the idea of becoming a military psychologist, and those students most likely obtained their undergraduate degrees under some form of military sponsorship.

The level of education needed for a career depends on the particular area of specialization and the employment setting. At a minimum, a master's degree is required for conducting research at a laboratory and working in support roles at military schools, bases, and government headquarters. A Ph.D. is more likely to be needed for academic positions and clinical practice at hospitals or clinics. For all jobs, the Ph.D. opens up a wider range of opportunities for advancement and independence in choosing a career direction. The Ph.D. also confers a certain level of status that is often valuable in dealing with military personnel.

The most appropriate type of education also depends on the particular area of specialization. Most students entering graduate training immediately after obtaining a bachelor's degree have not yet set their career sights on becoming a military psychologist. Some graduate students serve as research assistants to a professor who is conducting research under military sponsorship. Often military psychology becomes an attractive option for an individual seeking employment after obtaining an advanced degree, because a particular military institution is interested in the student's area of specialization.

Preparation for a research career in military psychology may be obtained at most universities that have accredited graduate-level psychology programs. Those interested in working in a particular military research laboratory should review available descriptions of work conducted at the laboratory and tailor their research specialization to coincide with emerging areas of interest. It may be possible to develop a doctoral dissertation using existing data from a military laboratory or by collecting data on military subjects.

The path for entry into the field can also involve joining the military. The Army, Navy, and Air Force offer research and clinical opportunities for uniformed psychologists. Students may enter one of the service academies with the goal of a research subspecialty in psychology. They may also join the military through other officer commissioning programs such as the Reserve Officer Training Corps (during college), Officer Candidate School (after college graduation), or direct commission as an officer (on receipt of a Ph.D.)

For more than 50 years, commissioned officers trained as clinical or counseling psychologists have served in the military. As a means of obtaining uniformed clinical psychologists, the military offers internships in basic clinical skills, the provision of acute care and short-term treatment, and the handling of combat crises. Some psychologists provide their required educational support payback service to the military and then return to civilian life; others choose to complete a 15-year or longer career in the military. Rath and Norton (1991) have provided an excellent description of the education and training of uniformed military clinical psychologists.

The attributes contributing to success in military psychology, although somewhat dependent on the particular area of work, are shaped by the nature of military organizations. A critical skill is the ability to communicate well both orally and in writing. Given that a military tradition of presenting information is the "briefing" using visual aids, most researchers will need to communicate technical details of their project, as well as its applicability to military problems, in non-technical terms understandable to a lay audience. Also, because most research laboratories publish their findings in technical reports, good writing skills are helpful for advancement.

One additional requirement for most military psychologists is the need to interact with operating forces. Typical peacetime interactions involve test instrument administration, training evaluation, and clinical support to troops and their families on military bases. Studies addressing combat or its aftermath could involve on-site assessment of battlefield performance or evaluation of combat stress. Other psychologists may be assigned to operational, highly classified special forces units.

Work Settings for Military Psychologists

Military psychologists often work in a broader range of settings than would be the case for most other psychological disciplines. We typically think of researchers as being employed in laboratories or in academia, and clinicians performing their services in hospitals or clinics. Because of the large number of bases, schools, offices and other sites under military jurisdiction, however, there are opportunities for assignment at many different locations in the United States and abroad. For example, the Walter Reed Army Institute of Research, headquartered in Washington, DC, has field offices both in the United States and in Heidelberg, Germany. It is not unusual for civilian military psychologists to serve at several geographical locations during their careers; it is expected of uniformed psychologists. Temporary assignments to serve the troops in combat zones, develop studies, collect data, present research findings, and so forth, are commonplace. Table 1 displays the six major types of settings in which military psychologists are located.

Table 1

Military Psychology Work Settings

Major Settings	Examples of Locations
Research facilities	Military laboratories and field units; contractor offices
Educational facilities	Colleges and universities; military educational institutions
Medical centers, hospitals and clinics	Military hospitals; outpatient clinics; mental health centers; drug treatment centers; prisons
Military schools and bases	Service training schools; military bases in the U.S.
Military deployments overseas	Military overseas bases and small missions; combat zones; military hospitals
Military organization offices	The Pentagon; service headquarters commands

Research Facilities

By far the largest number of civilian military psychologists work at large government laboratories or smaller facilities, performing the wide range of research outlined earlier. Laboratories can employ from a handful to hundreds of individuals. In some settings psychologists are the predominant professional group; in others, a few psychologists may be an integral part of multidisciplinary teams. Several laboratories are located on bases with operational military units which facilitates interactions with military personnel and data collection. Other organizations occupy leased commercial space.

A considerable number of other researchers are located at contractor offices across the United States, some of which may be co-located with government offices. A few contract organizations are similar to service laboratories in having a long history of military psychological research. Whereas in past years many contractors were solely devoted to DoD-supported research, recent reductions in defense spending reductions have spurred contractors to seek a mix of clients, including DoD, other government agencies, and private sector organizations.

Educational Facilities

Psychologists at many universities perform research under contract to military organizations. In some instances a faculty member and a small number of graduate students will work alone under a grant or contract. At the other extreme, some universities have set up institutes dedicated entirely to government-sponsored research. Many of these programs have existed for many years and support specific military needs for research and product development. Clinical research psychologists may find themselves at schools of medicine performing interdisciplinary medical research.

The military is also unique in having its own educational institutions such as military academies; command, staff, and war colleges; graduate education facilities (e.g., the Naval Postgraduate School, Monterey, CA); and special military colleges and schools (e.g., the Air Force Institute of Technology, Wright-Patterson AFB, OH). Psychologists as faculty members in these institutions teach classes, perform research, consult, and hold staff positions.

Medical Centers, Hospitals, and Clinics

Clinical psychologists serve in the widest variety of settings, among which are military hospitals, outpatient clinics, mental health centers, day care centers, prisons and shipboard duty. The military has major medical centers with modern facilities where the psychologist is a member of a team of medical professionals. Psychologists may also serve at smaller clinics and hospitals and at bases in isolated locations where they may have to function without much additional support. One of the characteristics of the uniformed psychologist's job is that there is generally a change of geographical location every 2 to 3 years.

Military Schools and Bases

The military operates a vast system for training enlisted and officer personnel. Military psychologists perform both research and educational support tasks at many entry-level and advanced training schools. Clinical psychologists serve as advisors to military base commanders concerning troop health and readiness. Psychologists also function as members of special forces and military intelligence units.

Military Deployments Overseas

Some military psychologists may spend significant time in overseas field settings, perhaps even in hostile environments such as Kuwait during the Gulf War. Clinical psychologists often go wherever there are concentrations of troops. They may be assigned to special operations units to assist in preparing troops for combat, or to help evaluate soldier stress during humanitarian missions (e.g., Operation Restore Hope in Somalia). Researchers may be assigned to temporary duty overseas to collect data from U.S. operating forces.

Military Organization Offices

A final group of military psychologists holds positions within the operational and headquarters commands of each of the services and the Office of the Secretary of Defense. Some of these individuals are responsible for budgeting, planning, and monitoring the research programs of military laboratories and contractors. Still others are involved in developing policy by providing information and guidance to military decision makers.

Professional Linkages

One of the most advantageous features of a career in military psychology is the opportunity to establish linkages with professionals performing similar work in other countries. In addition to interactions through professional societies, there are unique military-sponsored organizations for information exchange. For example, some researchers have had the opportunity to participate in NATO and other cross-national working groups on technical subjects of interest to a wide range of nations. An annual open meeting of interest to many is the International Military Testing Association that has met not only in the United States, but in Canada, Europe and Australia. The U.S. Air Force sponsors a biannual Psychology in Defense Symposium at the Air Force Academy which is attended by an international audience of psychologists.

Military psychologists have the opportunity to join national and local professional organizations that reflect their specific research interests. They can also publish the results of their research in a host of journals that cover the diverse areas of interest within the field.

The primary identification for many military clinicians and researchers is Division 19—the Division of Military Psychology—of the American Psychological Association (APA). Division 19 offers an affiliate membership status for students and for psychologists who are not APA members. It is common for military psychologists to belong to other APA divisions, such as Experimental Psychology (Division 3); Evaluation, Measurement and Statistics (Division 5): Clinical (Division 12); the Society for Industrial and Organizational Psychologists (Division 14); Applied Experimental and Engineering (Division 21); Health (Division 38); and Family (Division 43). Many are members of other professional organizations such as the American Psychological Society, the Human Factors Society, the Inter-University Seminar and the Academy of Management.

The Division of Military Psychology publishes its own quarterly journal, *Military Psychology,* which features original behavioral science research and scholarly integration of research findings performed in a military setting. *Military Psychology* has published contributions from a number of countries and has featured special issues on topics of particular interest to the military research community: team processes, training and performance; women in the navy; stimulants to ameliorate sleep loss during sustained operations; military service and the life-course perspective; and military occupational analysis. Other special issues scheduled for publication in 1997 and 1998 include the enhanced computer-administered test (ECAT) battery, the impact of chemical protective clothing on performance, and equal opportunity research in the U.S. military.

Range of Opportunities and Typical Careers

The possibilities for a rewarding career are great because military psychology encompasses so many different types of job opportunities. It is possible to work in a relatively narrow area of specialization throughout a career, to move across various research areas, or to pursue research management opportunities. It is not unusual for researchers to remain in Civil Service for their entire career, either within one laboratory or in a mix of military laboratory and headquarters jobs. Others select a combination of environments including academia, government employment, and working for a government contractor. Military researchers may function as members of a multidisciplinary team of scientists, either within the same laboratory or across organizational settings.

Uniformed psychologists can devote a sizable portion of their career in uniform or spend just a few years in military service. Some maintain a military connection after a short period of active duty by serving in the reserve forces. Many retire from the military in their forties and pursue a rewarding second career.

A career as a military psychologist is perhaps best understood through examples. Following are two examples of possible careers within the field.

Laboratory Research Psychologist

A typical psychologist working at a government research laboratory would begin a civil service career immediately after obtaining a master's or doctoral degree. The first assignment would probably find the researcher working within his or her graduate area of specialization as a member of a research team. Early in the career, the researcher would be given increasingly greater independence, including the development of research designs and planning for new projects. Project responsibilities might include interaction with other service laboratories and monitoring contractual programs performed by researchers within academia and private contractor organizations. Other responsibilities would be added over time, such as justifying the relevance of the work to military problems and obtaining support from military sponsors.

The researcher would be expected to publish findings in laboratory technical reports or in refereed journals. Other significant opportunities for professional growth include presentations at professional meetings and interactions with colleagues in the United States and other countries. Some researchers might become adjunct professors at local universities with responsibilities that could include part-time teaching and overseeing student thesis preparation.

Continued career growth would be reflected by promotions to head teams and larger organizational segments within the laboratory. Although increased managerial responsibility is an expected progression, some laboratories offer a limited number of dual career paths to accommodate personnel who wish to remain scientists. Tours of duty at other laboratories and at headquarters sites are a distinct possibility.

On retirement from the civil service, researchers might continue to pursue professional interests by maintaining contact with their laboratories in an emeritus status. Or, using their accumulated expertise and experience, they might pursue a second career in academia or with a research contractor organization.

Uniformed Psychologist

Uniformed psychologists typically fall into either a research or a clinical track. I will focus on the latter because the career of the uniformed research psychologist is very much like that of the government civil servant described above. The following 20-year career description is adapted from Rath and Norton (1991).

A uniformed clinical psychologist would serve a 1-year internship at one of several U.S. military sites after completing 4 years of graduate school. With receipt of the doctoral degree, the clinician would be assigned for 3 years to a hospital on a military base in the continental United States. This would likely be followed by another 3-year assignment to a military hospital overseas. At this point the officer might attend a 6-month officer development course and then go to a 1-year fellowship. This could be followed by assignment as a director of a small residential alcohol rehabilitation center or as the staff psychologist for a large military unit. The next assignment might be as a faculty member at a service academy or other military school.

At this point the officer would probably attend a 9-month command and staff course. The remainder of the career could be spent in a combination of certain nontraditional assignments (such as in survival schools or medical intelligence) or in a hospital setting as a chief psychologist. In addition to the training mentioned above, the officer would be expected to attend other professional leadership and specialized military education in order to advance in rank and responsibility.

On retirement from the military at a young age (mid-40s), the clinical psychologist would be likely to open a private practice, work in a clinic or hospital setting, or serve as a consultant to government and industry. Many clinicians maintain their association with the military in some capacity during fruitful second careers.

Employment Outlook and Financial Compensation

Even before the Cold War era officially ended in the late 1980s, the military started downsizing. This trend continued through the mid-1990s. Military research facilities suffered a corresponding reduction in funding during this time, accompanied by hiring freezes, reductions in the number of research and support personnel, and reductions in funding for university and private research contracts.

It is unlikely in the near future that the military psychological research field will reach the high level of funding and personnel staffing it attained by the early 1980s. A changing world situation, however, would definitely influence the level and type of support required from military psychologists. Research with clear application to solving military problems stands the best chance of being funded in the future; for example, research from which the findings or products can directly improve the performance of personnel in combat by providing immediate feedback to field commanders and information for policy makers. Given government emphasis on cost reduction, research that promises to reduce the costs of military operations will also be looked upon favorably.

The opportunities for uniformed clinical internships has also decreased and will probably decline another 15% to 20% as military downsizing continues. In contrast, there may be future opportunities for post-doctoral fellowship training. The uniformed research community will also face numerical reductions and restrictions in assignment opportunities. One potential growth area is that of special forces operations.

In general, the salary scale for military psychologists is aligned to government pay scales. Those entering Civil Service employment with a master's degree but without prior experience can generally expect to receive a salary at the GS-9 level. In 1996 that starting annual salary was $30,658. The GS-11 level (starting at $37,094 in 1996) is generally given to entering researchers with a Ph.D. Individuals with an exceptional work history could be hired at a higher level, such as GS-12 for a new Ph.D. (starting at $44,458 in 1996). The high end of the government salary scale (GS-15) is more than $95,000 in 1996, and a few psychologists in the Senior Executive Service make more than $120,000.

Salaries at research contractor organizations heavily involved in government research tend to approximate the Civil Service scale, although there is greater flexibility in setting actual dollar amounts. Other entry-level salaries, such as at universities and with contractors primarily engaged in industrial/organizational research, are defined by the particular institutions and organizations.

Uniformed psychologists receive a package of pay and benefits that depends on the particular service program and rank at which an officer is commissioned. Psychologists going through a military academy start at the lowest entry level, whereas clinical psychologists entering to serve an internship start at a slightly higher rank.

Advantages and Disadvantages of a Career in Military Psychology

A compelling reason to enter the field for many military psychologists is the security of government employment and the unique benefits of the Civil Service and military retirement systems. While reductions in force and changes in the government retirement system have diminished these advantages somewhat, government service still promises reasonable stability and the opportunity to make the types of contributions indicated earlier. There are additional reasons why employment in military psychology is attractive.

Given the military's need and penchant for the most current battlefield and management technologies, much military research is at the cutting edge of science. Military laboratories offer a psychologist the unique opportunity to conduct research without collateral requirements to teach or consult. Laboratory personnel can establish a career path to include increasing research management responsibilities and possible service in decision making roles within government. Uniformed psychologists have unusual opportunities to perform research, to provide clinical services in a unique environment, or to consult on matters of international importance.

There are also advantages to employment as a military contractor or as an academic doing research for the military on a contractual basis. Military contractors enjoy a wider range of research opportunities and the likelihood of mixing military research with other public and private sector work. Academic psychologists have the opportunity for independence in research focus and can tailor the direction to fit military needs and interests. In their research they often have access to a pool of research subjects that is not easily available elsewhere. A Science Directorate article in the April, 1995 APA Monitor stated: "Each year, 50 scientists at the Air Force's Armstrong Laboratory Lackland Test Facility in San Antonio, Texas, and 10 to 15 university researchers working on grants from the Air Force, have access to more than 35,000 Air Force recruits for four hours of their basic training." Whether the task is to evaluate training methodologies, conduct field experiments, collect survey data, or administer experimental test instruments, the opportunities for performing research on significant samples of relevant subjects are unparalleled. The military also maintains large-scale computer data bases on its population, going back more than 30 years, that are available for cross-sectional and longitudinal analyses.

Most military research has important applications in the private sector as well. Joint government-industry undertakings are becoming commonplace. Military issues and technologies cross national boundaries, and the international community of military researchers shares information at military and professional conferences and during exchange visits.

The major disadvantage of a career focus on military psychology is the uncertainty of government funding in the future. Opportunities may be more limited for laboratory researchers, uniformed psychologists, and contractors. Although salary levels are generally competitive with the private sector, Civil Service and military pay scales have more clearly defined upper limits than those for psychologists performing research or consulting in the private sector.

Why I Chose Military Psychology

The field of military psychology has provided me with support, friendships, and a full and rewarding professional career. While I was completing my Ph.D. at the University of Maryland, I started working for the Army Research Institute in Washington, DC. I had the opportunity to conduct interesting research in areas of testing and recruit assignment, and I also received invaluable assistance from my fellow psychologists and the ARI in completing my dissertation. After 4 years I moved to the Navy and spent 24 years in research, first in the Washington, DC area, at the Bureau of Naval Personnel and then in San Diego, CA, at the NPRDC. During my Washington, DC, employment I accepted an invitation to be a visiting professor for 1 year at the Naval Postgraduate School in Monterey, CA. This started a long term love affair with Monterey that eventually led to my spending the last 3 years of my civil service career there as a senior scientist with a newly formed DoD organization, the Defense Personnel Security Research Center.

During my career I was given increasing responsibility that culminated in my assignment as the director of the Manpower and Personnel Laboratory at the NPRDC. I am proudest of the applications of our research, such as the selection system for the U.S. Naval Academy, that have made a difference to the Navy and the lives of military personnel. I was able to help initiate a major new DoD program in the emerging technology of computerized adaptive testing, and I served as the officer-in-charge of the program. Since my retirement from civil service at age 55, I have been heading a small contractor research group in Monterey, CA conducting fascinating research into aspects of personnel security, counter-intelligence and espionage.

Military psychology has allowed me the opportunity to present and publish papers and to participate actively in my profession. A major career highlight was helping to create the APA journal *Military Psychology* that was first published in 1989. I have served as editor since that time.

My work and associations have crossed international boundaries. Several research programs have involved travel to different countries and joint projects with psychologists in these countries. Many of the manuscripts that I review for *Military Psychology* are submitted by authors residing outside of the United States. Articles and special issues in *Military Psychology* reflect some of the current research emphases. A good source of information on research programs within the service laboratories can be obtained from their documents such as annual reports. The laboratories also publish technical reports on the results of their research as well as bibliographies of these reports. The References contain selected recent publications that cover in greater depth various aspects of work in military psychology.

References

Defense Technical Information Center (1995). *Directory of Researchers*. MATRIS Office, DTIC-AM, San Diego.

Dillon, R. F., & Pellegrino, J. W. (Eds.) (1989). *Testing: Theoretical and applied perspectives*. New York: Praeger.

Driskell, J. E., & Olmstead, B. (1989). Psychology and the military: Research applications and trends. *American Psychologist, 44*, 43–54.

Ellis, H. A. (Ed). (1986). *Military contributions to instructional technology*. New York: Praeger.

Gal, R., & Mangelsdorff, A. D. (Eds). (1991). *Handbook of military psychology*. New York: Wiley.

Hunt, J. G., & Blair, H. D. (Eds.) (1985*). Leadership on the future battlefield*. New York: Pergamon-Brassey.

McGuire, F. (1990). *Psychology aweigh! A history of clinical psychology in the United States Navy*. Washington, DC: American Psychological Association.

O'Brien, T. G., & Charlton, S. G. (Eds.) (1995). *Handbook of human factors testing and evaluation*. New Jersey: Lawrence Erlbaum Associates, Inc.

Rath, F. H., Jr., & Norton, F. E., Jr. (1991). Education and training: Professional and paraprofessional. In R. Gal & A. D. Mangelsdorff, A. D. (Eds.) *Handbook of military psychology* (pp. 593–606). New York: Praeger.

Rumsey, M. G., Walker, C. B., & Harris, J. H. (Eds.) (1994). *Personnel selection and classification*. New Jersey: Lawrence Erlbaum Associates, Inc.

Sands, W. A., & McBride, J. R. (Eds.) (In press). *Computerized adaptive testing in the Department of Defense*. New Jersey: Lawrence Erlbaum Associates, Inc.

Solomon, Z. (1993). *Combat stress reaction: The enduring toll of war*. New York Plenum.

Taylor, H. L., & Alluisi, E. A. (1994). Military psychology. In *Encyclopedia of human behavior*, Volume 3 (pp. 191–201). New York: Academic Press, Inc.

Wigdor, A. K., & Green, B. F., Jr. (Eds.) (1991*). Performance assessment for the workplace*. Washington: National Academy Press.

Wiskoff, M. F., & Rampton, G. L. (Eds.) (1989). *Military personnel measurement: Testing, assignment, evaluation*. New York: Praeger.

Zeidner, J. (Ed.) (1986). *Human productivity and enhancement:* Volume 1: Training and human factors in systems design. New York: Praeger.

Zeidner, J. (Ed.) (1987). *Human productivity and enhancement:* Volume 2: Organization, personnel, and decision making. New York: Praeger.

Zeidner, J., & Drucker, A. (1983). *Behavioral science in the Army: A corporate history of the Army Research Institute*. Alexandria, VA: U.S. Army Research Institute for the Behavioral and Social Sciences.

14

The Future of Military Psychology

Christopher Cronin, Ph.D.

* Introduction * Advances in Technology * Budget Reductions and
the Military Drawdown * Political Changes * Changes in Personnel
Demographics * The Future of Military Psychology * Summary
* Conclusion * References

Introduction

This chapter will tackle the rather ambitious task of predicting how military psychology will develop in the future. The military holds a position of high visibility in society due to its political importance. Public, political, economic, and social pressures can all have a major impact on the military environment. Recent events ranging from the fall of the Berlin Wall to the military's sexual harassment cases all impact upon military psychology. These recent events also caution us from making predictions about the future; who could have predicted the fall of the Berlin Wall and subsequent expansion of NATO?

With this caveat in mind, this chapter will identify four trends which will influence military psychology in the coming years: (1) advances in technology, (2) budget reductions and the military drawdown, (3) recent political changes, and (4) changes in personnel demographics. The chapter will also describe the recent growth in the field of military psychology and includes comments on the future growth of military psychology. This chapter draws heavily from the preceding chapters regarding future directions.

Advances in Technology

The past several decades have witnessed an exponential growth in technology, both in terms of capability and availability. Personal computers were first mass-marketed as recently as the early 1980s. Many of our current laptop computers are more powerful than the large mainframe computers of the 1960s and early 70s.

With the use of computers, psychology has changed into a high-tech field (Wiggins, 1994). Psychologists of the future will need to feel at ease with technology. Computers will continue to play a significant role in the work of psychologists, both in the military and private sectors. Although technological advances will impact upon all branches of psychology, it is likely that these developments will have a significant impact on the work of military psychologists. Despite recent budget reductions, the military has a wealth of resources to draw upon. Additionally, technological changes which increase combat readiness and reduce costs will have a high priority in the military environment.

In addition to the obvious role computers play in research (data collection and analysis, and the dissemination of results), computers will undoubtedly improve selection and classification, human factors engineering, training, diagnosis and treatment, and information warfare.

Perhaps the most obvious area where computers will have a significant impact is the area of selection and classification. We can anticipate a major increase in computerized assisted testing and assessment services. Computerized adaptive testing (CAT) is becoming more common in both the military and the private sector. CAT technology has been applied to the ASVAB (Wiskoff & Schratz, 1989) and the Educational Testing Service now offers a computerized version of the Graduate Record Exam (GRE). The Graduate Management Admission Test (GMAT) is already available on computer (since Fall, 1997) and it is anticipated that the Scholastic Assessment Test (SAT) will be computerized in most urban areas by the year 2003.

Computerized adaptive testing

In CAT, test items to be administered are selected based on estimates of an examinee's ability calculated from his or her responses to previously administered items. If an examinee misses an item, the next item selected for administration is easier; if an item is correctly answered, the next item is more difficult. In the future, CAT technology will move beyond multiple-choice questions and become a more multidimensional test. Respondents will need to plot graphs, work on spreadsheets and draw diagrams to accompany their answers.

CAT technology has already demonstrated several benefits, including improved testing efficiency and increased acceptance of test taking by recruits. Future areas of research will include using computer-assisted testing to understand individual cognitive strategies as well as evaluating traditional psychometric constructs, such as validity and reliability. For an overview of computer-assisted and computerized adaptive testing in the military, see Steege & Fritscher (1991).

In the future, psychologists involved in selection and classification may also be using biological tests of intelligence, cognition, interests, attitudes, personality styles and predispositions (Matarazzo, 1992). Such biological measures will assess individual differences in brain functions at the neuromolecular, neurophysiologic and neurochemical levels. We can also anticipate human-computer interfaces and computer-mediated teletesting (Schlosser, 1991).

Psychologists are also expanding the use of computers in survey research. Computerized surveys on diversity issues have been found to have advantages over the traditional paper-and-pencil methodology (Edwards, Rosenfeld, Booth-Kewley & Thomas, 1996). Other paper-and-pencil selection tests have been automated with comparable success (e.g., Griffin & Koonce, 1996) as well as diagnostic tests (Roper, Ben-Porath, & Butcher, 1995). Computers have also been successfully used in training (Ricci, Salas & Cannon-Bowers, 1996).

A byproduct of the use of computers has been the development of statistical techniques used in the behavioral sciences. Computers have increased the use of sophisticated statistical procedures, such as structural equation modeling (SEM) and multidimensional scaling

(MDS). Psychologists will need polished statistical skills (e.g., Medsker, Williams & Holahan, 1994). Statistical skills are necessary for psychologists involved in research as well as work in applied settings. A thorough understanding of statistical procedures is as essential for interpreting a longitudinal study on leadership as for understanding a clinical assessment battery or an empirical study on the efficacy of treatments for PTSD.

Virtual reality will play a large part in the work of military psychologists. Clinical/counseling psychologists will be able to treat individuals with various disorders through the use of virtual reality (Wiggins, 1994). For example, clinicians can help clients work through phobias, such as the fear of heights (acrophobia), in the safety of the therapist's office by simulating high places. Virtual reality will allow researchers to study human behavior under a wide variety of simulated environments and training conditions. Simulator networking (SIMNET; Alluisi, 1991) will allow for multiple international teams to interactively conduct war-games in real-time across geographic and political boundaries (Oswalt, 1993).

Advances in technology will create new challenges for the developers of human-machine systems. It is reasonable to predict that future advances in technology will eventually outpace human capabilities (e.g. Lovesey, 1995). Military psychologists will need to identify human capabilities and limitations to develop effective, user-friendly human-machine systems. With the increases in technology, psychologists will be challenged to help humans attend to and process the wealth of available information. Computers which can detect and respond to human behavioral, physiologic, and cognitive changes will be further refined and introduced into human-machine systems. Military psychologists will continue to lead the way in research on cognition and information processing.

Advances in technology will also affect recruitment and training needs. Personnel will need extensive training to operate and maintain technologically advanced weapon and communication systems. The military will need to compete with the private sector in recruiting and retaining highly qualified personnel. Frequent training and updating of skills will be necessary to maintain combat readiness. Sophisticated weapon-systems will rely more on the use of teams for effective operation. The study of teams and teamwork will continue to be an important focus of research. Research on the identification of the KSAs required for teamwork will continue to progress (Borman, Hanson & Hedge, 1997).

Finally, the expanding Internet needs to be included in any predictions regarding technological advances in the future. Psychologists will have ready access to a much larger quantity (and quality) of information than ever before. There is no doubt that the information superhighway will greatly enhance scientists' ability to conduct research and communicate their results. However, as Bloom discusses in Chapter 8, the Internet will also be used as a tool in information warfare (cyberwarfare) and for access to sensitive government and private information. Psychologists engaged in IW will need to make full use of this tool while protecting their own databases from infiltration.

Students wishing to pursue a career in psychology will need to embrace technological advances during their academic careers. Graduates with a solid understanding of computer technology and statistics, including but not limited to use of the internet, electronic databases, multimedia, statistical software (such as the Statistical Package for the Social Sciences; SPSS) and basic computer operating systems will be sought after by the military.

Budget Reductions and the Military Drawdown

The end of the Cold War and reorganization of the former Soviet Union has produce reductions in the defense budgets of the majority of western nations. These budget reductions and the military drawdown continue with the DoD recently calling for further base closures. Cost concerns will impact on military psychologists in a number of ways.

The future will see a greater emphasis on applied as opposed to basic or theoretical research. Although the distinction between the two types of research is often blurred, basic research attempts to identify the underlying principles or processes of a particular behavior whereas applied research attempts to solve practical problems, often using principles discovered by basic research. The reduction in research funding will also generate more multi-national cooperative projects as research costs continue to rise and defense budgets are trimmed (Krueger, Chapter 2, this text).

The reduction in forces has also witnessed an increase in the number and frequency of deployments. The military has recently initiated one-year hardship tours (unaccompanied by family members) for service members stationed in Europe. Forced-separation has a significant impact on retention in the military. The challenge facing military psychologists is to develop ways to reduce the stress of family separation for the deployed service member as well as the member's spouse and children (Bartone, Chapter 7, this text).

Retention will become more critical for the military in the coming years. The number of eligible recruits has decreased, pitting the military against the private sector to recruit and retain personnel. Driskell and Olmstead (1989) identify three ways in which the military will need to respond to this competition for manpower. First, issues such as quality of life and family concerns will play a role in retention. Second, the military will need to increase training to make better use of the available human resources. Third, user-friendly systems which are easier to operate and maintain will need to be developed.

Budget constraints will also affect the number of authorized billets for psychologists in the Armed Services. The civilian health care system has moved toward a system of managed care, with the emphasis placed on the role of primary care. Psychologists in the private sector are only gradually finding their way through the changes brought about by managed care. Recent state and national legislation regulating the managed care industry suggests that the role of psychologists will continue to change. Similarly, the military's Medical Health Services System has adopted the Tri-care health care system for active military personnel, family members, and retirees. Military psychologists involved in clinical work will be finding their way through this new system just as their civilian counterparts grapple with the changes imposed by managed care (Kelleher, Talcott, Haddock & Freeman, 1996).

Base closures and downsizing will unquestionably account for the loss of some authorized billets for uniformed psychologists and fewer training opportunities. However, in addition to their role as treatment providers (Page, 1996), psychologists are playing a larger role in nontraditional health care areas. Professional psychologists' ultimate concern is with problems involving human coping skills and effective human behavior (Fox, 1994). Psychologists will expand their professional roles beyond that of health care providers to human behavior specialist. As discussed in the previous chapters, military psychologists now function as organizational consultants (Johnson, Chapter 10, this text) and as members of disaster response teams and on deployments for peacekeeping and humanitarian operations

(Bartone, Chapter 7, this text). Military psychologists, with their expert knowledge of psycho-social factors and human behavior, will work more in the area of prevention, helping service members maintain healthy lifestyles (e.g. LEAN program) and preventing service-related disorders, such as PTSD (e.g. Armfield, 1994; McCarroll, Ursano, Fullerton & Lundy, 1995).

The push for prescription privileges will help psychologists maintain a strong presence in the health care field. As discussed by Sexton (Chapter 12, this text) many civilian clinical psychologists support prescription privileges. Although the DoD PDP may come to an end due to pressures from special interest groups, it is unlikely that the issue of prescription privileges for psychologists will go away. Rather, military psychologists have paved the way for psychologists in the private sector. States will soon be introducing legislation for prescription privileges and training programs have been established. It is likely that this action will facilitate a change in how and where professional psychologists are trained and credentialed. Scott (1991) has suggested that the teaching of clinical psychologists will eventually move out of academic psychology departments and others have called for the Psy.D., as opposed to the Ph.D., degree for all practitioners (Shapiro & Wiggins, 1994). Students planning careers as clinical psychologists are advised to take the appropriate preparatory courses for these forthcoming changes at the undergraduate level. Suitable courses include chemistry, biology, anatomy, physiology, psychopharmacology and physiological psychology. Indeed, many graduate programs in psychology already recommend a strong foundation in the biological sciences.

Political Changes

Recent political changes such as the fall of the Berlin Wall and restructuring of the former Soviet Union will also affect the military's mission. These recent changes have already brought about the proposed expansion of NATO to include former Warsaw Pact countries. These changes will affect military psychologists in a variety of ways, ranging from basic research to training.

U.S. Forces will undoubtedly train and deploy with units from these countries. Military psychologists can play an active role in facilitating intercultural teamwork among nations. The understanding of different cultural organizational and leadership styles will help ensure mission success. Troops stationed in these host nations will benefit from intercultural training which addresses the former "adversary" status of these countries. Training needs must also address the potential future adversaries in the event of war (Schwalbe, 1993). Psychologists will need to consider the environmental and psychological challenges to training and combat in the potential "hotspots".

The last few years have seen a change in the type of military operations in which U.S. Forces function. There has been an increase in humanitarian (e.g., Operation Provide Promise, Croatia), peacekeeping (e.g., Operation Restore Hope, Somalia) and disaster response operations, often using multi-national forces under the auspices of the United Nations. The U.S. government has recently re-affirmed the role of the United Nations as demonstrated by the payment of overdue dues. Military psychologists can play a role in ensuring the success of these joint operations through intercultural training. Additionally, psychologists will need to identify stressors unique to these types of operations in which personnel trained for combat are asked to perform an extended humanitarian mission (Bartone, Chapter 7, this text).

The area of cultural diversity and gender issues will also develop an international perspective as U.S. troops deploy with other nations whose troops do not have the cultural or gender diversity found in American units. For example, troops from nations which restrict military service to males only may experience difficulty interacting appropriately with a female officer.

Military psychologists will also enjoy increased collaboration with their international colleagues. These efforts will be in terms of research as well as the provision of psychological services. Advances in technology and joint international operations will increase the opportunity for psychologists to conduct large scale, international research projects. Psychologists already function with their international peers on a number of research and clinical assignments (e.g. Krueger, Chapter 2, this text; Bartone, Chapter 7, this text). The military environment is comparable to a large research laboratory which will soon take on an international quality.

Changes in Personnel Demographics

Several changes in the nation's demographics will affect the military. As baby-boomers age, there will be an increased need for health care services for elderly veterans. There has also been an increase in the number of female troops entering into traditional "male soldiering" jobs. As mentioned above, the retention of highly qualified troops will be critical for the military. Troops who remain in the military tend to eventually marry and raise families, increasing the number of dependents for whom the military provides support. The divorce rate over the past several decades has added a new type of service member to the demographic checklists, the single-parent. Finally, the ethnic diversity of troops entering the military has changed and it is likely that the white male service member may find himself among the minority in the future. Each of these changes will be discussed in the following paragraphs.

The increase in the number of elderly will create new priorities for health care systems. This trend is also true for the mental health services provided to older adults in the Veterans Affairs (VA) system (Van Stone & Goldstein, 1993). Military psychologists will need to focus on treatment, training and research related to the care of elderly veterans.

The increase in the number of women entering the military will also pose some distinctive challenges. Recent court decisions have cleared the way for females to enroll in military academies and it is likely that the courts will decide on other gender-related issues, such as combat exclusion. Gender issues (e.g., pregnancy, child care, single-parenthood, combat exclusion) must be addressed to increase recruitment and retention rates among female service members (Dansby, Chapter 9, this text; St. Pierre, 1991). Related to VA health care, there is an increased need for providing quality health care for female veterans (Weiss, 1995). Gender differences, such as leadership styles (e.g., Lipman-Blumen, 1992) and human-machine interface will need to be explored as more women are promoted into positions of leadership and traditional male assignments.

Regarding the increase in military dependents (a service member's spouse and children), research suggests psychologists can play a larger role in preparing families for the stresses associated with military service (Bartone, Chapter 7, this text; Hobfoll, Spielberger, Breznitz, Figley et al., 1991; Rosen & Durand, 1995). Preparation of dependents regarding stresses related to military deployments can increase retention, which will reduce training costs.

Psychologists are also asked to provide psychological services to an increasing number of dependents (Ballenger, Chapter 11, this text).

Despite some difficulties, the military has been successful at integrating an ethnically diverse population. The military of the future will continued to be challenged to accommodate a culturally diverse group, including African-Americans, Hispanics and Whites, but also increasing numbers of Asian-Americans, Arab-Americans and the offspring of interracial marriages. The controversy surrounding homosexuals in the military continues and will need to be resolved along with other issues related to sexual activity, such as adultery and intimate relations between officers and enlisted personnel. Recent incidents involving high ranking officers which were reported in the press have prompted at least one member of congress to introduce legislation to change the laws of the Uniformed Code of Military Justice regarding sexual activity between consenting adults. (Note: As discussed at the beginning of this chapter, these incidents serve as an excellent example of how the military environment is subjected to public and social pressures due to its position in society.)

Regarding diversity issues, military psychologists can play a number of significant roles. Policy decisions should be based on sound scientific research as opposed to opinion or tradition. Psychologists can conduct studies to determine how any policy change might influence the military's mission. Psychologists will be involved in the formulation and implementation of manpower policies along with monitoring their effect on combat readiness.

The Future of Military Psychology

Psychology has enjoyed a very prosperous period during its symbiotic relationship with the military (Driskell & Olmstead, 1989). There is every reason to predict that this relationship will continue to flourish for many years to come. The preceding chapters have illustrated the variety of duties military psychologists perform. Their role continues to expand as psychologists apply their skills as human behavior specialists to the many issues confronted by the nation's largest employer. DoD's budget and manpower pool dictate an efficient management of resources. Military psychologists play a pivotal role in maintaining combat readiness while keeping costs down.

The previous chapters described the significant contributions psychology has achieved in the military environment. However, these accomplishments have been realized without much fanfare or public recognition. Very few people outside of the profession understand what is included in the specialty of military psychology. Indeed, it safe to say that a large percentage of psychologists and students of psychology are not fully aware of the role military psychology has played in the development of the discipline. This author was unable to identify any introductory textbooks which listed military psychology as a specialty and encountered a fair amount of misunderstanding when discussing the field with students and colleagues.

However, several recent developments suggest that military psychology will receive increased recognition in the coming years. There has been a recent trend for military psychologists to publish more of their work in the professional literature. An encyclopedic source on military psychology was recently published and the Division of Military Psychology is doing well and has established its own official publication. Finally, a recent APA book on careers in psychology includes a chapter on careers in military psychology.

275

Much of the work of military psychologists has not been published in professional journals or books. Rather, these studies appeared as technical reports which could be accessed using computer-assisted searches through the Defense Technical Information Center and the National Technical Information Service (Dansby, Chapter 9). Now, military psychologists have a professional journal in which to report their work. *Military Psychology*, the official journal of the APA's Division 19, was founded in 1989. Currently, the journal is doing very well and is financially stable (Wiskoff, 1995). An increase in the number of submissions from abroad reflects the high visibility the journal enjoys in the international military psychology community.

Other than a few narrowly focused books and numerous journal articles scattered throughout the literature, it was difficult to locate a single, comprehensive source reporting on the current work of military psychologists. In 1991, the *Handbook of Military Psychology* (Gal & Mangelsdorff, 1991) was published. This extensive text provides a detailed overview of military psychology and has helped other professionals become better acquainted with the field.

Perhaps as a result of these recent additions to the professional literature, membership in Division 19 has steadily increased. Currently, the Division has approximately 500 members. Many of the members and affiliates are professionals involved in the work of military psychology. Hopefully more students will become aware of the profession and join Division 19 as student affiliates.

Few students have the opportunity to learn about the field of military psychology. As stated above, this author was unable to locate any introductory psychology books which listed military psychology as a specialty. Yet, it is hoped that this will soon change. Recently, APA published a book on careers in psychology. Included was a chapter by Wiskoff (1997) on careers in military psychology, which served as the basis for Chapter 13 of this book. At the graduate level, the Uniformed Services University of the Health Sciences in Bethesda, Maryland has approximately 12 students enrolled in their doctoral program and recently awarded the first Ph.D. in Military Clinical Psychology. It is a sincere hope of this author that the addition of this text to the literature will make information about the field of military psychology more accessible to undergraduates and other professionals not currently involved in the specialty.

Summary

This chapter identified four trends which will affect the direction of military psychology in the coming years. These trends are: (1) advances in technology, (2) budget reductions and the military drawdown, (3) political changes, and (4) changes in personnel demographics. Under each of the four areas, specific examples were provided. The chapter then discussed recent changes in the profession of military psychology which point to the discipline's continued growth.

Conclusion

The purpose of the current text is to familiarize students to the diverse field of military psychology. It is hoped that the preceding chapters have inspired an interest in this profession. Readers who wish to learn more about the field are encouraged to participate in the profession as members of the Division and as participants at national and international meetings. The recruitment of highly qualified, dedicated professionals will contribute to the growth of the profession.

References

Alluisi, E. A. (1991). The development of technology for collective training: SIMNET, a case history. *Human Factors, 33*(3), 343–362.

Armfield, F. (1994). Preventing post-traumatic stress disorder resulting from military operations. *Military Medicine, 159*(12), 739–746.

Borman, W. C., Hanson, M. A., & Hedge, J.W. (1997). Personnel selection. In J. T. Spence, J. M. Darley & D. J. Foss (Eds). *Annual Review of Psychology*. Palto Alto: Annual Reviews, Inc.

Driskell, J. E. & Olmstead, B. (1989). Psychology and the military. *American Psychologist, 44*(1), 43–54.

Edwards, J. E., Rosenfeld, P., Booth-Kewley, S., & Thomas, M. D. (1996). Methodological issues in Navy surveys. *Military Psychology, 8*(4), 309–324.

Fox, R. E. (1994). Training professional psychologists for the twenty-first century. *American Psychologist, 49*(3), 200–206.

Schwalbe S. R. (1993). War gaming: In need of context. Special Issue: Military simulation/gaming. *Simulation and Gaming, 24*(3), 314–320.

Griffin, G. R. & Koonce, J. M. (1996). Review of psychomotor skills in pilot selection research of the U.S. military services. *International Journal of Aviation Psychology, 6*(2), 125–147.

Hobfoll, S. E., Spielberger, C. D., Breznitz, S., Figley, C., Folkman, S., Michenbaum, D., Lepper-Green, B., Milgram, N. A., Sander, I., Sarason, I., & van der Kolk, B. (1991). War-related stress: Addressing the stress of war and other traumatic events. *American Psychologist, 46*(8), 848–855.

Kelleher, W. J., Talcott, W. G., Haddock, K. C. & Freeman, K. R. (1996). Military psychology in the age of managed care: The Wilford Hall model. *Applied and Preventive Psychology, 5*(2), 101–110.

Lipman-Blumen, J. (1992). Connective leadership: Female leadership styles in the 21st-century workplace. Special Issue: Women in the workplace: Toward true integration. *Sociological Perspectives, 35*(1), 183–203.

Lovesey, E. (1995). Information flow between cockpit and aircrew. *Ergonomics, 38*(3), 558–564.

Matarazzo, J. D. (1992). Psychological testing and assessment in the 21st century. *American Psychologist, 47*(8), 1007–1018.

McCarroll, J. E., Ursano, R. J., Fullerton, C. S. & Lundy, A. (1995). Anticipatory stress of handling human remains from the Persian Gulf War: Predictors of intrusion and avoidance. *Journal of Nervous and Mental Disease, 181*(11), 698–703.

Medsker, G. J., Williams, L. J. & Holahan, P. J. (1994). A review of current practices for evaluating causal models in organizational behavior and human resources management research. *Journal of Management, 20*(2), 439–464.

Oswalt, I. (1993). Current applications, trends, and organization in U.S. military simulation and gaming. *Simulation and Gaming, 24*(2), 153–189.

Page, G. D. (1996). Clinical psychology in the military: Developments and issues. *Clinical Psychology Review, 16*(5), 383–396.

Ricci, K. E., Salas, E., & Cannon-Bowers, J. A. (1996). Do computer-based games facilitate knowledge acquisition and retention? *Military Psychology, 8*(4), 295–307.

Roper, B. L., Ben-Porath, Y. S. & Butcher, J. N. (1995). Comparability and Validity of Computerized Adaptive Testing with the MMPI 2. *Journal of Personality Assessment, 65*(2), 358–371.

Rosen, L. N. & Durand, D. B. (1995). The family factor and retention among married soldiers deployed in Operation Desert Storm. *Military Psychology, 7*(4), 221–234.

Schlosser, B. (1991). The future of psychology and technology in assessment. *Social Science Computer Review, 9*(4), 575–592.

Schwalbe, S. R. (1993). War gaming: In need of content. *Simulation and Gaming, 24*(3), 314–320.

Scott, T. R. (1991). A personal view of the future of psychology departments. *American Psychologist, 46*(9), 975–976.

Shapiro, A. E. & Wiggins, J. G. (1994). A PsyD Degree for every practitioner. *American Psychologist, 49*(3), 207–210.

St. Pierre, M. (1991). Accession and retention of minorities: Implications for the future. Special Issue: Racial, ethnic, and gender issues in the military. *International Journal of Intercultural Relations, 15*(4), 469–489.

Steege, F. W. & Fritscher, W. (1991). Psychological assessment and military personnel management. In R. Gal & A.D. Mangelsdorff (Eds.). *Handbook of military psychology.* New York: John Wiley & Sons.

Van Stone, W. W. & Goldstein, M. Z. (1993). Mental health services for older adults in the VA system. *Hospital and Community Psychiatry, 44*(9), 828–830.

Weiss, T. W. (1995), Improvements in VA health services for women veterans. *Women and Health, 23*(2), 1–12.

Wiggins, J. G. (1994). Would you want your child to be a psychologist? *American Psychologist, 49*(6), 485–492.

Wiskoff, M. F. (1995). Report on the journal of Military Psychology. *The Military Psychologist, 12*(1), 18.

Wiskoff, M. F. (1997). Defense of the Nation: Military Psychologists. In: Robert Sternberg (Ed.) *Career Paths in Psychology: Where Your Degree Can Take You,* (pp. 245–268). Washington, DC: American Psychological Association.

Wiskoff, M. F. & Schratz, M. K. (1989). Computerized adaptive testing of a vocational aptitude battery. In R. F. Dillon & J. W. Pellegrino (Eds.). *Testing: Theoretical and applied perspectives*, pp. 69–96. New York: Praeger.

Appendix

---

* Application for the Division of Military Psychology:
(Division 19 of the American Psychological Association)
* Web Sites for Psychology

Military Psychology Division (19) of the American Psychological Association

Military Psychology is a microcosm of all psychology disciplines. It was among the charter specialty divisions established in the American Psychological Association in 1946. Military psychologists' areas of work and study include: selection, classification, and assignment; training, human factors engineering; environmental stressors; leadership and team effectiveness; survey research; individual and group behavior; clinical practice and research. Work settings include: research facilities; educational facilities; medical centers and clinics; military schools and bases; overseas deployments; and operational policy offices.

What Are the Benefits of Division 19 Membership?

Division 19 members and affiliates receive the quarterly journal, **Military Psychology** and the biannual newsletter, **The Military Psychologist**. In addition, membership in the Division entitles you to vote on Division 19 matters and propose Division 19 papers, workshops, and symposia for presentation at APA conventions. Another benefit is the Division's diversity. Division 19 is about equally divided between clinical and research psychologists guaranteeing a stimulating exchange of ideas and perspectives in our journals and conference sessions. The small size (around 500 members) is also an advantage if you are looking for camaraderie and easy access to colleagues who share your research and/or practice interests.

How Do You Join Division 19?

There are four categories of membership in Division 19: **Members**, **Dues Exempt Members**, **Affiliates**, and **Student Affiliates**.

APA members, fellows and associates are accepted as Division 19 members. Long-term, dues-exempt APA members can become Division 19 Dues Exempt Members, but do need to pay a nominal fee to cover publication printing and mailing costs. Individuals who are not APA members may become Division 19 Affiliates and students may become Student Affiliates.

Yearly Fees:

Members	$25	(APA Members)
Student Affiliates	$19	(regardless of APA affiliation)
Affiliates	$30	(Non-APA Members)

To join, complete the application below, tear it out from the book and mail to the address indicated along with a check in US dollars for the appropriate amount made out to Division 19.

———————————————————————————————————————

APPLICATION FOR THE DIVISION OF MILITARY PSYCHOLOGY
(DIVISION 19) OF THE AMERICAN PSYCHOLOGICAL ASSOCIATION

Full Name _____ Date _____

Title (circle one):   Dr.   Mr.   Ms.   Mrs.   Other _____

Mailing Address: _____

_____

_____

Phone (w)_____(h)_____(fax)_____

E-Mail _____

Position Title _____

Organization _____

Division status you are applying for: Present APA status

___Member $25   __ Member

                                                 __ Associate

                                                 __ Fellow

__Student Affiliate $19   __ Student Affiliate

__Affiliate $30   __ Non-member

NOTE: Only paid APA members/associates/fellows can be Division 19 "members", and only *full time* students are eligible to be student affiliates. All others can join as affiliates.

If you belong to APA, please list your APA member number _____

and other divisions: _____

College or University	*Degree*	*Date*	*Field*
_____	_____	_____	_____
_____	_____	_____	_____
_____	_____	_____	_____

Return this form along with a check made out to "Division 19" to
Dr. Brian Waters
HumRRO, Suite 400
66 Canal Center Plaza
Alexandria, VA 22314

Phone: (703) 706-5647
Fax: (703) 548-5574
E-mail: bwaters@mail.humrro.org

Web Sites Review

Below is a list of web sites on the Internet related to psychology. This list is not comprehensive and additional sites will undoubtedly be developed. A useful guide to the internet is *Psychology on the Internet: A Student's Guide* by Andrew T. Stull (1997), published by Prentice Hall.

http://www.apa.org APA's web page, which provides updates on the Association's activities, public information items, excerpts from publications, employment opportunities, and membership information. APA's PsychNET offers information for both the public and members. Students can read the complete editorial text and classified ads of the *APA Monitor* online and find information on accredited graduate programs. Journal tables of contents and information about ordering APA books is also available. There is also information available on the APA's 49 subspecialty divisions. The search engine (called PsychCrawler, a free psychological information searching tool available on PsychNET) indexes other sites including the National Institute of Mental Health and the Substance Abuse and Mental Health Services Administration (SAMHSA).

http://www.gasou.edu/psychweb/psychweb.html This Web Site can be very helpful to both faculty and students. Just some of the features include: a self-quiz for Introduction to Psychology, a Hypertext Research Paper (which illustrates the APA style and includes links to explanations of the style), tips for preparing for the Psychology Graduate Record Exam, a "Find Anything" section, and links to over 700 psychology departments around the country. This is a good place for students to begin to explore the Internet's contributions to psychology.

http://psych.hanover.edu:80/APS/ This website offers information about the American Psychological Society, links to academic psychology department home pages, teaching resources, research sources, other psychological societies, and mental health resources. The site also provides a link to Internet searches. Fun for both faculty and students is the link to Today in Psychology's History which allows one to select a date (day and month) and then view the important events in the history of psychology which took place on that date.

http://www.pr.erau.edu/~security The International Bulletin of Political Psychology (IBPP) is a weekly electronic journal which frequently features articles on the psychology of military and national security issues.

http://dticam.dtic.dla.mil/ MATRIS Home Page: A Department of Defense directory of individuals and organizations who perform and manage current people-related research and development for the DoD.

http://www.anu.edu.au/psychology/PiP/pip.htm Positions in Psychology maintains an up-to-date listing of academic and clinical position openings in psychology worldwide. Each position announcement includes the location of the position, posting and closing dates, a link to the job description, contact information and usually a link to the posting agency's home page. You can register to have new announcements sent directly to you as they are posted. A valuable resource for job hunters and students wishing to identify career opportunities.

Glossary

Abilities The capacity to carry out physical and mental acts required by a job's tasks.

Acceleration To increase in speed, a change in velocity, and/or the rate of change of velocity with respect to time. Sustained acceleration means the change in velocity, usually a rapid one, continues long enough that our bodily organs respond to the "thrust" of the change. Measures of acceleration are usually in terms of 'g' forces or levels of gravitational pull experienced during the velocity change.

Active Straining Maneuver Military pilots use a muscle control maneuver consisting of shrugging the shoulders up around the neck and toward the head, and pulling the head downward to the body, (maximum shrugging) and slowly and forcefully exhaling through partially closed mouth while tensing all skeletal muscles. This maneuver helps the pilot to withstand excessive g-forces during aerial maneuvers involving rapid accelerations.

Adjustment Disorder A normal emotional and/or behavioral disturbance in response to a clear stressor. The disturbance is expected to resolve quickly.

Adjustment Reaction Transitory behavior resulting from an event in the recent past. An example could be the reduction in academic performance by an adolescent who has moved from one school to another and has left behind good friends and a familiar environment.

Administrative Discipline The use of administrative actions, rather than courts-martial or nonjudicial punishment, to enforce discipline. For example, individuals who violate rules may be discharged from the service, reassigned, or subjected to other administrative actions.

Administrative Separation Termination of an active duty member from the military due to a nonmedical incompatibility between the member and the military.

Advanced Development Research (6.3) Part of DoD's Program for RDT&E (see below). Tends to examine particular application of well developed psychological phenomenon with intention of eventually implementing it into operations. For example, verifying the dose response use of drug treatment regimens for application of treatment protocols.

Affirmative Duty A legal requirement to behave in a particular manner under given circumstances.

Aircrew Coordination Training A type of team training that was patterned after cockpit resource management training programs and designed to increase an aircrew's awareness of the need for coordination in the cockpit.

Air Force Specialty Code (AFSC) Categorization of a 'job' in the Air Force. For example, 6153b is an officer Behavioral Scientist AFSC.

American Psychological Association (APA) Founded in 1892, the APA is the major psychological organization in the United States with over 125,000 members. Membership consists of professionals and student affiliates interested in the science and practice of psychology. The organization consists of 49 Divisions, one of which is the Division of Military Psychology, Division 19.

Anthropomorphic Ascribing human characteristics to nonhuman entities—animate and inanimate.

Antidepressant medications Drugs used to treat depression; a drug which elevates mood.

Antimanic medications Lithium, some anticonvulsant medications, and others used to treat extreme elation and the corresponding behavioral disturbance.

Antipsychotic medications Drugs used to treat psychosis, which is a severe mental disorder manifested by hallucinations and significant disturbances in thinking.

APA Approval Approval of a doctoral program or internship setting by the American Psychological Association. APA approval signifies that the program has met criteria for excellence in training. Of late, APA approved training is required for licensure in an increasing number of states. Students can read about graduate training programs in psychology and whether they are APA approved in the book, *Graduate Study in Psychology*, published by the APA.

Apnea A sleep disorder in which the brain fails to send a signal to the breathing muscles to initiate respiration, or in obstructive apnea in which air cannot flow into or out of the person's nose or mouth although efforts to breathe continue.

Apparata The means by which something is accomplished.

Applied Psychology The professional practice of psychology for a practical purpose; usually involves the provision of psychological services such as therapy, assessment or consultation. The APA currently recognizes four areas of applied psychology: clinical, counseling, industrial/organization, and school.

Aptitude A combination of abilities that indicate an individual's potential to learn or develop proficiency in a particular area of education or training.

Armed Forces Qualification Test (AFQT) Joint-Service enlisted selection test composite comprised of Word Knowledge, Paragraph Comprehension, Arithmetic Reasoning, and Mathematics Knowledge test scores from the ASVAB.

Armed Services Vocational Aptitude Battery (ASVAB) A general aptitude test used by the armed services to aid in the classification of recruits.

Assertiveness Training A type of team training designed to teach the use of assertive communication when providing feedback to other team members, stating and maintaining opinions, offering potential solutions, initiating action, and offering and requesting assistance or backup when needed.

Assignment Personnel management action directing an individual to a specific unit at a specific location.

Ataxia Loss or lack of muscular coordination.

Attention Deficit Hyperactivity Disorder (ADHD) is a constellation of symptoms which include excessive psychomotor functioning and the inability to sustain attention on a task.

Audible Frequency Range The sound frequency range from about 100 to 8500 cycles per second (hertz). The practical limits of human speech intelligibility are probably in the ranges of from 200 to 6100 Hz.

Basic Research (6.1) Part of DoD's Program for RDT&E (see below). In psychological research, tends to look for basic mechanisms (often employing animal models, college students, young soldiers) to get at a basic explanation of how or why some human behavior pattern works the way it does. For example, why will low level white lighting in an aircraft instrument panel not disturb a pilot's dark adaptation?

Behavioral Approach A focus on what leaders did that identified two major behavioral dimensions: Consideration (mutual trust, support and respect) and initiation of structure (direction and control toward task accomplishment).

Behavioral Change The modification of individuals' behaviors to meet institutional standards, irrespective of whether the individuals' attitudes or values are modified. For example, the military's main focus in deterring racism is to ensure that all members behave in a non-racist manner, regardless of their personal attitudes toward other races. Certainly, attitude change is desirable and may be a product of behavioral change, but the goal is to have behaviors meet the standard.

Billet Duty assignment; employment or jobs within the military.

Blast Overpressure Wave The blast wave emitting large amounts of air pressure and usually accompanied by loud impulse noise, e.g., firing artillery, bombs, etc. If blast overpressure is high enough, those exposed can suffer not only hearing loss, but lesions in the air containing organs of the body, such as the lungs. Some artillery forces, e.g., Russians, dig a trench to lay in while pulling the lanyard to trigger large artillery pieces so that resultant blast goes over the top of them instead of hitting them directly.

Boredom With regards to stress on deployment, the lack of meaningful work.

Carboxyhemoglobin An accumulation of carbon usually from carbon monoxide (from fires, or even cigarettes) in the blood stream as it adheres to the hemoglobin, oxygen carrying red blood cells, and affects our performance and gradually our health because it does not permit our lungs from taking in enough oxygen.

Chemical Protective Clothing (CPC) Gas mask, protective hood and clothing garments worn over normal utility uniform to provide protection from chemical-biological warfare agents on the battlefield. Most CPC has a charcoal lining to absorb chemicals to keep them away from the skin. The bulky clothing looks like snowpants and makes manipulation of arms and legs more difficult, and retains body heat.

Circadian Rhythms Rhythmic biological cycles occurring at approximately 24-hour intervals such as body temperature, mental alertness, etc. Night shift workers must be aware of the effects of circadian rhythms on performance.

Classification Set of procedures used to match persons to a single job or cluster of jobs.

Clinical Training The practical component of training for psychologists which focuses on actual assessment and therapy skills with clients.

Clinical Psychology The branch of psychology that specializes in the application of clinical methods to persons suffering from behavioral disorders.

Cluster Grouping of a number of jobs into a set with relatively similar KSAs for job incumbents.

Cognitive Ability to use thought process to know or to become aware of and to use judgement.

Cognitive Assessment Measurement of mental abilities, generally through performance on an aptitude test or number of tests (e.g., AFQT).

Cohesion The quality of hanging together. When applied to social groups, (e.g., unit cohesion) refers to mutually shared feelings of belonging to a group and 'sticking together.'

Combat Fatigue Mental, physical, and/or emotional stress suffered as a result of combat operations.

Combat Readiness The ability to fight, on demand, anywhere.

Composite Score A score that combines two or more component scores statistically combined to produce a single measure.

Computer-adaptive Testing (CAT) Computerized testing in which test items to be administered are selected based on estimates of an examinee's ability calculated from his or her responses to previously administered items.

Conduct Disorder Behavior that violates the rights of others in a variety of settings; often a precursor to the development of an anti-social personality disorder.

Confidentiality A right to express one's self privately to another and to have the expression kept private unless the statements are such that the recipient has a legal and/or ethical responsibility to breach the privacy of the statement. A mental health client has the right to have information disclosed to a psychologist kept private.

Contingency Theories of Leadership Theories which suggest that leadership effectiveness is determined by leader characteristics, follower characteristics and situational factors such as task complexity.

Continuous Operations Teams of workers operate in work shifts to spell one another in the workplace and thereby give the organization the capability to sustain operations continuously around the clock. Important in military context in that modern military units plan to conduct operations around the clock, taking advantage of night vision systems to function and fight at night, and thereby causing large portions of their personnel staff to become "night fighters" and requiring close adherence to sleep management and sleep discipline for the entire force.

Contrast Sensitivity Visual perception ability that permits us to detect differences in contrast of objects in our visual field, and thus to pick out or detect visual stimuli from the background in different lighting situations.

Correlation The association, not the causal relationship, among variables; the degree to which the values of variables vary together.

Countermeasure Specifically, a measure taken to counter another measure. For example, the use of planned 45 minute naps to counter operator fatigue or the development of unit cohesion to counter the negative effects of stress.

Credentialing Permission given by a designated body, usually in a hospital, to an individual to practice a profession; to be given the right to exercise certain powers.

Critical Incident Stress Debriefing A crisis mental health intervention in which persons having experienced a traumatic event are encouraged to fully discuss and process the experience with others shortly after it occurs.

Cross Rotational Requirements Training components or duties which remain consistent across the various rotations or settings occupied by an intern or fellow during training.

Cross-Training A type of team training in which team members rotate positions in order to develop an understanding of the basic knowledge necessary to successfully perform the task, duties, and/or positions of the other team members.

Cultural Assimilation Model A training model designed to increase intercultural sensitivity by using simulation, role play, computer-based training, or other techniques to allow members of one group (e.g., the dominant group) to gain insight into the behavior, motivations, cultural meanings, etc., of another group (e.g., the nondominant group). For example, a white drill sergeant may respond to a scenario depicting interactions involving black soldiers' behavior. If the drill sergeant fails to select the culturally sensitive alternative, an explanation is provided and the sergeant is allowed to choose another alternative. This continues until the preferred alternative is selected. The preferred alternative is accompanied by an explanation as to why it is preferred.

Cultural Diversity A term used to indicate differences among individuals in cultural background, or sometimes to indicate the statistical level of such differences (as in "cultural diversity is high in the Army"). Though the term may be used expansively to include myriad differences (e.g., regional, educational), in the military it is generally applied to differences based on race, national origin (or ethnicity), color, gender, religion, disability, and/or age.

Cut Score A binary (Go-No Go) decision point in which all persons scoring equal to or above 'pass' and all scoring below 'fail.'

Cyclothymic Syndrome A mood disorder characterized by alternating periods of elevated mood and activity (hypomanic) and periods of inactivity and depression. As discussed in Chapter 9, sometimes occurs at moderately high altitudes where living with lower partial pressure of oxygen is believed to be the major contributor.

Dark Adaptation A process of rod (sensor cells in the retina of the eye) adjustment whereby the retina becomes over a million times more sensitive to dim light. Complete adaptation requires approximately four hours; however, most of the process is completed within 20–25 minutes.

De Novo To arise anew.

Deconstructionism The concept of unmasking cultural assumptions about other assumptions, concepts, beliefs, or human activities; the cultural assumptions being so pervasive as to be almost hidden.

Depression A mood disorder characterized by persistent feelings of sadness and despair and loss of interest in previous sources of pleasure.

Discrimination In general, the act of distinguishing differences; in diversity issues, this term usually indicates actions or policies that have a differential adverse impact on one group (or more) as compared to others.

Disparity Statistical inequity between compared groups in the rate of some action (either positive or negative) affecting both groups. Typically, when diversity issues are considered, disparity indicates the occurrence for one group (e.g., percentage of black soldiers who are incarcerated) is statistically different from the rate for another group (e.g., percentage of white soldiers who are incarcerated). Since disparity may occur for many reasons, it does not necessarily indicate illegal discrimination has occurred, though the presence of a disparity may lead to a determination of whether discrimination has occurred.

Division of Military Psychology, Division 19 See Appendix.

Doctoral Dissertation A major original research project completed by a doctoral student, often as the last step in doctoral training.

Double Agent A sense that one is simultaneously serving two opposing sides.

Dual Roles Simultaneously occupying multiple, often incompatible, functions or duties.

Dyspnea A sense of difficulty breathing often associated with heart or lung disease.

Empiricism The study of knowledge through observation.

Engineering Development (6.4) Part of DoD's Program for RDT&E (see below). Usually for hardware systems, and thus may involve human engineers in verification that their man-machine interactions actually work in field testing, and identifying any alterations needed in operator procedures or in design modifications.

Engineering Psychologists Psychologists who do research on human factors design issues usually regarding how human operators acquire information and handle it in operation of equipment. Engineering psychologists design and conduct experiments with human operators using equipment systems; their particular design skills are best used in designing experiments with people.

Epistemology The study of how people know what they know.

Ergonomics The science of fitting the individual to the job and the job to the individual.

Ergonomic Design The field of ergonomics incorporates knowledge of body biomechanics, physiology, anthropometry, and other human engineering factors into a user oriented design approach to making equipment of jobs more functional, easier to accomplish, more comfortable, and less threatening in terms of body stresses and biomechanical risks to the body's joints.

Ethics The rules or standards governing the conduct of a person or the members of a profession. Psychologists adhere to the *Ethical Principles of Psychologists and Code of Conduct*. Providers of psychological services also adhere to the *General Guidelines for Providers of Psychological Services*; both publications are available free of charge from the APA. See the APA web site in the Appendix.

Etiology The study of causes, origins or reasons as in the medical study of causes of disease. For example, trauma is an etiological factor in the development of Post-traumatic Stress Disorder.

Exploratory Research (6.2) Part of DoD's Program for RDT&E (see below). Usually directed at answering some applied question for a military question. For example, should countermeasures to operator fatigue include planned nap taking of less than 45 minutes so as to prevent slow wave sleep and the risk of sleep inertia upon awakening?

False-Negative Rejection of an applicant who would have succeeded in training and on the job if given the opportunity (Type I error).

False-Positive Acceptance of an applicant who fails to meet performance standards in training or on the job (Type II error).

Fellowship Advanced education and training beyond the doctoral degree, usually consists of a one or two year period of intensive clinical training in a specialty area.

G Suit An overgarment with pumping mechanisms built into it and worn over an aviator's flight suit.

G Vectors The three linear vectors (Gx, Gy, Gz) of "tug" or "pull" on the body due to acceleration are normally referred to in terms of the amount of 'g' or gravitational effect on the body. Three dimensional effects on the body are readily apparent: Gx is the direction thrust from the chest to the back, Gx is the tendency to lean sideways or laterally, and Gz is the longitudinal vector from feet to head.

Garrison Environment A military post, in particular, a relatively permanent installation.

Gender-Stereotyped Jobs Jobs that have historically been considered by society (or some reference group) to be more appropriate for one gender or the other. For example, nurse and secretary have traditionally been considered jobs that are more appropriate for women, while firefighter (usually called fireman) and infantry soldier (i.e. infantryman) have been considered more appropriate for men.

Generalist Psychologist A psychologist competent in evaluating and treating a broad range of clinical disorders in a broad range of client types.

Guided Team Self-Correction The process whereby a leader or instructor imposes structure on a team's feedback in order to facilitate the team's ability to diagnose important team processes and develop effective solutions to these problems.

Gynopomorphic Ascribing women's characteristics to men and to non-human entities—animate and inanimate.

Hardiness A personality syndrome that is marked by commitment, challenge, and control and that is associated with strong stress resistance.

Hermeneutics The study of different approaches to interpretation.

Host Nation The nation or country in which United States military facilities are located. Examples include Germany, Panama, Korea, and Japan.

Human Factors Engineering A branch of psychology and engineering that addresses issues related to machine design, systems design, working conditions, skills and work efficiency.

Human Factors Studies Research conducted by behavioral scientists and engineers with a goal of defining equipment or systems design issues of concern to human operators, e.g. design for ease of use, less likely to lead to accidents, to maximize productivity, to take advantage of what we know about the capabilities and limitations of human operators in use of systems of equipment.

Hydrogen Cyanide (HCN) A gas produced by the combustion of burnable construction materials. Breathing even modest amounts can be fatal.

Hypostatized Constructions Hypothetical constructs (see definition below) that are wrongly assumed to be real and to have material existence.

Hypothermia When body temperature gets very low, many of our physiological systems begin to shut down or respond differently and present an immediate risk to health and ability to function.

Hypothetical Construct A concept that helps us understand the world and obtain meaning from it, though the concept may not physically exist or exist in some absolute sense.

Hypoxia (hypoxemic) A condition caused by a deficiency in the amount of oxygen that reaches body tissues; as lungs are deprived of sufficient oxygen, ultimately the brain, working muscles and other parts of the body are too. Oxygen shortages cause profound effects on sensory processes, mental capacity, and work physiology; judgement is adversely affected and life threatening mistakes are more common. Prolonged hypoxia can result in coma.

Hysteresis Levels The failure of a property that has been changed by an external agent to return to its original value when the change agent is removed. For example, the viscosity of motor oil may be different after undergoing rapid changes in temperature.

Immersion Foot A disease of the feet encountered when they are exposed to moisture or dampness for considerable time and the skin begins to deteriorate in much the same way as it does with trench foot. See below for a definition of trench foot.

Impulse Noise Noise created as if by a single shot, burst, loud bang, gunfire, artillery fire, etc. Repeated exposure to impulse noise causes hearing loss. Generally, the hearing loss is in the noise frequency octave bands correlating to the type of noise one encounters.

Indelicate Experiments Experiments having the rigors of principles of good experimental design, but for one or more reasons, (e.g., not enough data to fill all needed data cells), ending up giving research insights, but often not sufficient data to make valid statistical inferences.

Indoctrination To instruct and initiate new military members with military doctrine or principles.

Information A set of facts or ideas gained through experience, investigation or practice. More generally, anything from which people derive meaning (see Chapter 8 for more details).

Information Warfare The collection, protection, transmission, and modification of information to help achieve political objectives.

Informed Consent Consent for psychological procedures or treatment which is based on full information about the possible risks and benefits of the procedures.

Inoculation Approach When applied to military discipline issues, a method of discouraging individuals (usually recruits) from engaging in actions that may result in their being disciplined by educating them in advance as to rules of behavior and allowing them to experience vicarious examples of discipline-generating interactions in a "safe" (e.g., training) environment. By experiencing the troublesome interactions in a mild form (such as a video presentation, role play, etc.), individuals might learn to avoid behaviors that would result in punishment.

Inoculations Serums, a vaccine, or antigenic substances which are injected into the body, especially to produce or boost immunity to a specific disease.

Institutional Discrimination A form of discrimination in which institutional rules, policies, or practices (whether legal or illegal) have a differential adverse impact (e.g., reduced opportunities for promotion, retention, access to benefits) for one demographic group (or more) as compared to others. For example, restrictions on women entering direct combat specialties may prevent women from obtaining experience that would help them gain promotion to the highest military grades. At this time, such restrictions are required by Congress and are legal, despite possible adverse impact on women.

Interactionism This perspective emphasizes the interaction of situation and person variables in determining human behavior.

Internship The final year of doctoral training in clinical and counseling psychology in which the student practices full time in a supervised setting.

Job-Person Match Classification action which aligns job requirements with applicant KSAs.

KSA Knowledges, Skills and Abilities (see individual definitions).

Knowledges The degree to which employees have mastered a technical body of material directly involved in the performance of a job.

Leader Match A self-programmed instruction book developed by Fiedler and Chemers for leadership training.

Leadership To provide purpose, direction and motivation to subordinates in order to accomplish a mission.

LEAN An Army weight loss program which emphasizes healthy **L**ifestyles, **E**motions and **E**xercise, **A**ttitudes and **N**utrition.

Least Preferred Co-worker (LPC) A scale developed by Fiedler that measures whether a leader is task or relationship oriented.

Line Community Those military groups assigned to combat-ready billets and most directly tasked with defense of the country.

Machiavellianism An emotionally cool, analytical approach to manipulating people for achieving one's objectives.

Man-machine Interface The design features associated with human operators interacting with equipment systems in the sense that they use controls and displays for communicating directions and information with the machinery or equipment. Operators require feedback telling them the machine is going to behave the way they instructed it to. The "interface" between the operator and the machine is of design concern to human factors specialists, engineering psychologists, human engineers, and ergonomists.

Material Life Cycle Management Military planners manage the life of engineering and material systems from the time of identifying the requirement for some new piece of equipment, through the program 6 RDT&E sequence (see below), to procurement and then use, maintenance and eventual destruction or abandonment of a system as it is gradually replaced by newer systems.

Materialist A theory of knowledge based on assumptions that reality comprises real, physical entities.

Medical Board A formal discharge from military service as a result of medical disability directly related to military service.

Metabolic Heat The internal heat energy our body generates through activity.

Micro-environment Usually used to refer to individual encapsulation of individuals in some confining, restrictive enclosure. Used in Chapter 9 to refer to wearing chemical protective clothing designed to keep chemicals off the body, but in so doing place the body into an enclosed micro-environment of its own.

Military Occupational Specialty (MOS) Army or Marine Corps job classifications. For example, MOS 19B is an enlisted MOS for infantrymen in the Army.

Military Psychology The study and application of psychological principles and methods to the military environment.

Mind A hypothetical construct (see definition above) describing some nonmaterial aspect of a person associated with that person's thoughts, feelings, motives, and behavior.

Misnorming The use of incorrect information to establish norms (standards) on standardized tests, such as the Armed Services Vocational Aptitude Battery (ASVAB). The ASVAB was inadvertently misnormed due to statistical errors between 1976–1980. During this time, nearly 360,000 individuals whose tests scores would have been too low if the test had been normed properly entered military service.

Morale Term used to describe the mental fitness and motivation of individuals and groups.

Motion Sickness A feeling of nausea and eventually vomiting induced by forms of motion which rotate the head simultaneously in more than one plane (e.g., sea sickness, train sickness, air sickness). The trouble originates in the labyrinth of the inner ear where the organs of equilibrium are located. See Chapter 9 for more details.

Multifactor Leadership Questionnaire (MLQ) A questionnaire designed by Bass and Avolio to measure components of transactional and transformational leadership.

Neurolinguistic Programming A theory of influencing people based on an assumed compatibility between characteristics of language and communication on the one hand and the human nervous system on the other.

Night Fighters Military personnel expected to adjust their schedules to do most of their work at night, in effect becoming night-shift workers. Their circadian rhythms rarely fully adjust to night work, most due to other demands during daylight hours and frequently lose large amounts of needed sleep. Night fighters must understand circadian rhythm physiology better than most others because they are most dramatically affected by its implications.

Nonjudicial Punishment (NJP) A form of military discipline that is administered by a commander or other designated authority to punish relatively minor offenses at the option of the offending party in lieu of a court-martial. NJP is intermediate in severity between minor corrective discipline (e.g., extra duty, short-term restriction) and court-martial. NJP does not involve incarceration, but may result in reduction in rank, fines, restriction to base, or extra duty.

Norm A representative standard for a given group; usually expressed as a numerical value such as a score or percentile.

Nonsequitur An inference that does not follow from a previous one.

Normalizing Conveying that one's experience or distress is normal and appropriate.

Nurse Practitioners Nurses who have been given specialized training in an area of healthcare, permitting them to perform procedures beyond the range of most nurses and with less physician oversight.

Operational Training Military training which occurs in the military environment.

Organizational Culture A system of shared meaning held by members that distinguishes the organization from other organizations.

Organizational Effectiveness A general field of research and practice dealing with how to improve the management, communications, job, and human resources climates in organizations in order to make them more effective in accomplishing their goals; also used to indicate the degree to which such effectiveness has been obtained (e.g., "the 33rd Fighter Squadron demonstrates a high degree of organizational effectiveness").

Otolith Organs Particles of the inner ear (from French: ear stone) that interact with the labyrinth and affect our equilibrium and therefore play a role in the development of motion sickness.

Partial Pressure of Oxygen in the Air At higher altitudes, air molecules are spread farther apart due to less gravitational pull on that air. Consequently, the pressure of oxygen on air breathed at high altitudes is less creating a need to breathe in more air to obtain the required amount of oxygen.

Pay Back Years of required military service following a period of training. For psychologists, a period of three years payback currently follows one year of formal training.

Personality An individual's unique constellation of consistent behavioral traits.

Personality Disorder A developmental and life-long pattern of behavior which is rigid and causes problems in relationships and occupational performance.

Pharmacotherapeutic Interventions Treatment using medications.

Political Objective The control of a finite resource in a world of infinite need for that resource.

Pool Group of candidates from whom selections are made.

Post Doctoral Training which occurs following completion of the doctoral degree.

Post Traumatic Stress Disorder (PTSD) An anxiety disorder resulting from exposure to traumatic situations such as combat, natural disasters, personal attack, sexual assault or other terrifying experiences.

Power The ability and will one has to resolve the disparity between who one is and who one wants to be, what one has and what one wants to have, the real and ideal.

Practicum The clinical work with patients which follows the classroom training.

Prescription Privileges Having the authority to prescribe medications. This authority is granted by a legal or regulatory body.

Prescribing Psychologists' Register One of the first and largest training programs in psycho-pharmacology for psychologists.

Prevention The use of measures designed to reduce the likelihood of the development of an unwanted or pathological condition. For example, the structuring of the work environment to increase commitment to the mission among troops. A sense of commitment has been shown to reduce the negative effects of catastrophic stress.

Privileged Communication Communication that is legally protected and the recipient of the communication cannot be compelled to release the content of the communication by judicial authority. Communication between lawyer and client; priest and penitent are examples.

Procrustean Constraint An analogy to a mythical Greek giant who stretched or shortened his captives to make them fit his beds.

Psychiatrist A physician with additional training to treat mental, emotional and/or behavioral disturbances.

Psychiatry The specialized branch of medicine dealing with the diagnosis, treatment and prevention of mental disorders.

Psychoanalysis A theory of human psychology based on assumptions that there is continuous unconscious conflict at the root of dysfunctional thought, feelings, motives, and behaviors.

Psychological Autopsy An assessment made to determine the state of mind of the deceased prior to death.

Psychological Evaluation An assessment of an individual's current mental health functioning.

Psychological Test A general concept including diagnostic, personality, mental-ability, achievement and special-aptitude tests.

Psychometrics The specialized branch of psychology involved in mental testing.

Psychomotor Performance Operator performance which involves mental processes in carrying out manual manipulation of some control(s) which necessitates use of our motor functions. For example, driving a vehicle or flying an airplane.

Psychopathology The branch of psychology concerned with the study of mental disorders.

Psychopharmacology The use of medications to influence mental, emotional and/or behavioral problems. The science of drugs used to treat psychiatric disturbance, including their composition, uses, and effects.

Psychopharmacology Demonstration Project (PDP) The PDP is the Department of Defense's fellowship to train psychologists to prescribe psychotropic medications.

Psychotherapy The application of specialized techniques (e.g., systematic desensitization, directive counseling, etc.) by a trained specialist (e.g., clinical psychologists, psychiatrist) to the treatment of mental disorders or the problems of everyday adjustment.

Psychotropics Chemical substances that modify mental, emotional, or behavioral functioning.

Doctor of Psychology (Psy.D.) A doctorate with an emphasis on applied clinical work versus scholarship and teaching.

Quod Erat Demonstrandum Q.E.D. Latin for "which was to be demonstrated".

Racial Separatism A desire to have separate activities, facilities, etc., for different races. For example, when this desire is high one may see people separating themselves by racial groups in dining facilities, clubs, etc., on military installations. On college campuses, this may be reflected in the desire for separate dormitories and student centers for different racial groups.

Rating Navy job classification. For example, RM is an enlisted rating for radiomen.

Rationalists A theory of knowledge based on reasoning as opposed to systematic observation.

Reliability Degree of consistency between two tests of the same thing; consistency of scores if a person took a measure many times.

Research, Development, Test and Evaluation (RDT&E) Management oversight program at all levels of the U.S. Department of Defense to govern the planning, programming, budgeting and accomplishment of major and minor scale research and development projects for all facets of the U.S. military services. See also Basic Research (6.1), Exploratory Research (6.2), Advanced Development Research (6.3), and Engineering Development (6.4).

Response A general term widely used in psychology, frequently with qualifiers. In general, any behavior, whether overt or covert, in response to a stimulus.

Rotations Specific practice setting within an internship or fellowship program. The intern or fellow rotates among settings to gain a range of experience.

Screen As a noun, denotes score point below which candidate is rejected; as a verb, denotes rejecting of candidate.

SEAR The Navy's Survival, Evasion and Resistance school.

Selection Method for identifying 'hired' individuals from a pool of applicants for a job.

Selection Ratio The ratio of the number of persons selected for a job divided by the number of job applicants.

Self-fulfilling Prophecy The finding that expectations of the leader actually bring about that behavior on the part of subordinates.

Sexual Harassment A form of gender discrimination in which the victim is subjected to unwelcome behaviors of a sexual nature. There are two basic types: quid pro quo, in which sexual favors are demanded in return for some job benefit (such as retention, promotion, etc.), and hostile environment, in which sexually related behaviors establish an environment that is hostile, intimidating, or offensive or interferes with a person's work.

Sisyphean Task An analogy to a mythical Greek King of ancient Corinth who was condemned forever to roll a huge stone up a hill in Hades, i.e., Hell, only to have it roll down again on nearing the top.

Situational Leadership Theory A Leadership theory developed by Hersey and Blanchard that focuses on diagnosis of followers' ability and motivation as the key determinants of leadership; behavior and effectiveness.

Skills The capacity to perform tasks requiring the use of tools, equipment, and machinery.

Sleep Management Military forces are becoming smarter about managing their rest and sleep schedules during sustained and continuous operations much as they manage any other logistical supply like fuel, food, and ammunition.

Social Support Various types of aid and backing provided by one's social networks.

Speech Intelligibility For human operators, equipment design decisions (human engineering) may be based on measurements of speech intelligibility through one of two procedures: 1) noise measurement and calculation of an articulation index (AI) indicating how much trouble operators are likely to have communicating in and around the noisy environment, or, b) having an audiologist or trained technician conduct intelligibility testing with participants listening and responding to various spoken sounds and then calculating from the tests a speech intelligibility index.

Stigma Disgrace or shame associated with seeking mental health services.

Stimulant therapy The use of medicines which adjust mental processes through excitation.

Stimulus Any detectable input from the environment.

Stress Events or forces in the environment, outside the person, as opposed to subjective, internal responses, which threaten or are perceived to threaten one's well-being; stress can influence performance, social adjustment, and health.

Survivor Assistance Officer (SAO) Army officers assigned to assist family members of deceased service members who died in the line of duty.

Sustained Operations Term which is more individually based (as opposed to group) implying that a person is expected to work non-stop for an extended period of time and may not know when to expect rest, sleep or relief on the sustained performance tasks until the mission is completed. A real problem for those who sustain work longer than 18 or 24 or 36 hours without rest or sleep.

Synovial Joint Fluids Transparent albuminous fluid that lubricates bodily joints.

Synergistic Two or more items or objects interacting in such a way that their combined effects are different (preferably better) than either or both acting alone or independently.

T Scores Normalized scores such that a T score of 50 is the mean and a T score of 60 is one standard deviation above the mean at the 84th percentile; a T score of 40 is one standard deviation below the mean at the 16th percentile on a Gaussian (normal-probability) curve.

Taskwork Skills Skills associated with the technical aspects of the job such as the execution of the task or the mission itself (e.g., pilot's stick and rudder skills, knowledge of equipment).

Team A distinguishable set of two or more people who interact dynamically, inter-dependently, and adaptively toward a common goal/objective/mission, who have each been assigned specific roles or functions to perform, and who have a limited life-span of membership.

Team Adaptation and Coordination Training A team training strategy that provides team members with specific information on how best to optimize preplanning session, idle periods, information dissemination, information anticipation, and redistribution of workload.

Team Leader Training Training designed to teach leaders of teams to develop a "team climate" in their briefings by critiquing themselves and requesting feedback on their own performance which in turn encourages the team members to admit their own mistakes and offer suggestions to others.

Teamwork Skills The knowledges, skills, and attitudes required to work effectively with others in pursuit of a common goal (e.g., backup behavior, mutual performance monitoring, error correction).

Thematic Apperception Test (TAT) A projective tool that requires the patient to develop a story about what is occurring in the stimulus picture presented.

Theory X and Theory Y The assumption that employees dislike work, are lazy and must be coerced to perform versus the assumption that employees like work, are creative and seek responsibility.

Time-of-usefulness-consciousness (TUC) The time between rapid decompression of an aircraft cabin when oxygen supply is lost and the brain goes into an unconsciousness state or coma. The TUC differs as a function of how high an altitude the decompression occurs and the access to supplemental oxygen.

Total Quality Management A management philosophy that emphasizes customer satisfaction through continuous improvement.

Trait Personality, social, physical or intellectual characteristics that may differentiate effective from ineffective leaders.

Transactional Leadership Leadership that emphasizes the clarification of role and task requirements with the promise of rewards for performance.

Transformational Leadership Leadership that inspires followers and has an extraordinary effect on performance.

Treatment The application of measures designed to alleviate or reduce an unwanted condition. For example, a service member suffering from post-traumatic stress disorder may seek counseling to relieve the condition.

Trench Foot A disease of the feet caused by continued dampness and cold. Characterized by discoloration, weakness, and sometime gangrene. A potential war stopper if most of the soldiers experience this disabling foot problem—a lesson relearned in many winter wars.

Triage Determining which injured persons are most likely to respond to treatment and what type of treatment; deciding where a patient should be referred after the initial provider evaluates the patient.

Uniformed Services University of the Health Sciences (USUHS) University located in Bethesda, Maryland. USUHS offers a Ph.D. training program in Military Clinical Psychology. For information on this program, contact the Program Administrator, Department of Medical and Clinical Psychology, Uniformed Services University of the Health Sciences, 4301 Jones Bridge Road, Bethesda, Maryland 20814-4799, USA.

Validation Verification that scores used in selection correlate well with criterion performance, i.e., training or job performance.

Validity The extent to which a test measures what it is claimed to measure for the people tested; validity has different connotations for various types of tests.

Vestibular Function A division of the acoustic nerve which conducts stimuli related to bodily equilibrium to the brain. The vestibular function helps us to maintain balance and body equilibrium and can be upset with motion conditions that lead to onset of motion sickness.

Vibration Rapid linear motion about an equilibrium position. Military vibration in all sorts of vehicles can affect performance, especially fine motor performance if one is doing tasks while his/her body is being subjected to vibration. Excessive vibration or sustained exposure can lead to health consequences. Design engineers attempt to dampen out vibration of operating platforms.

Vigilance Visual watchfulness. Any job task that requires an operator to spend a considerable amount of time watching for small changes in signals or targets is usually referred to as a vigilance task. For example, watching a radar screen or looking for the enemy while on sentry duty.

Whole-Person 'Holistic' evaluation process in which a variety of predictors are combined to assess job candidates, including tests scores, interviews, exercises, grades, recommendations, etc.

Index